T0137143

Studies in Fuzziness and Soft Computing

Volume 372

The series "Studies in Fuzziness and Soft Computing" contains publications on various topics in the area of soft computing, which include fuzzy sets, rough sets, neural networks, evolutionary computation, probabilistic and evidential reasoning, multi-valued logic, and related fields. The publications within "Studies in Fuzziness and Soft Computing" are primarily monographs and edited volumes. They cover significant recent developments in the field, both of a foundational and applicable character. An important feature of the series is its short publication time and world-wide distribution. This permits a rapid and broad dissemination of research results.

More information about this series at http://www.springer.com/series/2941

Said Melliani · Oscar Castillo
Editors

Recent Advances in Intuitionistic Fuzzy Logic Systems

Theoretical Aspects and Applications

Editors
Said Melliani
Department of Mathematics
Université Sultan Moulay Slimane
Beni Mellal, Morocco

Oscar Castillo
Division of Graduate Studies and Research
Tijuana Institute of Technology
Tijuana, Baja California, Mexico

ISSN 1434-9922 ISSN 1860-0808 (electronic)
Studies in Fuzziness and Soft Computing
ISBN 978-3-030-13208-8 ISBN 978-3-030-02155-9 (eBook)
https://doi.org/10.1007/978-3-030-02155-9

This Springer imprint is published by the registered company Springer Nature Switzerland AG
The registered company address is: Gewerbestrasse 11, 6330 Cham, Switzerland

Preface

ICIFSMAS 2018, second edition of international conference on intuitionistic fuzzy sets and its applications, was held at Al Akhawayn University, Ifrane, Morocco, during 11–13 April 2018. Following the tradition of the predecessor, this meeting gathered researchers around topics in present recent progress and new trends in intuitionistic fuzzy sets theory and its applications. A total number of 89 participants from several countries have attended the conference in Al Akhawayn University, Ifrane (AUI). The main goal of the event is to introduce Moroccan faculty and graduate students to important ideas in the mainstream of intuitionistic fuzzy sets theory. A secondary goal is for Moroccan mathematicians to open channels of communication with specialists from around the globe and eventually begin collaborative research projects. The audience was multidisciplinary allowing the participants to exchange diversified ideas and to show the wide attraction of intuitionistic fuzzy sets theory and its applications. There were two kinds of lectures: keynote talks of one hour presented by K. Atanassov (Bulgarian Academy of Sciences, Sofia) and Oscar Castillo (Tijuana Institute of Technology, Mexico), and then, there were three main sessions:

- Session 1: Intuitionistic Fuzzy Sets
- Session 2: Partial Differential Equations
- Session 3: Numeric and Informatics

The organizing committee was constituted by Said Melliani (Chair) from Moulay Slimane University, Beni Mellal; Bouchaib Falah from Al Akhawayn University, Ifrane; and Elhoussine Azroul from Sidi Ben Abdellah University of Fes. We are particularly indebted to our plenary speakers: Vassia Atanassova, Yubka Doukovska, Olympia Roeva, Dafina Zoteva, all from Bulgarian Academy of Sciences, Sofia, Bulgaria. Thanks are also due to the presenters of contributed papers, as well as everyone who attended for making the event a success. According to the evaluations of the scientific committee, there were several excellent talks presented by invited speakers.

The selected papers of ICIFSMAS 2018 were significantly revised and extended and are now presented in this volume in the Springer series of Studies in Fuzziness and Soft Computing. With the publication of these proceedings, we hope that a wider mathematical audience will benefit from the seminar research achievements and new contributions to the field of intuitionistic fuzzy sets theory and its applications.

Ifrane, Morocco
April 2018

Said Melliani
Oscar Castillo
Al Akhawayn University

Contents

Construction of a Topological Degree Theory in Generalized Sobolev Spaces

Mustapha Ait Hammou and Elhoussine Azroul

Abstract In this paper, we construct an integer-valued degree function in a suitable classes of mappings of monotone type, using a complementary system formed of Generalized Sobolev Spaces in which the variable exponent $p \in \mathscr{P}^{log}(\Omega)$ satisfy $1 < p'^{-} \leq p'^{+} \leq \infty$, where $\Omega \subset \mathbb{R}^N$ is open and bounded. This kind of spaces are not reflexives.

1 Introduction

Topological degree theory is one of the most effective tools in solving nonlinear equations.

Brouwer had published a degree theory in 1912 for continuous maps defined in finite dimensional Euclidean space [1]. Leray and Schauder developed the degree theory for compact operators in infinite dimensional Banach spaces [2]. Since then numerous generalizations and applications have been investigated in various ways of approach (see e.g. [3–6]). Browder introduced a topological degree for nonlinear operators of monotone type in reflexive Banach spaces [7, 8]. The theory was constructed later by Berkovits and Mustonen by using the Leray-Schauder degree [9–11] which can be applied to partial differential operators of general divergence form.

The purpose of this article is to generalize this theory to Sobolev spaces with variable exponent in the case where these spaces are not reflexives, exactly in the case where the variable exponent p satisfies $1 < p'^{-} \leq p'^{+} \leq \infty$ where p' is the dual variable exponent of p. We will construct this theory for appropriate classes of monotone mappings using a complementary system formed of Generalized Sobolev spaces.

M. Ait Hammou · E. Azroul (✉)
Sultan Moulay Slimane University, BP 523 Beni Mellal, Morocco
e-mail: elhoussine.azroul@usmba.ac.ma

M. Ait Hammou
e-mail: mus.aithammou@usmba.ac.ma

S. Melliani and O. Castillo (eds.), *Recent Advances in Intuitionistic Fuzzy Logic Systems*, Studies in Fuzziness and Soft Computing 372,
https://doi.org/10.1007/978-3-030-02155-9_1

The paper is divided into three parts. In the second section, we introduce some preliminary definitions and results concerning the generalized Lebesgue and Sobolev spaces, we construct a complementary system of these spaces and we present some classes of monotone cartography. The third section is dedicated to the construction of degree theory in generalized Sobolev spaces.

2 Preliminary Definitions and Results

In the sequel, we consider a natural number $N \geq 1$ and an open and bounded domain $\Omega \subset \mathbb{R}^N$ with segment property.

2.1 Generalized Lebesgue Spaces

We define $\mathscr{P}(\Omega)$ to be the set of all measurable function: $p : \Omega \to [1, +\infty]$. Functions $p \in \mathscr{P}(\Omega)$ are called variable exponents on Ω. We define $p^- = \operatorname{ess}\inf_{\Omega} p$ and $p^+ = \operatorname{ess}\sup_{\Omega} p$.

If $p \in \mathscr{P}(\Omega)$, then we define $p' \in \mathscr{P}(\Omega)$ by $\frac{1}{p(x)} + \frac{1}{p'(x)} = 1$, where $\frac{1}{\infty} := 0$. The function p' is called the dual variable exponent of p.

We say that a function $\alpha : \Omega \to \mathbb{R}$ is *locally log-Hölder continuous* on Ω if there exists $c_1 > 0$ such that

$$|\alpha(x) - \alpha(y)| \leq \frac{c_1}{log(e + 1/|x - y|)}$$

for all $x, y \in \Omega$. We say that α satisfies the *log-Hölder decay* condition if there exist $\alpha_\infty \in \mathbb{R}$ and a constant $c_2 > 0$ such that

$$|\alpha(x) - \alpha_\infty| \leq \frac{c_2}{log(e + |x|)}$$

for all $x \in \Omega$. We say that α is *globally log-Hölder continuous* in Ω if it is locally log-Hölder continuous and satisfies the log-Hölder decay condition.

We define the following class of variable exponents

$$\mathscr{P}^{log}(\Omega) := \{p \in \mathscr{P}(\Omega) : \frac{1}{p} \text{ is globally log-Hölder continuous}\}.$$

We can deduce that $p \in \mathscr{P}^{log}(\Omega)$ if and only if $p' \in \mathscr{P}^{log}(\Omega)$.

For $t \geq 0$, $x \in \Omega$ and $1 \leq p < \infty$ we define

$$\varphi_{p(x)}(t) := t^{p(x)}$$

Moreover we set

$$\varphi_{\infty}(t) := \infty.\chi_{(1,\infty)}(t) = \begin{cases} 0 & \text{if } t \in [0,1] \quad , \\ \infty & \text{if } t \in (1,\infty) \ . \end{cases}$$

We will use $t^{p(x)}$ as an abbreviation for $\varphi_{p(x)}(t)$, also in the case $p = \infty$. Similarly, $t^{\frac{1}{p(x)}}$ will denote the inverse function $\varphi_{p(x)}^{-1}(t)$; note that in case $p = \infty$ we have $t^{\frac{1}{\infty}} = \varphi_{\infty}^{-1}(t) = \chi_{(0,\infty)}(t)$.
For any variable exponent $p(\cdot)$ and any measurable function u, we define the modular

$$\rho_{p(\cdot)}(u) = \int_{\Omega} |u(x)|^{p(x)}\, dx,$$

and we define the variable exponent Lebesgue space

$$L^{p(\cdot)}(\Omega) = \{u;\ u : \Omega \to \mathbb{R} \text{ is measurable and } \rho_{p(\cdot)}(\lambda u) < \infty \text{ for some } \lambda > 0\}$$

equipped with the norm, called the Luxemburg norm,

$$||u||_{p(\cdot)} = \inf\{\lambda > 0/\rho_{p(.)}(\frac{u}{\lambda}) \leq 1\}.$$

It is a Banach space ([12, Theorem 3.2.7]). The space $E^{p(\cdot)}(\Omega)$ is the closure of the space $L^{\infty}(\Omega)$ with respect to the Luxemburg norm.

Theorem 1 [12] *Let $p(\cdot)$ and $q(\cdot)$ be the exponent and $\Omega \subset \mathbb{R}^N$ open and bounded. Then*

(i) $E^{p(\cdot)}(\Omega) \subset L^{p(\cdot)}(\Omega)$,
(ii) $E^{p(\cdot)}(\Omega) = L^{p(\cdot)}(\Omega)$ *iff* $p^+ < \infty$,
(iii) $E^{p(\cdot)}(\Omega)$ *is separable,*
(iv) $(E^{p(\cdot)}(\Omega))^* = L^{p'(\cdot)}(\Omega)$,
(v) $L^{p(\cdot)}(\Omega)$ *is reflexive iff* $1 < p^- \leq p(x) \leq p^+ < \infty$.

We say that a sequence $\{u_n\} \subset L^{p(\cdot)}(\Omega)$ converges to $u \in L^{p(\cdot)}(\Omega)$ *in the modular sense*, denote $u_n \to u(mod)$ in $L^{p(\cdot)}$, if there exists $\lambda > 0$ such that

$$\rho_{p(\cdot)}(\frac{u_n - u}{\lambda}) \to 0, \text{ when } n \to \infty.$$

Let X and Y be arbitrary Banach spaces with bilinear bicontinuous pairing $\langle .,.\rangle_{X,Y}$. We say that a sequence $u_n \subset X$ converges to $u \in X$ with respect the topology $\sigma(X, Y)$, denote $u_n \to u(\sigma(X, Y))$ in X, if $\langle u_n, v\rangle \to \langle u, v\rangle$ for all $v \in Y$. When $Y^* \cong X$, we denote only $u_n \rightharpoonup u$ in X.

Theorem 2 [12, 13] *In any Generalized Lebesgue space* $L^{p(\cdot)}(\Omega)$

(i) norm convergence implies modular convergence,
(ii) norm convergence and modular convergence are equivalent iff $p^+ < \infty$,
(iii) modular convergence implies $\sigma(L^{p(\cdot)}, L^{p'(\cdot)})$ *convergence.*

2.2 Generalized Sobolev Spaces and Complementary System

Definition 1 Let Y and Z be Banach spaces in duality with respect with to a continuous pairing $\langle ., .\rangle$ and let Y_0 and Z_0 be closed subspaces of Y and Z respectively. Then the quadruple $\begin{pmatrix} Y & Z \\ Y_0 & Z_0 \end{pmatrix}$ is called *a complementary system* if, by means of $\langle ., .\rangle$, $Y_0^* \cong Z$ and $Z_0^* \cong Y$.

An example of a complementary system is

$$\begin{pmatrix} L^{p(\cdot)}(\Omega) & L^{p'(\cdot)}(\Omega) \\ E^{p(\cdot)}(\Omega) & E^{p'(\cdot)}(\Omega) \end{pmatrix}$$

The following lemma gives an important method by which from a complementary system $\begin{pmatrix} Y & Z \\ Y_0 & Z_0 \end{pmatrix}$ and a closed subspace E of Y, one can construct a new complementary system $\begin{pmatrix} E & F \\ E_0 & F_0 \end{pmatrix}$. Define $E_0 = E \cap Y_0$, $F = Z/E_0^{\perp}$ and $F_0 = \{z + E_0^{\perp}; z \in Z_0\} \subset F$, where $\perp$ denotes the orthogonal in the duality (Y, Z), i.e. $E_0^{\perp} = \{z \in Z; \langle y, z\rangle = 0 \text{ for all } y \in E_0\}$.

Lemma 1 [14, Lemma 1.2] *The pairing* $\langle ., .\rangle$ *between* Y *and* Z *induces a pairing between* E *and* F *if and only if* E_0 *is* $\sigma(Y, Z)$ *dense in* E. *In this case,* $\begin{pmatrix} E & F \\ E_0 & F_0 \end{pmatrix}$ *is a complementary system if* E *is* $\sigma(Y, Z_0)$ *closed, and conversely, when* Z_0 *is complete,* E *is* $\sigma(Y, Z_0)$ *closed if* $\begin{pmatrix} E & F \\ E_0 & F_0 \end{pmatrix}$ *is a complementary system.*

Next, let $p \in \mathscr{P}(\Omega)$ and $m \in \mathbb{N}$.
We define the spaces

$$W^{m,p(.)}(\Omega) = \{u \in L^{p(.)}(\Omega) : D^{\alpha}u \in L^{p(.)}(\Omega),\ |\alpha| \le m\},$$

$$H^{m,p(.)}(\Omega) = \{u \in E^{p(.)}(\Omega) : D^{\alpha}u \in E^{p(.)}(\Omega),\ |\alpha| \le m\}$$

with the norm

$$||u||_{m,p(.)} = \sum_{|\alpha|\leq m} ||D^\alpha u||_{p(.)}.$$

The spaces $W^{m,p(.)}(\Omega)$ and $H^{m,p(.)}(\Omega)$ are Banach spaces (see [12]).
We say that a sequence $\{u_n\} \subset W^{m,p(.)}(\Omega)$ converges to $u \in W^{m,p(.)}(\Omega)$ *in the modular sense*, denote $u_n \to u(mod)$ in $W^{m,p(.)}$, if there exists $\lambda > 0$ such that

$$\rho_{p(.)}(\frac{D^\alpha u_n - D^\alpha u}{\lambda}) \to 0, \text{ when } n \to \infty,$$

for $|\alpha| \leq m$.
The Sobolev space $W_0^{m,p(.)}(\Omega)$ with zero boundary values is the closure of the set of $W^{m,p(.)}(\Omega)$-functions with compact support, i.e.

$$\{u \in W^{m,p(.)}(\Omega) : u = u\chi_K \text{ for a compact } K \subset \Omega\}$$

in $W^{m,p(.)}(\Omega)$.
The space $W^{m,p(.)}(\Omega)$ will always be identified to a subspace of the product $\Pi_{|\alpha|\leq m} L^{p(.)} = \Pi L^{p(.)}$; this subspace is $\sigma(\Pi L^{p(.)}, \Pi E^{p'(.)})$ closed and $W_0^{m,p(.)}(\Omega)$ will be the $\sigma(\Pi L^{p(.)}, \Pi E^{p'(.)})$ closure of $\mathcal{D}(\Omega) = \bigcap_{m=1}^{\infty} C_0^m(\Omega)$ in $W^{m,p(.)}(\Omega)$.
The (norm) closure of $\mathcal{D}(\Omega)$ in the space $W^{m,p(.)}(\Omega)$ (or in $\Pi L^{p(.)}$) is denoted by $H_0^{m,p(.)}(\Omega)$.
If $p \in \mathcal{P}^{log}(\Omega)$ is bounded, then $W_0^{m,p(.)}(\Omega) = H_0^{m,p(.)}(\Omega)$ [12, Corollary 11.2.4].
The space $W_0^{m,p(.)}(\Omega)$ is a Banach space, which is separable if p is bounded, and reflexive and uniformly convex if $1 < p^- \leq p^+ < \infty$ [12, Theorem 8.1.13].
Let $p \in \mathcal{P}^{log}(\Omega)$ satisfy $1 < p'^- \leq p'^+ \leq \infty$. We denote the dual spaces of Sobolev spaces $W_0^{m,p(.)}(\Omega)$ and $H_0^{m,p(.)}(\Omega)$ as follows

$$W^{-m,p'(.)}(\Omega) := (W_0^{m,p(.)}(\Omega))^* \text{ and } H^{-m,p'(.)}(\Omega) := (H_0^{m,p(.)}(\Omega))^*.$$

Proposition 1 [12, Proposition 12.3.2] *Let $\Omega \subset \mathbb{R}^N$ be a domain, let $p \in \mathcal{P}^{log}(\Omega)$ satisfy $1 < p'^- \leq p'^+ \leq \infty$ and let $m \in \mathbb{N}$. For each $F \in W^{-m,p'(.)}(\Omega)$ there exists $f_\alpha \in L^{p'(.)}(\Omega)$, $|\alpha| \leq m$, such that*

$$\langle F, u\rangle = \sum_{|\alpha|\leq m} \int_\Omega f_\alpha D^\alpha u \, dx$$

for all $W_0^{m,p(.)}(\Omega)$. Moreover,

$$\|F\|_{-m,p'(\cdot)} \approx \sum_{|\alpha|\leq m} \|f_\alpha\|_{p'(\cdot)}.$$

We can write an analogue proposition for $H^{-m,p'(.)}(\Omega)$ and then

$$W^{-m,p'(.)}(\Omega) = \{F \in \mathscr{D}'(\Omega) : F = \sum_{|\alpha|\leq m} (-1)^{|\alpha|} D^{\alpha} f_{\alpha}, \text{ where } f_{\alpha} \in L^{p'(.)}(\Omega)\},$$

$$H^{-m,p'(.)}(\Omega) = \{F \in \mathscr{D}'(\Omega) : F = \sum_{|\alpha|\leq m} (-1)^{|\alpha|} D^{\alpha} f_{\alpha}, \text{ where } f_{\alpha} \in E^{p'(.)}(\Omega)\}.$$

By Lemma 1, the quadruple

$$\begin{pmatrix} W_0^{m,p(.)}(\Omega) & W^{-m,p'(.)}(\Omega) \\ H_0^{m,p(.)}(\Omega) & H^{-m,p'(.)}(\Omega) \end{pmatrix}$$

forms a complementary system.
We say that a sequence $\{u_n\} \subset W^{-m,p'(.)}(\Omega)$ converges to $u \in W^{-m,p'(.)}(\Omega)$ *in the modular sense*, denote $u_n \to u(mod)$ in $W^{-m,p'(.)}$, if u_n and u have representations

$$u_n = \sum_{|\alpha|\leq m} (-1)^{|\alpha|} D^{\alpha} g_{\alpha}^{(n)}, \quad u = \sum_{|\alpha|\leq m} (-1)^{|\alpha|} D^{\alpha} g_{\alpha},$$

such that $g_{\alpha}^{(n)}, g_{\alpha} \in L^{p'(.)}(\Omega)$ and $g_{\alpha}^{(n)} \to g_{\alpha}(mod)$ in $L^{p'(.)}$ for all $|\alpha| \leq m$.

Let A be a subset of a Generalized Sobolev Space Y. We denote by $\bar{A}^{mod}$ the sequential modular closure of A,i.e.

$$\bar{A}^{mod} = \{u \in Y / \text{ there exists } \{u_n\} \subset A \text{ such that } u_n \to u(mod) \text{ in } Y\}.$$

2.3 Some Classes of Mappings of Monotone Type

Let

$$\begin{pmatrix} Y & Z \\ Y_0 & Z_0 \end{pmatrix} = \begin{pmatrix} W_0^{m,p(.)}(\Omega) & W^{-m,p'(.)}(\Omega) \\ H_0^{m,p(.)}(\Omega) & H^{-m,p'(.)}(\Omega) \end{pmatrix}$$

be a complementary system formed of Generalized Sobolev Spaces in which $\Omega \subset \mathbb{R}^N$ is open, bounded and satisfies the segment property and $p \in \mathscr{P}^{log}(\Omega)$ satisfy $1 < p'^{-} \leq p'^{+} \leq \infty$. We consider mappings $F : D_F \to Z$ which satisfy the following conditions:

(i) $Y_0 \subset D_F \subset Y$,
(ii) F is *finitely continuous*, i.e., the restriction of the mapping F to any finite dimensional subspace $X \subset Y_0$ is continuous from the topology of X to the weak topology of Z.

We shall next define some classes of mappings of monotone type:

(i) F is *bounded*, denote $F \in (BD)$, if the set $F(A) \subset Z$ is bounded when $A \subset D_F$ is bounded.
(ii) F is *strongly quasibounded*, denote $F \in (QB)$, if the conditions $\{u_n\} \subset D_F$ bounded and $\langle F(u_n), u_n - \bar{u}\rangle$ is bounded from above for some $\bar{u} \in Y_0$ imply that $\{F(u_n)\}$ is bounded in Z.
(iii) F is *continuous*, denote $F \in (CONT)$, if the conditions $\{u_n\} \subset D_F$, $u \in D_F$ and $\|u_n - u\|_Y \to 0$
imply that $\|F(u_n) - F(u)\|_Z \to 0$.
(iv) F is of the class (S_+), denote $F \in (S_+)$, if the conditions $\{u_n\} \subset D_F$, $u_n \rightharpoonup u \in Y$ in Y and $limsup_{n\to\infty}\langle F(u_n), u_n - u\rangle \leq 0$ imply that $u \in D_F$, and $\|u_n - u\|_Y \to 0$.
(v) F is *semicontinuous* if the conditions $\{u_n\} \subset D_F$, $u \in D_F$ and $u_n \to u$ imply that $F(u_n) \rightharpoonup F(u)$.
(vi) F is *pseudomonotone*, $F \in (PM)$, if the conditions $\{u_n\} \subset D_F$, $u_n \rightharpoonup u$ in Y, $F(u_n) \rightharpoonup \chi$ in Z and $limsup_{n\to\infty}\langle F(u_n), u_n\rangle \leq \langle \chi, u\rangle$
imply that $u \in D_F$, $\chi = F(u)$ and $\langle F(u_n), u_n\rangle \to \langle F(u), u\rangle$.
(vii) F is of the class (MOD), denote $F \in (MOD)$, if the conditions $\{u_n\} \subset D_F$, $u_n \rightharpoonup u$ in Y, $F(u_n) \rightharpoonup \chi$ in Z and $limsup_{n\to\infty}\langle F(u_n), u_n\rangle \leq \langle \chi, u\rangle$
imply that $u \in D_F$, $\chi = F(u)$ and there exists a subsequence $\{u_{n'}\}$ such that $u_{n'} \to u$ (mod) in Y and $F(u_{n'}) \to F(u)$ (mod) in Z.

3 Degree Theory in Generalized Sobolev Spaces

3.1 An Outline of Brouwers Degree Theory

Theorem 3 *Let $X = \mathbb{R}^n = Y$ for a given positive integer n. For bounded open subsets G of X, consider continuous mappings $f : \bar{G} \to Y$ and points y_0 in Y such that $y_0 \notin f(\partial G)$. Then to each such triple (f, G, y_0), there corresponds an integer $d(f, G, y_0)$ having the following properties:*

(a) *Existence: if $d(f, G, y_0) \neq 0$, then $y_0 \in f(G) =$,*
(b) *Additivity: if $f : \bar{G} \to Y$ is a continuous map with G a bounded open set in X and G_1, G_2 are a pair of disjoint open subsets of G such that $y_0 \notin f(\bar{G} \setminus (G_1 \cup G_2))$, then*

$$d(f, G, y_0) = d(f, G_1, y_0) + d(f, G_2, y_0),$$

(c) *Invariance under homotopy: Let G be a bounded open set in X, and consider a continuous homotopy $\{f_t : 0 \leq t \leq 1\}$ of maps of $\bar{G}$ into Y. $\{y_t : 0 \leq t \leq 1\}$ be a continuous curve in Y such that $y_t \notin f_t(\partial G)$ for any $t \in [0, 1]$, then $d(f_t, G, y_t)$ is constant in t on* $[0, 1]$,

(d) *Normalization: If* f_0 *is the identity map of* X *onto* Y*, then for every bounded open* G *and* $y_0 \in f_0(G)$ *then*

$$d(f_0, G, y_0) = 1.$$

Theorem 4 *The degree function* $d(f, G, y0)$ *is uniquely determined by the four conditions of Theorem 3.*

Remark 1 Theorem 3 is an appropriately formalized version of the properties of the classical Brouwer degree. Theorem 4 contains an observation made independently in 1972 and 1973 by Fuhrer [15] and Amann and Weiss [16], respectively.

3.2 Construction of a Degree Function in Generalized Sobolev Spaces

Let

$$\begin{pmatrix} Y & Z \\ Y_0 & Z_0 \end{pmatrix} = \begin{pmatrix} W_0^{m,p(.)}(\Omega) & W^{-m,p'(.)}(\Omega) \\ H_0^{m,p(.)}(\Omega) & H^{-m,p'(.)}(\Omega) \end{pmatrix}$$

be a complementary system formed of Generalized Sobolev Spaces in which $\Omega \subset \mathbb{R}^N$ is open, bounded and satisfies the segment property and $p \in \mathcal{P}^{log}(\Omega)$ satisfy $1 < p'^- \leq p'^+ \leq \infty$. We define the class $\mathcal{F}$ of *admissible mappings* and the class $\mathcal{H}$ of *admissible homotopies* as follows:
$F : D_F \subset Y \to Z$ belongs to $\mathcal{F}$, if

(a) F is a strongly quasibounded mapping of the class (MOD).
$F : D_F \subset Y \to Z$ belongs to $\mathcal{F}^a$, if $F \in \mathcal{F}$ and the following conditions hold:
(b) if $\{u_n\} \subset D_F$ is bounded , $t_n \to 0^+$ and $\langle t_n F(u_n), u_n - \bar{u}\rangle$ is bounded from above for some $\bar{u} \in Y$, then $\{t_n F(u_n)\} \subset Z$ is bounded,
(c) if $\{u_n\} \subset D_F, u_n \to u \in Y$ for $\sigma(Y, Z_0), t_n \to 0^+, t_n F(u_n) \to \chi \in Z$ for $\sigma(Z, Y_0)$ and $limsup\langle t_n F(u_n), u_n\rangle \leq \langle \chi, u\rangle$, then $\langle t_n F(u_n), u_n\rangle \to \langle \chi, u\rangle$,
(d) if $\{u_n\} \subset D_F$, $u_n \to u$ (mod) in Y, $t_n \to 0^+$, $t_n F(u_n) \to \chi \in Z(\sigma(Z, Y_0))$ in Z and $limsup\langle t_n F(u_n), u_n\rangle \leq \langle \chi, u\rangle$, then $t_n F(u_n) \to 0$ (mod) in Z.

The homotopy $H : D_H \to Z$ belongs to $\mathcal{H}$, if H is a strongly quasibounded homotopy of the class (MOD).

Lemma 2 *If* $F, G \in \mathcal{F}^a$*, then* $H(t, u) = tF(u) + (1 - t)G(u)$ *belongs to* $\mathcal{H}$ *with*

$$D_{H_t} = \begin{cases} D_F \cap D_G, & if\, 0 < t < 1 \\ D_G, & if\, t = 0 \\ D_F, & if\, t = 1. \end{cases}$$

Proof **Step 1**
Since F and G are finitely continuous, the homotopy H is finitely continuous from the norm topology of $[0, 1] \times Y$ to the weak topology of Z.
Step 2
We shall prove that H is strongly quasibounded.
Assume $\{t_n\} \subset [0, 1]$, $\{u_n\} \subset D_{H_{t_n}}$ and $\langle H(t_n, u_n), u_n - \bar{u}\rangle$ is bounded from above for some $\bar{u} \in Y_0$. It follows that $\langle t_{n'}F(u_{n'}), u_{n'} - \bar{u}\rangle$ or $\langle (1 - t_{n'})G(u_{n'}), u_{n'} - \bar{u}\rangle$ is bounded from above for a subsequence.
We may also suppose that $\langle t_{n'}F(u_{n'}), u_{n'} - \bar{u}\rangle$ is bounded from above. By condition **(b)** and the fact that F is strongly quasibounded, the sequence $\{t_{n'}F(u_{n'})\}$ is bounded in Z. Consequently, $\langle t_{n'}F(u_{n'}), u_{n'} - \bar{u}\rangle$ is bounded implying that $\langle (1 - t_{n'})G(u_{n'}), u_{n'} - \bar{u}\rangle$ is bounded from above. Therefore $\{(1 - t_{n'})G(u_{n'})\}$ is also bounded in Z and hence $\{H(t_{n'}, u_{n'})\}$ is bounded in Z.
By contradiction argument, $\{H(t_n, u_n)\}$ is bounded in Z.
Step 3
We shall next prove that H is a homotopy of the class *(MOD)*.
Assume that $t_n \subset [0, 1]$, $t_n \to t \in [0, 1]$, $\{u_n\} \subset D_{H_{t_n}}$, $u_n \rightharpoonup u$ in Y, $H(t_n, u_n) \rightharpoonup \chi$ in Z and $limsup_{n\to\infty}\langle H(t_n, u_n), u_n\rangle \le \langle \chi, u\rangle$.
Deducing as above, $\{t_{n'}F(u_{n'})$ and $\{(1 - t_{n'})G(u_{n'})\}$ are bounded in Z for some subsequence. We may assume that $t_{n'}F(u_{n'}) \rightharpoonup \chi_1$ and $(1 - t_{n'})G(u_{n'}) \rightharpoonup \chi_2$ in Z. It is clear that $\chi = \chi_1 + \chi_2$. We may assume that $limsup_{n\to\infty}\langle t_{n'}F(u_{n'}), u_{n'}\rangle \le \langle \chi_1, u\rangle$ or $limsup_{n\to\infty}\langle (1 - t_{n'})G(u_{n'}), u_{n'}\rangle \le \langle \chi_2, u\rangle$. Suppose, for example, that

$$limsup_{n\to\infty}\langle t_{n'}F(u_{n'}), u_{n'}\rangle \le \langle \chi_1, u\rangle.$$

By condition **(c)** and the fact that F belongs to the class *(MOD)*, we have $\langle t_{n'}F(u_{n'}), u_{n'}\rangle \to \langle \chi_1, u\rangle$. Hence

$$limsup_{n\to\infty}\langle (1 - t_{n'})G(u_{n'}), u_{n'}\rangle \le \langle \chi_2, u\rangle.$$

If $0 < t < 1$, then $F(u_{n'}) \rightharpoonup \frac{\chi_1}{t}$ in Z, $G(u_{n'}) \rightharpoonup \frac{\chi_2}{1-t}$ in Z, $\langle F(u_{n'}), u_{n'}\rangle \to \langle \frac{1}{t}\chi_1, u\rangle$ and $limsup_{n\to\infty}\langle G(u_{n'}), u_{n'}\rangle \le \langle \frac{1}{1-t}\chi_2, u\rangle$.
Therefore $u \in D_F \cap D_G$, $\frac{\chi_1}{t} = F(u)$, $\frac{\chi_2}{1-t} = G(u)$, $u_{n'} \to u(mod)$ in Y and $F(u_{n'}) \to F(u)$, $G(u_{n'}) \to G(u)(mod)$ in Z. Hence $u \in D_{H_t}$ and $H(t_{n'}, u_{n'}) \to H(t, u)(mod)$ in Z. If $t = 0$, then $G(u_{n'}) \rightharpoonup \frac{\chi_2}{1-t}$ in Z and $limsup_{n\to\infty}\langle G(u_{n'}), u_{n'}\rangle \le \langle \frac{1}{1-t}\chi_2, u\rangle$.
Therefore $u \in D_G$, $u_{n'} \to u(mod)$ in Y and $G(u_{n'} \to G(u)(mod)$ in Z. Moreover, by condition **(d)**, we have $t_{n'}F(u_{n'}) \to 0$ in the modular sense. Hence $u \in D_{H_0}$ and $H(t_{n'}, u_{n'}) \to H(t, u)(mod)$ in Z. If $t = 1$, we make an analogue deduction to obtain $u \in D_F = D_{H_1}$ and $u_{n'} \to u(mod)$ in Y and $F(u_{n'}) \to F(u)(mod)$ in Z.
Consequently, H is a homotopy of the class *(MOD)*.

Our aim in this subsection is to construct an integer-valued degree function $d(F, G, f)$ for $F \in \mathscr{F}$, $G \subset Y_0$ open and bounded in Y_0, $f \in Z_0$ and $f \notin F(\overline{\partial_{Y_0} G}^{mod})$ $\cap \overline{F(\partial_{Y_0} G}^{mod}$ satisfying the following conditions:

(C_1) Existence: if $d(F, G, f) \neq 0$, then $f \in F(\bar{G}^{mod}) \cap \overline{F(G)}^{mod}$,
(C_2) Additivity: if $G_1, G_2 \subset G$ are open and bounded, $G_1 \cap G_2 = \emptyset$ and $f \notin \overline{F(\bar{G} \setminus (G_1 \cup G_2)}^{mod}) \cap \overline{F(\bar{G} \setminus (G_1 \cup G_2))}^{mod}$,
then

$$d(F, G, f) = d(F, G_1, f) + d(F, G_2, f),$$

(C_3) Homotopy invariance: if $H \in \mathscr{H}$, $f \in Z_0$ and $f \notin H([0, 1] \times \overline{\partial_{Y_0} G}^{mod}) \cap \overline{H([0, 1] \times \partial_{Y_0} G)}^{mod}$, then

$$d(H(t, \cdot), G, f) = \textit{constant} \text{ for all } t \in [0, 1],$$

(C_4) Normalization: There exists a normalising map $K \in \mathscr{F}^a$ such that if $f \in Z_0$, $f \notin K(\overline{\partial_{Y_0} G}^{mod}) \cap \overline{K(\partial_{Y_0} G)}^{mod}$ and $f \in K(G)$, then

$$d(K, G, f) = 1.$$

Remark 2 We shall always assume in the applications that
$1 < p^- \leq p(\cdot) \leq p^+ < \infty$. This restriction means that instead of modular closure we have norm closures. In these cases the corresponding degree theories can be formulated as fellows:
$F \in \mathscr{F}$, $G \subset Y$ open and bounded in Y, $f \in Z_0$ and $f \notin F(\partial_Y G)$

(c_1) Existence: if $d(F, G, f) \neq 0$, then $f \in F(G)$,
(c_2) Additivity: if $G_1, G_2 \subset G$ are open and bounded, $G_1 \cap G_2 = \emptyset$ and $f \notin F(\bar{G} \setminus (G_1 \cup G_2))$, then

$$d(F, G, f) = d(F, G_1, f) + d(F, G_2, f),$$

(c_3) Homotopy invariance: if $H \in \mathscr{H}$, $f \in Z_0$ and $f \notin H([0, 1] \times \partial_Y G)$, then

$$d(H(t, \cdot), G, f) = \textit{constant} \text{ for all } t \in [0, 1],$$

(c_4) Normalization: There exists a normalising map $K \in \mathscr{F}^a$ such that if $f \in Z_0$, $f \notin K(\partial_Y G)$ and $f \in K(G)$, then

$$d(K, G, f) = 1.$$

For the construction of such degree, we need the following:

Definition 2 Let Λ be the set of all finite dimensional subspaces of Y_0.
Denote
$Y_0 = X_\lambda \oplus Y_\lambda$ for all $X_\lambda \in \Lambda$, where Y_λ is the closed complement of X_λ ([17, p. 157]),
$A_\lambda = A \cap X_\lambda$, when $A \subset Y$,
$P_\lambda : Y_0 \to X_\lambda$ the projection map,
$P_\lambda^* : X_\lambda \to Z$, $\langle P_\lambda^*(u), v\rangle = \langle u, P_\lambda(v)\rangle$ for all $u \in X_\lambda$ and $v \in Y_0$.

For the natural injection $\phi_\lambda : X_\lambda \to Y_0$, we define
$\phi_\lambda^* : Z \to X_\lambda$, $\langle \phi_\lambda^*(u), v\rangle = \langle u, \phi_\lambda(v)\rangle$ for all $u \in Z$ and $v \in X_\lambda$.
If $F : Y_0 \to Z$, then $F_\lambda : X_\lambda \to X_\lambda$, $F_\lambda(x) = \phi_\lambda^*(F(\phi_\lambda(x)))$.

Let d_n be the Brouwer degree for continuous maps from $\mathbb{R}^n$ to $\mathbb{R}^n$.

Lemma 3 [7] *Let $G \subset \mathbb{R}^n$ be open and bounded and $F : \bar{G} \subset \mathbb{R}^n \to \mathbb{R}^n$ continuous such that $0 \notin F(\partial G)$. Define a mapping*

$$F' : \bar{G} \times [-1, 1]^m \subset \mathbb{R}^{n+m} \to \mathbb{R}^{n+m}$$

$$F'(x, y) = (F(x), y) \text{ for all } x \in \bar{G}, y \in [-1, 1]^m.$$

Then

$$d_n(F, G, 0) = d_{n+m}(F', G \times (-1, 1)^m, 0).$$

Let d_λ be the Brouwer degree in the space X_λ.

Lemma 4 *Let $G \subset Y_0$ be open and bounded in Y_0, $X_\lambda \subset X_\mu \in \Lambda$, such that $G \cap X_\lambda \neq \emptyset$ and $F \in (MOD)$. If $d_\lambda(F_\lambda, G_\lambda, 0) \neq d_\mu(F_\mu, G_\mu, 0)$ or one of the degrees is not defined, then there exists $u \in \partial_{Y_0} G$ such that*

$$\langle F(u), u\rangle \leq 0 \text{ and } \langle F(u), v\rangle = 0 \text{ for all } v \in X_\lambda.$$

Proof If one of the degrees $d_\lambda(F_\lambda, G_\lambda, 0)$ and $d_\mu(F_\mu, G_\mu, 0)$ is not defined, there exists $u \in \partial_{X_\lambda} G_\lambda \subset \partial_{Y_0} G$ such that $F_\lambda(u) = 0$ or $u \in \partial_{X_\mu} G_\mu \subset \partial_{Y_0} G$ such that $F_\mu(u) = 0$. In both cases the proof is complete.
Otherwise we define a continuous mapping $S : X\mu \to X_\mu$,

$$S(x) = \phi_\lambda^*(F(P_\lambda(x))) + x - P_\lambda(x.)$$

By Lemma 3 we have

$$d_\lambda(F_\lambda, G_\lambda, 0) = d_\mu(S, G_\lambda \times (-1, 1)^m, 0),$$

where $m = dimX_\mu - dimX_\lambda$. If $S(x) = 0$ for some $x \in G_\mu$, then $x - P_\lambda(x) = 0$, implying $x \in G_\lambda \times (-1, 1)^m$. If $S(x) = 0$ for some $x \in G_\lambda \times (-1, 1)^m$ then $x - P_\lambda(x) = 0$, which means that $x \in G_\lambda \subset G_\mu$. By the excision property of the Brouwer degree, we have

$$d_\mu(S, G_\mu, 0) = d_\mu(S, G_\lambda \times (-1, 1)^m, 0).$$

Hence

$$d_\mu(F_\mu, G_\mu, 0) \neq d_\mu(S, G_\mu, 0).$$

Define another mapping $S' : X\mu \to X_\mu$,

$$S'(x) = \phi_\lambda^*(Fx)) + x - P_\lambda(x).$$

We shall prove that $d_\mu(S, G_\mu, 0) = d_\mu(S', G_\mu, 0)$. Consider the homotopy

$$H(t, u) = tS(u) + (1-t)S'(u) = u - P_\lambda(u) + \phi_\lambda^*[tF(P_\lambda(u)) + (1-t)F(u)].$$

If $H(t, u) = 0$ for some $0 \le t \le 1$ and $u \in \bar{G}_\mu^{X_\mu}$, then $u = P_\lambda(u) \in X_\lambda$ implying $\phi_\lambda^*(F(u)) = F_\lambda(u) = 0$ and therefore $u \in G_\lambda \subset G_\mu$. By homotopy invariance, we have

$$d_\mu(S, G_\mu, 0) = d_\mu(S', G_\mu, 0).$$

We can thus deduce that

$$d_\mu(F_\mu, G_\mu, 0) \neq d_\mu(S', G_\mu, 0).$$

Consider the homotopy $H : [0, 1] \times X_\mu \to X_\mu$,

$$H(t, u) = tS'(u) + (1-t)F_\mu(u).$$

By the homotopy invariance of the Brouwer degree, $H(t, u) = 0$ for some $u \in \partial_{X_\mu} G_\mu \subset \partial_Y G$ and $t \in (0, 1)$. Let $v \in X_\lambda$ be arbitrary, then

$$\begin{aligned}\langle H(t, u), v\rangle &= t\langle S'(u), v\rangle + (1-t)\langle F_\lambda(u), v\rangle \\ &= t\langle F(u), v\rangle + t\langle u - P_\lambda(u), v\rangle + (1-t)\langle F(u), v\rangle \\ &= t\langle F(u), v\rangle + (1-t)\langle F(u), v\rangle = 0,\end{aligned}$$

implying $\langle F(u), v\rangle = 0$. Moreover,

$$\begin{aligned}\langle H(t, u), u - P_\lambda(u)\rangle &= t\langle \phi_\lambda^*(F(u)), u - P_\lambda(u)\rangle + t\langle u - P_\lambda(u)), u - P_\lambda(u)\rangle \\ &\quad + (1-t)\langle F_\lambda(u), u - P_\lambda(u)\rangle \\ &= t||u - P_\lambda(u)||^2 + (1-t)\langle F(u), u\rangle = 0.\end{aligned}$$

Hence $\langle F(u), u\rangle \le 0$.

Lemma 5 *Let $H : D_H \to Z$ be a strongly quasibounded (MOD) homotopy and $A \subset Y_0$ closed and bounded. If*

$$0 \notin H([0, 1] \times \bar{A}^{mod}) \cap \overline{H([0, 1] \times A)}^{mod},$$

then there exists $X_\lambda \in \Lambda$ such that

$$0 \notin H_\mu([0, 1] \times A_\mu)\ for\ all\ X\mu \supset X_\lambda.$$

Proof By contradiction, suppose that

$$\forall X_\lambda \in \Lambda, \exists X\mu \supset X_\lambda, \exists (t_\mu, a_\mu) \in [0,1] \times A_\mu; H_\mu(t_\mu, a_\mu) = 0.$$

Define a set

$$V_\lambda = \{(t,a) \in [0,1] \times A \mid \langle H(t,a), a\rangle \leq 0 \text{ and } \langle H(t,a), v\rangle = 0 \text{ for all } v \in X_\lambda\},$$

which is non-empty for all λ. If $X_{\lambda_1}, X_{\lambda_2}, \ldots, X_{\lambda_n} \in \Lambda$ and if $\cup_{i=1}^n X_{\lambda_i} \subset X_\lambda$, then

$$V_\lambda \subset \cap_{i=1}^n V_{\lambda_i}.$$

Therefore the family $\{V_\lambda\}$ has the finite intersection property. Denote τ the topology $||.||_{\mathbb{R}} \times \sigma(Y, Z_0)$. The set $\overline{V_\lambda}^\tau$ is bounded and τ-closed, which implies by Alaoglu's Theorem that it is τ-compact. Hence

$$\cap \overline{V_\lambda}^\tau \neq \emptyset.$$

Choose$(t_0, u_0) \in \cap \overline{V_\lambda}^\tau$. The space Y_0 is separable, we may denote $Y_0 = \overline{\{y_1, y_2, \ldots\}}$. Define $X_{\lambda_i} = sp\{y_1, y_2, \ldots, y_i\}$, then we have $X_{\lambda_1} \subset X_{\lambda_2} \subset \ldots \subset X_{\lambda_i} \subset \ldots$. The space Z_0 is separable, then the τ-topology in the set $\overline{[0,1] \times A}^\tau$ is metrizable ([6, p. 782]. Denote d_τ this metric. For every $i = 1, 2, \ldots$, we have $(t_0, u_0) \in \overline{V_{\lambda_i}}^\tau$, then we can choose a sequence $\{(t_n^{(i)}, u_n^{(i)})\}_n \subset V_{\lambda_i}$ such that

$$(\forall i \in \mathbb{N}), d_\tau\{(t_n^{(i)}, u_n^{(i)}), (t_0, u_0)\} < \frac{1}{n}.$$

Therefore $(t_n^{(n)}, u_n^{(n)}) \to (t_0, u_0)$ with respect to τ and

$$\langle H(t_n^{(n)}, u_n^{(n)}), u_n^{(n)}\rangle \leq 0, \tag{1}$$

$$\langle H(t_n^{(n)}, u_n^{(n)}), v\rangle = 0 \text{ for all } v \in X_{\lambda_n}. \tag{2}$$

Since H is strongly quasibounded, the sequence $\{H(t_n^{(n)}, u_n^{(n)})\}$ is bounded in Z by (1). Therefore we can choose a subsequence $\{H(t_{n'}^{(n')}, u_{n'}^{(n')})\}$ such that $H(t_{n'}^{(n')}, u_{n'}^{(n')}) \rightharpoonup z \in Z$. Moreover, by (2), we have

$$\langle \chi, v\rangle = 0 \text{ for all } v \in \cup_{n=1}^\infty X_{\lambda_n}.$$

Because $\cup_{n=1}^\infty X_{\lambda_n}$ is norm-dense in Y_0, we have

$$\langle \chi, v\rangle = 0 \text{ for all } v \in Y_0.$$

Therefore $\chi = 0$ as an element of the dual space Z. Moreover, Y_0 is $\sigma(Y, Z)$-dense in Y, which implies that

$$\langle \chi, v \rangle = 0 \text{ for all } v \in Y.$$

Consequently, we have

$$limsup\langle H(t_{n'}^{(n')}, u_{n'}^{(n')}), u_{n'}^{(n')} \rangle \leq 0 = \langle \chi, u_0 \rangle.$$

Since H is a (MOD) homotopy, we have $u_0 \in D_{H_t}$, $u_{n'}^{(n')} \to u_0(mod)$ in Y and $H(t_{n'}^{(n')}, u_{n'}^{(n')}) \to H(t_0, u_0) = 0(mod)$ in Z for a subsequence. Therefor

$$0 \in H([0,1] \times \bar{A}^{mod}) \cap \overline{H([0,1] \times A)}^{mod},$$

which is a contradiction.

The next lemma proves that the degree d_μ will stabilize when we go to the limit.

Lemma 6 *Let $F \in (MOD)$ be a strongly quasibounded and $G \subset Y_0$ open and bounded in Y_0. If*

$$0 \notin F(\overline{\partial_{Y_0} G}^{mod}) \cap \overline{F(\partial_{Y_0} G)}^{mod},$$

then there exists $X_\lambda \in \Lambda$ such that

$$0 \notin F_\mu(\partial_{X_\mu} G_\mu) \text{ and } d_\mu(F_\mu, G_\mu, 0) = constant \text{ for every } X_\mu \supset X_\lambda.$$

Proof The first part follows immediately from Lemma 5.
For the second part, suppose, by contradiction, that

$$\forall X_\lambda \in \Lambda, \exists X_\mu \supset X_\lambda; d_\lambda(F_\lambda, G_\lambda, 0) \neq d_\mu(F_\mu, G_\mu, 0).$$

By Lemma 4, there exists $u_\lambda \in \partial_{Y_0} G$ such that

$$\langle F(u_\lambda), u_\lambda \rangle \leq 0 \text{ and } \langle F(u_\lambda), v \rangle = 0 \text{ for every } v \in X_\lambda.$$

Define a set

$$V_\lambda = u \in \partial_{Y_0} G \mid \langle F(u), u \rangle \leq 0 \text{ and } \langle F(u), v \rangle = 0 \text{ for all } v \in X_\lambda\},$$

which is non-empty. As in the proof of Lemma 5, we can deduce the existence of $u_0 \in \cap \overline{V_\lambda}^{\sigma(Y, Z_0)}$ and $\{u_n\} \subset \partial_{Y_0} G$ such that $u_n \to u_0 \in D_F(mod)$ in Y, $F(u_n) \to F(u_0)(mod)$ in Z for a subsequence and $F(u_0) = 0$.
Consequently, $0 \in F(\overline{\partial_{Y_0} G}^{mod}) \cap \overline{F(\partial_{Y_0} G)}^{mod}$, which is a contradiction.

We can now define a degree function in the complementary system

$$\begin{pmatrix} Y & Z \\ Y_0 & Z_0 \end{pmatrix} = \begin{pmatrix} W_0^{m,p(.)}(\Omega) & W^{-m,p'(.)}(\Omega) \\ H_0^{m,p(.)}(\Omega) & H^{-m,p'(.)}(\Omega) \end{pmatrix}$$

Definition 3 Let $F \in \mathscr{F}$, $G \subset Y_0$ open and bounded in Y_0, $f \in Z_0$ and $f \notin F(\overline{\partial_{Y_0} G}^{mod}) \cap \overline{F(\partial_{Y_0} G)}^{mod}$. We then define

$$d(F, G, f) = \lim_{\lambda} d_\lambda(F_\lambda - \phi_\lambda^*(f), G_\lambda, 0).$$

Theorem 5 *The mapping d in Definition 3 satisfies the conditions* $(C_1) - (C_3)$. *Any mapping* $K \in \mathscr{F}^a$ *satisfying*

$$\langle K(u), u\rangle > 0, \text{ when } u \neq 0, \text{ and } K(0) = 0$$

can be chosen as a normalising map.

Proof It is enough to prove the conditions $(C_1) - (C_3)$ for $f = 0$, because $F - f \in \mathscr{F}, H - f \in \mathscr{H}$ and

$$d(F, G, f) = \lim_{\lambda} d_\lambda(F_\lambda - \phi_\lambda^*(f), G_\lambda, 0) = \lim_{\lambda} d_\lambda((F - f)_\lambda, G_\lambda, 0 = d(F - f), G, 0).$$

(C_1) If $d(F, G, 0) \neq 0$, then there exists $X_\lambda \in \Lambda$ such that $d_\mu(F_\mu, G_\mu, 0) \neq 0$ for all $X_\mu \supset X_\lambda$. Choose a sequence $\{v_{\mu_n}\}$ such that $v_{\mu_n} \in G_{\mu_n}$, $F_{\mu_n}(v_{\mu_n}) = 0$, $\dim X_{\mu_n} \to \infty$ and $\cup X_{\mu_n}$ is dense in Y_0. Chose a subsequence $\{v_{\mu_{n'}}\}$ such that $v_{\mu_{n'}} \rightharpoonup v \in Y$. Since $\langle F(v_{\mu_{n'}}), v_{\mu_{n'}}\rangle = 0$ and F is strongly quasibounded, we have $F(v_{\mu_{n'}}) \rightharpoonup \chi \in Z$ for a subsequence. We immediately see that $\langle \chi, w\rangle = 0$ for all $w \in X_{\mu_n}$ for every n. Therefore, by density, $\langle \chi, w\rangle = 0$ for all $w \in Y_0$, and by the $\sigma(Y, Z)$-density of Y_0 in the space Y, $\langle \chi, v\rangle = 0$. Hence $\chi = 0$ and

$$limsup\langle F(v_{\mu_{n'}}), v_{\mu_{n'}}\rangle = 0 = \langle \chi, v\rangle,$$

implying $v_{\mu_{n'}} \to v \in D_F(mod)$ in Y and $F(v_{\mu_{n'}}) \to F(v)(mod)$ in Z for a further subsequence. Therefore $0 \in F(\bar{G}^{mod}) \cap \overline{F(G)}^{mod}$.
(C_2) If $G_1, G_2 \subset G$ are open and bounded in Y_0, $G_1 \cap G_2 = \emptyset$ and

$$0 \notin F(\overline{\bar{G} \setminus (G_1 \cup G_2)}^{mod}) \cap \overline{F(\bar{G} \setminus (G_1 \cup G_2))}^{mod},$$

then, by Lemma 6, there exists $X_\lambda \subset \Lambda$ such that

$$0 \notin F_\mu(\bar{G}_\mu \setminus (G_{1,\mu} \cup G_{2,\mu})) \text{ for all } X_\mu \supset X_\lambda.$$

Hence

$$\begin{aligned} d(F, G, 0) &= \lim_{\lambda} d_{\lambda}(F_{\lambda}, G_{\lambda}, 0) \\ &= \lim_{\lambda} [d_{\lambda}(F_{\lambda}, G_{1,\lambda}, 0) + d_{\lambda}(F_{\lambda}, G_{2,\lambda}, 0)] \\ &= d(F, G_1, 0) + d(F, G_2, 0). \end{aligned}$$

(C_3) Let $H \in \mathscr{H}$ and $G \subset Y_0$ be open and bounded in Y_0. Suppose that

$$0 \notin H([0, 1] \times \overline{\partial_{Y_0} G}^{mod}) \cap \overline{H([0, 1] \times \partial_{Y_0} G)}^{mod}.$$

By Lemma 5, there exists $X_{\lambda} \in \Lambda$ such that

$$0 \notin H_{\mu}([0, 1] \times \partial G_{\mu}) \text{ for all } X_{\mu} \supset X_{\lambda}.$$

Consequently,

$$d_{\mu}(H_{\mu}(t, .), G_{\mu}, 0) = \textit{constant} \text{ for all } t \in [0, 1], \text{ when } X_{\mu} \supset X_{\lambda}.$$

Let $t_1, t_2 \in [0, 1]$ be arbitrary. Then

$$d_{\mu}(H_{\mu}(t_1, .), G_{\mu}, 0) = d_{\mu}(H_{\mu}(t_2, .), G_{\mu}, 0),$$

and going to the limit we obtain

$$d(H(t_1, .), G, 0) = d(H(t_2, .), G, 0),$$

which means that $d(H(t, .), G, 0) = \textit{constant}$ for all $t \in [0, 1]$.
(C_4) Let $K \in \mathscr{F}^a$ a mapping satisfying

$$\langle K(u), u \rangle > 0, \text{ when } u \neq 0, \text{ and } K(0) = 0.$$

Suppose that $f \notin K(\overline{\partial_{Y_0} G}^{mod}) \cap \overline{K(\partial_{Y_0} G)}^{mod}$ and $f \in K(G)$. Let $u \in G$ be such that $K(u) = f$, and choose $X_{\lambda} \in \Lambda$ such that $u \in X_{\lambda}$. Then $\phi_{\mu}^{*}(f) \in K_{\mu}(G_{\mu})$ for all $X_{\mu} \supset X_{\lambda}$. Moreover, $\langle K_{\mu}(v), v \rangle > 0$ for every $v \in X_{\mu}$, $v \neq 0$. By the basic properties of the Brouwer degree, we have $d_{\mu}(K_{\mu}, G_{\mu}, \phi_{\mu}^{*}(f)) = 1$ for all $X_{\mu} \supset X_{\lambda}$.
Hence $d(K, G, f) = 1$.

3.3 Properties of the Degree Function

Using the conditions $(C_1) - (C_4)$ for the degree function, we can deduce some standard properties.

Proposition 2 *Let* $F, T \in \mathscr{F}^a$, $G \subset Y_0$ *open and bounded in* Y_0, $F/\partial_{Y_0} G = T/\partial_{Y_0} G$ *and* $f \in Z_0$.
If $1 < p^- \leq p(\cdot) \leq p^+ < \infty$ *and* $f \notin F(\partial_Y G)$, *then* $d(F, G, f) = d(T, G, f)$.

Proof Define an affine homotopy $H : D_H \to Z$,

$$H(t, u) = tF(u) + (1 - t)T(u),$$

which belongs ti the class $\mathscr{H}$ by Lemma 2. It is clear that

$$H([0, 1] \times \partial_{Y_0} G) = F(\partial_{Y_0} G).$$

Since $f \notin F(\partial_Y G)$, we have $f \notin H([0, 1] \times \partial_Y G)$. By homotopy invariance,

$$d(F, G, f) = d(T, G, f).$$

Proposition 3 *If* $F \in \mathscr{F}$ *and* $G \subset Y_0$ *is an open and bounded in* Y_0.
If $1 < p^- \leq p(\cdot) \leq p^+ < \infty$, *then* $d(F, G, .)$ *is constant on each open component in* Z_0 *of the open set* $Z_0 \setminus F(\partial_Y G)$.

Proof Let $\Delta \subset Z_0 \setminus F(\partial_Y G)$ be an open component in Z_0 and $f_1, f_2 \in \Delta$ arbitrary. Then there exists a continuous curve $y : [0, 1] \to Z_0$ such that $y(0) = f_1$, $y(1) = f_2$ and $y(t) \in \Delta$ for all $t \in [0, 1]$. Therefore $y(t) \notin F(\partial_Y G)$. We see immediately that $F(u) - y(t) \in \mathscr{H}$ and $0 \notin F(\partial G) \quad y([0, 1])$. By homotopy invariance, $d(F, G, y(0)) = d(F, G, y(1))$ and we have the proof.

Proposition 4 *Let* $F \in \mathscr{F}$, $G \subset Y_0$ *open and bounded in* Y_0 *and* $u_0 \in G$. *Define a mapping* $s : Y_0 \to Y_0$, $s(u) = u - u_0$. *If* $0 \notin F(\overline{\partial_{Y_0} G}^{mod}) \cap \overline{F(\partial_{Y_0} G)}^{mod}$, *then*

$$d(F, G, 0) = d(Fos^{-1}, s(G), 0).$$

Proof Choose $X_{\lambda_0} \in \Lambda$ such that $u_0 \in X_{\lambda_0}$. Now

$$d(F, G, 0) = \lim_{\lambda \geq \lambda_0} d_\lambda(F_\lambda, G_\lambda, 0).$$

By the properties of the Brouwer degree, we have

$$d_\lambda(F_\lambda, G_\lambda, 0) = d_\lambda(F_\lambda os^{-1}, s(G_\lambda), 0) - d_\lambda((Fos^{-1})_\lambda, (s(G))_\lambda), 0).$$

Moreover, it is easy to check that $Fos^{-1} \in \mathscr{F}$, $s(G) \subset Y_0$ is open and bounded in Y_0 and $0 \notin F(\overline{\partial_{Y_0} s(G)}^{mod}) \cap \overline{F(\partial_{Y_0} s(G))}^{mod}$. Therefore

$$d(F, G, 0) = \lim_{\lambda \geq \lambda_0} d_\lambda((Fos^{-1})_\lambda, (s(G))_\lambda), 0) = d(Fos^{-1}, s(G), 0).$$

References

1. L.E. Brouwer, Uber Abbildung von Mannigfaltigkeiten. Math. Ann. **71**, 97–115 (1912)
2. J. Leray, J. Schauder, Topologie et equationes fonctionnelles. Ann. Sci. Ec. Norm. Super. **51**, 45–78 (1934)
3. K. Deimling, *Nonlinar Functional Analysis* (Springer, Berlin, 1985)
4. I.V. Skrypnik, *Nonlinear Higher Order Elliptic Equations* (Naukova Dumka, Kiev, 1973). (in Russian)
5. I.V. Skrypnik, Methods for analysis of nonlinear elliptic boundary value problems. Am. Math. Soc. Transl., Ser. II, **139** (1994) (AMS, Providence)
6. E. Zeidler, *Nonlinear Functional Analysis and Its Applications I: Fixed-Point-Theorems* (Springer, New York, 1985)
7. F.E. Browder, Fixed point theory and nonlinear problems. Bull. Am. Math. Soc. **9**, 1–39 (1983)
8. F.E. Browder, Degree of mapping for nonlinear mappings of monotone type. Proc. Natl. Acad. Sci. USA **80**, 1771–1773 (1983)
9. J. Berkovits, On the degree theory for nonlinear mappings of monotone type. Ann. Acad. Sci. Fenn. Ser. A I Math. Dissertationes **58** (1986)
10. J. Berkovits, V. Mustonen, On topological degree for mappings of monotone type. Nonlinear Anal. **10**, 1373–1383 (1986)
11. J. Berkovits, V. Mustonen, Nonlinear mappings of monotone type I. Classification and degree theory. Preprint No 2/88, Mathematics, University of Oulu
12. L. Dingien, P. Harjulehto, P. Hästö, M. Ruzicka, *Lebesgue and Sobolev Spaces with Variable Exponents* (Springer, Berlin, 2011)
13. O. Kováčik, J. Rákosník, On spaces $L^{p(x)}$ and $W^{1,p(x)}$. Czechoslovak Math. J. **41**, 592–618 (1991)
14. J.P. Gossez, Nonlinear elliptic boundary value problems for equations with rapidly (or slowly) increasing coefficients. Trans. Am. Math. Soc. **190**, 163–205 (1974)
15. L. Fuhrer, Ein elementarer analytischer Beweis zur Eindeutigkeit des A bbildungsgrades im Rn. Math. Nachr. **54**, 259–267 (1972)
16. H. Amann, S. Weiss, On the uniqueness of the topological degree. Math. Z. **130**, 39–5 (1973)
17. L. Narici, E. Beckenstein, *Topological Vector Spaces* (Marcel Dekker Inc, New York, 1985)

Existence of Positive Solutions of Nonlinear Fractional Quadratic Differential Equations

K. Hilal, Y. Allaoui and K. Guida

Abstract In this work, we prove the existence as well as approximations of the positive solutions for an initial value problem of nonlinear fractional quadratic differential equations. We use some properties of the Mittag-Leffler functions and its relationship with fractional calculus. Also we obtain some results regarding the existence of positive solutions using the Dhage iterative method embodied in a recent hybrid fixed point theorem of Dhage in partially ordered normed linear spaces.

1 Introduction

Fractional differential equations have received increasing attention during recent years due to their application in various fields of science and engineering, such as viscoelasticity, electrochemistry, porous media and electromagnetism [1–4].
For more details on this theory and application, we refer the readers to Podlubny [5], Miller and Ross [6], Kilbas et al. [7] and Zhou [8].

Very recently, the study of existence and approximation of the solutions for the hybrid differential equations is initiated in Dhage [9] and Dhage et al. [10, 11] via hybrid fixed point theory.
Dhage and Dhage [9] discussed the following quadratic differential equations.

$$\begin{cases} \dfrac{d}{dt}\left[\dfrac{x(t)}{f(t,x(t))}\right] + \lambda\left[\dfrac{x(t)}{f(t,x(t))}\right] = g(t,x(t)), & t \in J = [t_0, t_0 + a] \\ x(t_0) = x_0 \in \mathbb{R}, \end{cases}$$

K. Hilal · Y. Allaoui (✉) · K. Guida
Sultan Moulay Slimane University, BP 523 Beni Mellal, Morocco
e-mail: youssefbenlarbi1990@gmail.com

K. Hilal
e-mail: hilal.khalid@yahoo.fr

K. Guida
e-mail: guida.karim@gmail.com

S. Melliani and O. Castillo (eds.), *Recent Advances in Intuitionistic Fuzzy Logic Systems*, Studies in Fuzziness and Soft Computing 372,
https://doi.org/10.1007/978-3-030-02155-9_2

for $\lambda \in \mathbb{R}^+$ where $f : J \times \mathbb{R} \longrightarrow \mathbb{R}^*$ and $g : J \times \mathbb{R} \longrightarrow \mathbb{R}$ are a continuous functions. They established the existence of positive solutions using the Dhage iterative method.

From this work, we develop the theory of fractional quadratic differential equations involving Caputo differential operators of order $0 < \alpha < 1$. We prove the existence of positive solutions of the following fractional quadratic differential equations (for short FQDE):

$$D^\alpha\left[\frac{x(t)}{f(t,x(t))}\right] - \lambda\left[\frac{x(t)}{f(t,x(t))}\right] = g(t,x(t)), \quad t \in J = [0,1] \tag{1}$$

$$x(0) = x_0 \in \mathbb{R}, \tag{2}$$

where D^α denotes the Caputo fractional derivative of order α, where $0 < \alpha < 1$, $\lambda > 0$, $f : J \times \mathbb{R} \longrightarrow \mathbb{R}^*$ is a continuous function and $g : J \times \mathbb{R} \longrightarrow \mathbb{R}$ is a L^1- *Caratheodory* function.

We note that the (FQDE) (1)–(2) with $\lambda = 0$ is considered by Hilal and Kajouni [12, 13].

The rest of this paper is organized as follows: In Sect. 2, we give some preliminaries which are used in the sequel. In Sect. 3 we prove our main results.

2 Preliminaries

In this section, we introduce some definitions and results which are used throughout this paper.

Definition 1 [14] Let $x \in C^n[0,\infty)$ and $n-1 < \alpha < n$, where $n \in \mathbb{N}^*$, the Caputo's derivative of order α for function $x : [0,\infty) \longrightarrow \mathbb{R}$ can be written as

$$D^\alpha x(t) = \frac{1}{\Gamma(n-\alpha)} \int_0^t (t-s)^{n-\alpha-1} x^{(n)}(s)ds. \tag{3}$$

Definition 2 [14] The fractional integral of order α is defined as

$$I^\alpha x(t) = \frac{1}{\Gamma(\alpha)} \int_0^t (t-s)^{\alpha-1} x(s)ds, \tag{4}$$

which is called the Riemann-Liouville integral.

We give a useful lemma which plays an important role in the fractional calculus.

Definition 3 [15, 16] Let $\alpha, \beta > 0$. The two-parameters Mittag-Leffler function $E_{\alpha,\beta}(t)$ is defined by the series expansion

$$E_{\alpha,\beta}(t) = \sum_{i=0}^{\infty} \frac{t^i}{\Gamma(\alpha i + \beta)}.$$

Especially, if $\beta = 1$, $E_{\alpha,1}(t)$ becomes the one-parameter Mittag-Leffler function $E_\alpha(t)$, i.e., $E_{\alpha,1}(t) = E_\alpha(t)$.

Lemma 1 [16] *Let $0 < \alpha < 1$, and $\lambda \in \mathbb{R}$. Then for any $t \in [0, T]$, we have*

(1) Let $0 < \alpha < 1$, and $K, U \in \mathbb{R}^{n\times n}$. Then for any $t \in [0, T]$, it has

$$\int_0^t (t-\tau)^{\alpha-1} E_{\alpha,\alpha}(-K(t-\tau)^\alpha) U (D^\alpha x)(\tau) d\tau =$$
$$Ux(t) - E_\alpha(\lambda^\alpha) Ux(0) - K \int_0^t (t-s)^{\alpha-1} E_{\alpha,\alpha}(-K(t-s)^\alpha) Ux(s) ds$$

(2)

$$D_0^\alpha \left[\int_0^t f(t-s) g(s) ds \right](t) = \int_0^t D_0^\alpha [f(t)](s) g(t-s) ds + g(t) \lim_{t\to 0^+} [{}_t I_{0^+}^{1-\alpha} f](t).$$

Lemma 2 [7, 15]

(a) $D_{a^+}^\alpha E_\alpha[\lambda(t-a)^\alpha](x) = \lambda E_\alpha[\lambda(x-a)^\alpha], \quad (Re(\alpha) > 0, \lambda \in \mathbb{C})$

(b) $\left(I_{a+}^{\alpha'} (t-a)^{\beta-1} E_{\mu,\beta}[\lambda(t-a)^\mu]\right)(x) = (x-a)^{\alpha'+\beta-1} E_{\mu,\alpha'+\beta}[\lambda(x-a)^\mu],$

with $\alpha' > 0$, $\beta > 0$ and $\mu > 0$.

(c) $\int_0^z t^{\beta-1} E_{\alpha,\beta}(\lambda t^\alpha) dt = z^\beta E_{\alpha,\beta+1}(\lambda z^\alpha)$.

(d) $|E_{\alpha,\beta}(z)| \le C_1 exp(\sigma |z|^\rho)$, *for all $\sigma > 1$ and* $\rho = \dfrac{1}{Re(\alpha)}$.

The following definitions are useful in the sequel.
We conserve the same definitions given in Dhage [9].
Let E denotes a partially ordered real-normed linear space with an order relation $\preceq$ and the norm $\| \ . \ \|$.

Definition 4 [9] A mapping $\mathscr{T} : E \longrightarrow E$ is called non decreasing if it preserves the order relation $\preceq$, that is if $x \preceq y$ implies $\mathscr{T}x \preceq \mathscr{T}y$ for all $x, y \in E$.

Definition 5 [9] A mapping $\mathscr{T} : E \longrightarrow E$ is called partially continuous at a point $a \in E$ if for $\varepsilon > 0$ there exists a $\delta > 0$ such that $\| \mathscr{T}x - \mathscr{T}a \| < \varepsilon$ whenever x is comparable to a and $\| x - a \| < \delta$. $\mathscr{T}$ called partially continuous on E if it is partially continuous at every point of it.It is clear that if $\mathscr{T}$ is partially continuous on E, then it is continuous on every chain C contained in E.

Definition 6 [9] A mapping $\mathscr{T} : E \longrightarrow E$ is called partially bounded if $\mathscr{T}(C)$ is bounded for every chain C in E. $\mathscr{T}$ is uniformly partially bounded if all chain $\mathscr{T}(C)$ in E are bounded by a unique constant.
$\mathscr{T}$ is called bounded if $\mathscr{T}(E)$ is a bounded subset of E.

Definition 7 [9] A mapping $\mathscr{T} : E \longrightarrow E$ is called partially compact if $\mathscr{T}(C)$ is a relatively compact subset of E for all totally ordered sets or chain C in E. $\mathscr{T}$ is uniformly partially compact if $\mathscr{T}(C)$ is a uniformly partially bounded and partially compact on E. $\mathscr{T}$ is called partially totally bounded if for any totally ordered and bounded subset C of E, $\mathscr{T}(E)$ is a relatively compact subset of E. If $\mathscr{T}$ is partially continuous and partially totally bounded, then it is called partially completely continuous on E.

Definition 8 [9] An upper semi-continuous and nondecreasing function $\psi : \mathbb{R}^+ \longrightarrow \mathbb{R}^+$ is called a $\mathscr{D}$ *– function*, provided $\psi(0) = 0$. Let $(E, \leq, \| \cdot \|)$ be a partially ordered normed linear space. A mapping $\mathscr{T} : E \longrightarrow E$ is called partially nonlinear $\mathscr{D}$ *– Lipschitz* if there exists a $\mathscr{D}$ *– function* $\psi : \mathbb{R}^+ \longrightarrow \mathbb{R}^+$ such that

$$\| \mathscr{T}x - \mathscr{T}y \| \leq \psi(\| x - y \|), \tag{5}$$

for all comparable elements $x, y \in E$. If $\psi(r) = kr$, $k > 0$, then $\mathscr{T}$ is called a partially Lipschitz with a Lipschitz constant k.

Let $(E, \leq, \| \cdot \|)$ be a partially ordered normed linear algebra. We denote

$$E^+ = \{x \in E \mid x \geq \theta, \textit{ where } \theta \textit{ is the zero element of } E\},$$

$$\mathscr{K} = \{E^+ \subset E \mid uv \in E^+ \textit{ for all } u, v \in E^+\},$$

and $\mathscr{P}_{ch}(E)$ is the set of all subsets of E.

The elements of the set $\mathscr{K}$ are called the positive vectors in E. The following lemma follows immediately from the definition of the set $\mathscr{K}$, which is oftentimes used in the hybrid fixed point theory of Banach algebras and applications to nonlinear differential and integral equations.

Lemma 3 [9] *If* $u_1, u_2, v_1, v_2 \in \mathscr{K}$ *are such that* $u_1 \leq v_1$ *and* $u_2 \leq v_2$ *then* $u_1u_2 \leq v_2v_2$.

Definition 9 [9] An operator $T : E \rightarrow E$ is said to be positive if the range $R(T)$ of T is such that $R(T) \subseteq \mathscr{K}$.

Theorem 1 [9] *Let* $(E, \leq, \| \cdot \|)$ *be a regular partially ordered complete normed linear algebra such that the order relation* $\leq$ *and the norm* $\| \cdot \|$ *in E are compatible in every compact chain of E. Let* $\mathscr{A}, \mathscr{B} \longrightarrow \mathscr{K}$ *be two nondecreasing operators such that*

(a) $\mathscr{A}$ *is partially bounded and partially nonlinear* $\mathscr{D}$ *– Lipschitz with* $\mathscr{D}$ *– function* $\psi_{\mathscr{A}}$,
(b) $\mathscr{B}$ *is partially continuous and uniformly partially compact,*
(c) $M\psi_{\mathscr{A}}(r) < r$, $r > 0$, *where* $M = \sup\{\| \mathscr{B}(C) \| : C \in \mathscr{P}_{ch}(E)\}$, *and*
(d) there exists an element $x_0 \in X$ *such that* $x_0 \leq \mathscr{A}x_0\mathscr{B}x_0$ *or* $x_0 \geq \mathscr{A}x_0\mathscr{B}x_0$

then the operator equation

$$\mathscr{A}x\mathscr{B}x = x$$

has a positive solution x^* *in E and the sequence* $\{x_n\}$ *of successive iterations defined by* $x_{n+1} = \mathscr{A}x_n\mathscr{B}x_n$, $n = 0, 1, \ldots$; *converges monotonically to* x^*.

Let $\mathscr{C}(J \times \mathbb{R}, \mathbb{R})$ be the class of functions $g : J \times \mathbb{R} \longrightarrow \mathbb{R}$ such that

(i) the map $t \longrightarrow g(t, x)$ is measurable for each $x \in \mathbb{R}$, and
(ii) the map $x \longrightarrow g(t, x)$ is continuous for each $t \in J$.

The class $\mathscr{C}(J \times \mathbb{R}, \mathbb{R})$ is called the Caratheodory class of functions on $J \times \mathbb{R}$ which are Lebesgue integrable when bounded by a Lebesgue integrable function on J.

$L^1(J, \mathbb{R})$ denotes the space of Lebesgue integrable real-valued functions on J equipped with the norm $\| \,.\, \|_{L^1}$ defined by

$$\| x \|_{L^1} = \int_0^1 | x(s) | \, ds$$

3 Main Results

Definition 10 By a solution of the FQDE, we mean a function $x \in C^1(J, \mathbb{R})$ that satisfies

(i) $t \longrightarrow \frac{x}{f(t,x)}$ is a continuously differentiable function for each $x \in \mathbb{R}$ and
(ii) x satisfies the equations (1), (2) on J.

We consider the FQDE in the space $C(J, \mathbb{R})$ of continuous real-valued functions defined on J. We define a norm $\| \,.\, \|$ and the order relation $\leq$ in $C(J, \mathbb{R})$ by

$$\| x \| = \sup_{t \in J} |x(t)|, \tag{6}$$

and

$$x \leq y \Leftrightarrow x(t) \leq y(t) \quad \text{for all} \quad t \in J. \tag{7}$$

Clearly, $C(J, \mathbb{R})$ is a Banach algebra with respect to above supremum norm and is also partially ordered with respect to the above partially order relation $\leq$. It is known that the partially ordered Banach algebra $C(J, \mathbb{R})$ has some nice properties with respect to the above order relation in it. The following lemma follows by an application of Arzela-Ascoli theorem.

Lemma 4 [9] *Let* $(C(J, \mathbb{R}), \leq, \| \,.\, \|)$ *be a partially ordered Banach space with the norm* $\| \,.\, \|$ *and the order relation* $\leq$ *defined by (6)–(7), respectively. Then,* $\| \,.\, \|$ *and* $\leq$ *are compatible in every partially compact subset of* $C(J, \mathbb{R})$.

Definition 11 A function $u \in C^1(J, \mathbb{R})$ is said to be a lower solution of the FQDE if the function $t \longrightarrow \frac{u(t)}{f(t,u(t))}$ is continuously differentiable and satisfies

$$D^\alpha \left[\frac{u(t)}{f(t, u(t))} \right] - \lambda \left[\frac{u(t)}{f(t, u(t))} \right] \leq g(t, u(t)), \quad t \in J$$
$$u(0) \leq x_0.$$

Similarly, a function $v \in C^1(J, \mathbb{R})$ is said to be upper solution of FQDE if satisfies the above property and inequalities with reverse sign.

Consider the following assumptions:

(A_0) The map $x \longrightarrow \frac{x}{f(t,x)}$ is increasing in $\mathbb{R}$ for each $t \in J$.
$(A_1) f$ defines a function $f : J \times \mathbb{R} \longrightarrow \mathbb{R}^+$.
(A_2) There exists a constant $M_f > 0$ such that $0 < f(t, x) < M_f$ for all $t \in J$ and $x \in \mathbb{R}$.
(A_3) There exists a $\mathscr{D}$ – *function* ϕ, such that

$$0 \leq f(t, x) - f(t, y) \leq \phi(x - y),$$

for all $t \in J$ and $x, y \in \mathbb{R}, x \geq y$.
(B_1) g defines a function $g : J \times \mathbb{R} \longrightarrow \mathbb{R}^+$.
(B_2) There exists a function $h \in L^1(J, \mathbb{R})$ such that $g(t, x) \leq h(t)$ for all $t \in J$ and $x \in \mathbb{R}$.
(B_3) $g(t, x)$ is nondecreasing in x for all $t \in J$.
(B_4) The FQDE has a lower solution $u \in C^1(J, \mathbb{R})$.

Lemma 5 *Suppose that hypothesis A_0 hold. Then a function $x \in C^1(J, \mathbb{R})$ is a solution of the FQDE*

$$D^\alpha \left[\frac{x(t)}{f(t, x(t))} \right] - \lambda \left[\frac{x(t)}{f(t, x(t))} \right] = g(t) \tag{8}$$
$$x(0) = x_0 \in \mathbb{R}, \tag{9}$$

if and only if it is a solution of the nonlinear integral equation

$$x(t) = f(t, x(t)) \left(\frac{x_0}{f(0, x_0)} E_\alpha(\lambda t^\alpha) + \int_0^t (t-s)^{\alpha-1} E_{\alpha,\alpha}(\lambda(t-s)^\alpha) g(s) ds \right),$$

for all $t \in J$.

Proof Consider the problem $D^\alpha \left[\frac{x(t)}{f(t,x(t))} \right] - \lambda \left[\frac{x(t)}{f(t,x(t))} \right] = g(t)$.
By the relations (4.1.65) and (4.1.66) in [7], the solution fo the integral equation is

$$\frac{x(t)}{f(t, x(t))} = \int_0^t (t-s)^{\alpha-1} E_{\alpha,\alpha}(\lambda(t-s)^\alpha) g(s) ds + \frac{x(0)}{f(0, x(0))} E_\alpha(\lambda t^\alpha). \tag{10}$$

According to the condition (2), we get

$$x(t) = f(t, x(t))\left(\frac{x_0}{f(0, x_0)}E_\alpha(\lambda t^\alpha) + \int_0^t (t-s)^{\alpha-1}E_{\alpha,\alpha}(\lambda(t-s)^\alpha)g(s)ds\right).$$

Conversely, if we have

$$x(t) = f(t, x(t))\left(\frac{x_0}{f(0, x_0)}E_\alpha(\lambda t^\alpha) + \int_0^t (t-s)^{\alpha-1}E_{\alpha,\alpha}(\lambda(t-s)^\alpha)g(s)ds\right),$$

then

$$D^\alpha\left(\frac{x(t)}{f(t, x(t))}\right) = D^\alpha\left(\int_0^t (t-s)^{\alpha-1}E_{\alpha,\alpha}(\lambda(t-s)^\alpha)g(s)ds\right) + D^\alpha\left(\frac{x_0}{f(0, x_0)}E_\alpha(\lambda t^\alpha)\right)$$

Substituting $x = t$ and $a = 0$ in the relation (a) in Lemma 2.2, we have

$$D^\alpha E_\alpha(\lambda t^\alpha) = \lambda E_\alpha(\lambda t^\alpha)$$

This implies that

$$D^\alpha(\frac{x_0}{f(0, x_0)}E_\alpha(\lambda t^\alpha)) = \frac{x_0}{f(0, x_0)}\lambda E_\alpha(\lambda t^\alpha)$$

Setting $F(u) := u^{\alpha-1}E_{\alpha,\alpha}(\lambda u^\alpha)$.
According to the relation (2) in Lemma 2.1, we have

$$D^\alpha\left[\int_0^t F(t-s)g(s)ds\right] = \int_0^t D^\alpha[s^{\alpha-1}E_{\alpha,\alpha}(\lambda s^\alpha)]g(t-s)ds] + g(t)\lim_{t\to 0^+} I^{1-\alpha}F(t)$$
$$= \int_0^t \lambda s^{\alpha-1}E_{\alpha,\alpha}(\lambda s^\alpha)g(t-s)ds] + g(t)\lim_{t\to 0^+} I^{1-\alpha}F(t)$$

Setting $t - s = u$, then we have

$$D^\alpha\left[\int_0^t F(t-s)g(s)ds\right](t) = \int_0^t \lambda(t-u)^{\alpha-1}E_{\alpha,\alpha}(\lambda(t-u)^\alpha)g(u)du + g(t)\lim_{t\to 0^+} I^{1-\alpha}F(t)$$

In the other hand, *For* $\alpha' = 1-\alpha$, $\mu = \beta = \alpha$, $a = 0$ and $t = x$, in the relation (b), Lemma 2.2, we have

$$I^{1-\alpha}t^{\alpha-1}E_{\alpha,\alpha}\left[\lambda t^\alpha\right] = E_{\alpha,1}\left[\lambda t^\alpha\right]$$

wish implies

$$g(t) \lim_{t \to 0^+} I^{1-\alpha} F(t) = g(t) \lim_{t \to 0^+} E_{\alpha,1} \left[\lambda t^\alpha\right] = g(t)$$

In conclusion

$$\begin{aligned} D^\alpha \left(\frac{x(t)}{f(t, x(t))}\right) &= \frac{x_0}{f(0, x_0)} \lambda E_\alpha(\lambda t^\alpha) \\ &+ \int_0^t \lambda (t-u)^{\alpha-1} E_{\alpha,\alpha}(\lambda (t-s)^\alpha) g(s) ds + g(t) \end{aligned}$$

$$D^\alpha \left(\frac{x(t)}{f(t, x(t))}\right) - \lambda \left(\frac{x(t)}{f(t, x(t))}\right) = g(t)$$

Finally, for $t = 0$ in the relation (10) and by the hypothesis (A_0) which gives us $x(0) = x_0$ we get

$$\frac{x(0)}{f(0, x(0))} = \frac{x_0}{f(0, x_0)},$$

This completes the proof.

Theorem 2 *Assume that hypothesis* $(A_0) - (A_3)$ *and* $(B_1) - (B_4)$ *hold. Furthermore, assume that*

$$\left(C_1 exp(2\lambda^\rho) \left(\| h \|_{L^1} + \left|\frac{x_0}{f(0, x_0)}\right|\right)\right) \phi(r) < r, r > 0,$$

then, the FQDE has a positive solution x^* *defined on* J *and the sequence* $\{x_n\}_{n=1}^\infty$ *of successive approximations defined by*

$$\begin{aligned} x_{n+1}(t) = [f(t, x_n(t)] &\left(\frac{x_0}{f(0, x_0)} E_\alpha(\lambda t^\alpha)\right. \\ &\left. + \int_0^t (t-s)^{\alpha-1} E_{\alpha,\alpha}(\lambda (t-s)^\alpha) g(s, x_n(s)) ds\right), \end{aligned} \tag{11}$$

for $t \in \mathbb{R}$, *where* $x_1 = u$, *converges monotonically to* x^*.

Proof Set $E = C(J, \mathbb{R})$, then, by Lemma 3.1, every compact chain in E possesses the compatibility property with respect to the norm $\| \ . \ \|$ and the order relation $\leq$ in E. By an application of Lemma 3.2, the FQDE (1), (2) is equivalent to the nonlinear integral equation

$$\begin{aligned} x(t) = [f(t, x(t)] &\left(\frac{x_0}{f(0, x_0)} E_\alpha(\lambda t^\alpha)\right. \\ &\left. + \int_0^t (t-s)^{\alpha-1} E_{\alpha,\alpha}(\lambda (t-s)^\alpha) g(s, x(s)) ds\right). \end{aligned} \tag{12}$$

Define two operators $\mathscr{A}$ and $\mathscr{B}$ on E by

$$(\mathscr{A}x)(t) = f(t, x(t)), \quad t \in J$$

$$(\mathscr{B}x)(t) = \frac{x_0}{f(0, x_0)} E_\alpha(\lambda t^\alpha) + \int_0^t (t-s)^{\alpha-1} E_{\alpha,\alpha}(\lambda(t-s)^\alpha) g(s, x(s)) ds$$

We shall show that the operators $\mathscr{A}$ and $\mathscr{B}$ satisfy all the conditions of the Theorem 2.1. This is achieved in the series of following steps.

Step I: $\mathscr{A}$ and $\mathscr{B}$ are nondecreasing on E.

Let $x, y \in E$ be such that $x \geq y$. Then by hypothesis (A_3), we obtain

$$(\mathscr{A}x)(t) = f(t, x(t)) \geq f(t, y(t)) = (\mathscr{A}y)(t)$$

for all $t \in J$. This shows that $\mathscr{A}$ is nondecreasing operator on E into E. Similarly using hypothesis (B_3), it is shown that the operator $\mathscr{B}$ is also nondecreasing on E into itself. Thus, $\mathscr{A}$ and $\mathscr{B}$ are nondecreasing positive operators on E into itself.

Step II: $\mathscr{A}$ is partially bounded and partially $\mathscr{D}$-Lipschitz on E.

Let $x \in E$ be arbitrary. Then by (A_2), $|(\mathscr{A}x)(t)| = f(t, x(t)) \leq M_f$, for all $t \in J$.

Taking supremum over t, we obtain $\|\mathscr{A}x\| \leq M_f$ and so, $\mathscr{A}$ is bounded. This further implies that A is partially bounded on E.

Let $x, y \in E$ be such that $x \geq y$. Then,

$$|(Ax)(t) - (Ay)(t)| = f(t, x(t)) - f(t, y(t)) \leq \phi(|x(t) - y(t)|) \leq \phi(\|x - y\|)$$

for all $t \in J$. Taking supremum over t, we obtain

$$\|\mathscr{A}x - \mathscr{A}y\| \leq \phi(\|x - y\|) \text{ for all } x, y \in E, \quad x \geq y$$

Hence, $\mathscr{A}$ is partially nonlinear $\mathscr{D}$-lipschitz on E which further implies that $\mathscr{A}$ is partially continuous on E.

Step III: $\mathscr{B}$ is partially continuous on E.

Let $\{x_n\}_{n\in\mathbb{N}}$ be a sequence in a chain C of E such that $x_n \longrightarrow x$. Then, by the Lebesgue dominated convergence theorem, for all $t \in J$.

$$\begin{aligned}\lim_{n\to\infty}(\mathscr{B}x_n)(t) &= \lim_{n\to\infty}\int_0^t (t-s)^{\alpha-1}E_{\alpha,\alpha}(\lambda(t-s)^\alpha)g(s,x_n(s))ds + \lim_{n\to\infty}\frac{x_0}{f(0,x_0)}E_\alpha(\lambda t^\alpha)\\ &= \int_0^t (t-s)^{\alpha-1}E_{\alpha,\alpha}(\lambda(t-s)^\alpha)\lim_{n\to\infty} g(s,x_n(s))ds + \frac{x_0}{f(0,x_0)}E_\alpha(\lambda t^\alpha)\\ &= \int_0^t (t-s)^{\alpha-1}E_{\alpha,\alpha}(\lambda(t-s)^\alpha)g(s,x(s))ds + \frac{x_0}{f(0,x_0)}E_\alpha(\lambda t^\alpha)\\ &= (\mathscr{B}x)(t).\end{aligned}$$

This shows that $\mathscr{B}x_n$ converges monotonically to $\mathscr{B}x$ pointwise on J.
Next, we will prove that $\{\mathscr{B}x_n\}$ is an equicontinuous sequence of functions in E. Let $t_1, t_2 \in J$ with $t_1 < t_2$.
Setting $\psi(t) = \int_0^t h(s)ds$. Then,

$$\begin{aligned}&|(\mathscr{B}x_n)(t_2) - (\mathscr{B}x_n)(t_1)| \leq\\ &\leq \left|\int_0^{t_2}(t_2-s)^{\alpha-1}E_{\alpha,\alpha}(\lambda(t_2-s)^\alpha)g(s,x_n(s))ds\right.\\ &\quad\left.-\int_0^{t_1}(t_1-s)^{\alpha-1}E_{\alpha,\alpha}(\lambda(t_1-s)^\alpha)g(s,x_n(s))ds\right|\\ &\quad+\left|E_\alpha(\lambda t_2^\alpha) - E_\alpha(\lambda t_1^\alpha)\right|\left|\frac{x_0}{f(0,x_0)}\right|\end{aligned}$$

$$\begin{aligned}&|(\mathscr{B}x_n)(t_2) - (\mathscr{B}x_n)(t_1)| \leq\\ &\leq \left|\int_0^{t_1}\left[(t_2-s)^{\alpha-1}E_{\alpha,\alpha}(\lambda(t_2-s)^\alpha) - (t_1-s)^{\alpha-1}E_{\alpha,\alpha}(\lambda(t_1-s)^\alpha)\right]g(s,x_n(s))ds\right|\\ &+\left|\int_{t_1}^{t_2}(t_2-s)^{\alpha-1}E_{\alpha,\alpha}(\lambda(t_2-s)^\alpha)g(s,x_n(s))ds\right|\\ &+\left|E_\alpha(\lambda t_2^\alpha) - E_\alpha(\lambda t_1^\alpha)\right|\left|\frac{x_0}{f(0,x_0)}\right|\end{aligned}$$

$$\begin{aligned}&|(\mathscr{B}x_n)(t_2) - (\mathscr{B}x_n)(t_1)| \leq\\ &\leq \left|\int_0^{t_1}\left|(t_2-s)^{\alpha-1}E_{\alpha,\alpha}(\lambda(t_2-s)^\alpha) - (t_1-s)^{\alpha-1}E_{\alpha,\alpha}(\lambda(t_1-s)^\alpha)\right||g(s,x_n(s))|\,ds\right|\\ &+\left|\int_{t_1}^{t_2}(t_2-s)^{\alpha-1}E_{\alpha,\alpha}(\lambda(t_2-s)^\alpha)\,|g(s,x_n(s))|\,ds\right| + \left|E_\alpha(\lambda t_2^\alpha) - E_\alpha(\lambda t_1^\alpha)\right|\left|\frac{x_0}{f(0,x_0)}\right|\end{aligned}$$

$$|(\mathscr{B}x_n)(t_2) - (\mathscr{B}x_n)(t_1)| \leq$$
$$\leq \| h \|_{L^1} \int_0^{t_1} (t_2 - s)^{\alpha-1} E_{\alpha,\alpha}(\lambda(t_2 - s)^\alpha) - (t_1 - s)^{\alpha-1} E_{\alpha,\alpha}(\lambda(t_1 - s)^\alpha) ds$$
$$+ C_1 exp(2\lambda^\rho) \int_{t_1}^{t_2} h(s)ds + \left|E_\alpha(\lambda t_2^\alpha) - E_\alpha(\lambda t_1^\alpha)\right| \left|\frac{x_0}{f(0, x_0)}\right|.$$

In the other hand, we have

$$\begin{aligned}\int_0^{t_1} (t_2 - s)^{\alpha-1} E_{\alpha,\alpha}(\lambda(t_2 - s)^\alpha) ds &= -\int_{t_2}^{t_2-t_1} u^{\alpha-1} E_{\alpha,\alpha}(\lambda(u)^\alpha) du \\ &= \int_{t_2-t_1}^{t_2} u^{\alpha-1} E_{\alpha,\alpha}(\lambda(u)^\alpha) du \\ &= \int_0^{t_2} u^{\alpha-1} E_{\alpha,\alpha}(\lambda(u)^\alpha) du \\ &\quad - \int_0^{t_2-t_1} u^{\alpha-1} E_{\alpha,\alpha}(\lambda(u)^\alpha) du\end{aligned}$$

According to the relation (c) in Lemma 2.2, with $\beta = \alpha$, we obtain

$$\int_0^{t_1} (t_2 - s)^{\alpha-1} E_{\alpha,\alpha}(\lambda(t_2 - s)^\alpha) ds =$$
$$t_2^\alpha E_{\alpha,\alpha+1}(\lambda t_2^\alpha) - (t_2 - t_1)^\alpha E_{\alpha,\alpha+1}(\lambda(t_2 - t_1)^\alpha \tag{13}$$

Similarly, we have:

$$\int_0^{t_1} (t_1 - s)^{\alpha-1} E_{\alpha,\alpha}(\lambda(t_1 - s)^\alpha) ds = t_1^\alpha E_{\alpha,\alpha+1}(\lambda t_1^\alpha). \tag{14}$$

Finally,

$$\begin{aligned}\left|(\mathscr{B}x_n)(t_2) - (\mathscr{B}x_n)(t_1)\right| &\leq \left|E_\alpha(\lambda t_2^\alpha) - E_\alpha(\lambda t_1^\alpha)\right| \left|\frac{x_0}{f(0, x_0)}\right| \\ &+ \| h \|_{L^1} \Big[t_2^\alpha E_{\alpha,\alpha+1}(\lambda t_2^\alpha) - t_1^\alpha E_{\alpha,\alpha+1}(\lambda t_1^\alpha) \\ &+ (t_2 - t_1)^\alpha E_{\alpha,\alpha+1}(\lambda(t_2 - t_1)^\alpha)\Big] + C_1 exp(2\lambda^\rho)|\psi(t_2) - \psi(t_1).|\end{aligned}$$

This implies that

$$|(\mathscr{B}x_n)(t_2) - (\mathscr{B}x_n)(t_1)| \to 0 \quad \text{as} \quad t_2 - t_1 \to 0$$

Uniformly for all n $\in \mathbb{N}$. This shows that the convergence $\mathscr{B}x_n \to Bx$ is uniform and hence $\mathscr{B}$ is partially continuous on E.

Step IV: $\mathscr{B}$ is uniformly partially compact operator on E.

Let C be an arbitrary chain in E. We prove that $\mathscr{B}(C)$ is a uniformly bounded and equicontinuous set in E. First, we prove that $\mathscr{B}(C)$ is uniformly bounded. Let $y \in \mathscr{B}(C)$ be any element. Then, there is an element $x \in C$, such that $y = \mathscr{B}x$. By hypothesis (B_2), for all $t \in J$,

$$\begin{aligned}|y(t)| &\leq \left|\int_0^t (t-s)^{\alpha-1} E_{\alpha,\alpha}(\lambda(t-s)^\alpha) g(s, x(s)) ds\right| + \left|\frac{x_0}{f(0, x_0)} E_\alpha(\lambda t^\alpha)\right| \\ &\leq C_1 exp(2\lambda^\rho) \int_0^1 h(s) ds + \left|\frac{x_0}{f(0, x_0)}\right| C_1 exp(2\lambda^\rho) \\ &\leq C_1 exp(2\lambda^\rho) \left(\| h \|_{L^1} + \left|\frac{x_0}{f(0, x_0)}\right|\right) = M, \end{aligned} \tag{15}$$

Thanks to the relation (d) in Lemma 2.2.
Taking supremum over t, we obtain

$$\| y \| = \| Bx \| \leq M \quad \forall y \in \mathscr{B}(C)$$

Hence, $\mathscr{B}(C)$ is a uniformly bounded subset of E. Moreover, $\| \mathscr{B}(C) \| \leq M$ for all chains C in E. Hence, $\mathscr{B}$ is a uniformly partially bounded operator on E.
Next, we will prove that $\mathscr{B}(C)$ is an equicontinuous set in E. Let $t_1, t_2 \in J$ with $t_1 < t_2$ and $\psi(t) = \int_0^t h(s) ds$. Then, for any $y \in \mathscr{B}(C)$, we have

$$\begin{aligned}\left|y(t_2) - y(t_1)\right| &\leq \left|\int_0^{t_2} (t_2-s)^{\alpha-1} E_{\alpha,\alpha}(\lambda(t_2-s)^\alpha) g(s, x(s)) ds\right. \\ &\quad \left. - \int_0^{t_1} (t_1-s)^{\alpha-1} E_{\alpha,\alpha}(\lambda(t_1-s)^\alpha) g(s, x(s)) ds\right| \\ &\quad + \left|E_\alpha(\lambda t_2^\alpha) - E_\alpha(\lambda t_1^\alpha)\right| \left|\frac{x_0}{f(0, x_0)}\right|\end{aligned}$$

$$\begin{aligned}\left|y(t_2) - y(t_1)\right| &\leq \left|\int_0^{t_1} \Big[(t_2-s)^{\alpha-1} E_{\alpha,\alpha}(\lambda(t_2-s)^\alpha)\right. \\ &\quad \left. -(t_1-s)^{\alpha-1} E_{\alpha,\alpha}(\lambda(t_1-s)^\alpha)\Big] g(s, x(s)) ds\right| \\ &\quad + \left|\int_{t_1}^{t_2} (t_2-s)^{\alpha-1} E_{\alpha,\alpha}(\lambda(t_2-s)^\alpha) g(s, x(s)) ds\right| \\ &\quad + \left|E_\alpha(\lambda t_2^\alpha) - E_\alpha(\lambda t_1^\alpha)\right| \left|\frac{x_0}{f(0, x_0)}\right|\end{aligned}$$

$$
\begin{aligned}
\left|y(t_2) - y(t_1)\right| \leq & \left|\int_0^{t_1} \left|(t_2 - s)^{\alpha-1} E_{\alpha,\alpha}(\lambda(t_2 - s)^{\alpha})\right.\right. \\
& \left.\left. -(t_1 - s)^{\alpha-1} E_{\alpha,\alpha}(\lambda(t_1 - s)^{\alpha})\right| |g(s, x(s))| \, ds\right| \\
& + \left|\int_{t_1}^{t_2} (t_2 - s)^{\alpha-1} E_{\alpha,\alpha}(\lambda(t_2 - s)^{\alpha}) \, |g(s, x(s))| \, ds\right| \\
& + |E_\alpha(\lambda t_2^\alpha) - E_\alpha(\lambda t_1^\alpha)| \left|\frac{x_0}{f(0, x_0)}\right|
\end{aligned}
$$

$$
\begin{aligned}
\left|y(t_2) - y(t_1)\right| \leq & \| h \|_{L^1} \int_0^{t_1} (t_2 - s)^{\alpha-1} E_{\alpha,\alpha}\Big(\lambda(t_2 - s)^{\alpha}\Big) - (t_1 - s)^{\alpha-1} E_{\alpha,\alpha}(\lambda(t_1 - s)^{\alpha}\Big) ds \\
& + C_1 exp(2\lambda^\rho) \int_{t_1}^{t_2} h(s) ds + \left|E_\alpha(\lambda t_2^\alpha) - E_\alpha(\lambda t_1^\alpha)\right| \left|\frac{x_0}{f(0, x_0)}\right|
\end{aligned}
$$

Finally, according to the relation (13) and (14), we have

$$
\begin{aligned}
|y(t_2) - y(t_1)| \leq & \left|E_\alpha(\lambda t_2^\alpha) - E_\alpha(\lambda t_1^\alpha)\right| \left|\frac{x_0}{f(0, x_0)}\right| \\
& + \| h \|_{L^1} \left[t_2^\alpha E_{\alpha,\alpha+1}(\lambda t_2^\alpha) - t_1^\alpha E_{\alpha,\alpha+1}(\lambda t_1^\alpha) + (t_2 - t_1)^\alpha E_{\alpha,\alpha+1}(\lambda(t_2 - t_1)^\alpha)\right] \\
& + C_1 exp(2\lambda^\rho) |\psi(t_2) - \psi(t_1)|
\end{aligned}
$$

This implies that

$$
|y(t_2) - y(t_1)| \to 0 \quad \text{as} \quad t_2 - t_1 \to 0
$$

uniformly for all $y \in \mathscr{B}(C)$. Hence $\mathscr{B}(C)$ is an equicontinuous subset of E. $\mathscr{B}(C)$ is a uniformly bounded and equicontinuous set of functions in E, so it is compact. Consequently, $\mathscr{B}$ is a uniformly partially compact operator on E into itself.

Step V: u satisfies the operator inequality $u \leq \mathscr{A} u \mathscr{B} u$.

By hypothesis (B_4), the FQDE (1),(2) has a lower solution u defined on J. Then, we have

$$
D^\alpha \left(\frac{u(t)}{f(t, u(t))}\right) - \lambda \left(\frac{u(t)}{f(t, u(t))}\right) \leq g(t, u(t)) \quad \forall t \in J, \quad u(0) \leq x_0
$$

Multiplying the above inequality by the integrating factor $E_{\alpha,\alpha}$, we obtain for all $t \in J$,

$$
\begin{aligned}
& \int_0^t (t - s)^{\alpha-1} E_{\alpha,\alpha}(\lambda(t - s)^\alpha) \left(D^\alpha \left(\frac{u(t)}{f(t, u(t))}\right) - \lambda \left(\frac{u(t)}{f(t, u(t))}\right)\right) ds \leq \\
& \quad \int_0^t (t - s)^{\alpha-1} E_{\alpha,\alpha}(\lambda(t - s)^\alpha) g(s, u(s)) ds
\end{aligned}
$$

Using the relation (1) in Lemma 2.1, with $n = 1$, $U = 1$ and $K = -\lambda$ one has

$$\int_0^t (t-s)^{\alpha-1} E_{\alpha,\alpha}(\lambda(t-s)^\alpha) D^\alpha u(s) ds =$$
$$u(t) - E_\alpha(\lambda t^\alpha) u(0) + \lambda \int_0^t (t-s)^{\alpha-1} E_{\alpha,\alpha}(\lambda(t-s)^\alpha) u(s) ds$$

Then we obtain:

$$\frac{u(t)}{f(t, u(t))} \le E_\alpha(\lambda t^\alpha) \frac{u(0)}{f(0, u(0))} + \int_0^t (t-s)^{\alpha-1} E_{\alpha,\alpha}(\lambda(t-s)^\alpha) g(s, u(s)) ds,$$

for all $t \in J$. From the definitions of the operators $\mathscr{A}$ and $\mathscr{B}$, it follows that $u(t) \le (\mathscr{A}u)(t)(\mathscr{B}u)(t)$, for all $t \in J$. Hence $u \le \mathscr{A}u\mathscr{B}u$.
Step VI: The $\mathscr{D}$-function ϕ satisfies the growth condition $M\phi_A(r)$, $r > 0$.

The $\mathscr{D}$-function ϕ of the operator $\mathscr{A}$ satisfies the inequality given in hypothesis (c) of Theorem 2.1. From the estimate (15), it follows that

$$M\phi_A(r) = \left(C_1 exp(2|\lambda|^\rho) \left(\| h \|_{L^1} + \left| \frac{x_0}{f(0, x_0)} \right| \right) \right) \phi(r) < r, \quad \forall r > 0$$

Thus, $\mathscr{A}$ and $\mathscr{B}$ satisfy all the conditions of Theorem 1 and we apply it to conclude that the operator equation $\mathscr{A}x\mathscr{B}x = x$ has a positive solution. Consequently, the integral equation (12) and the FQDE has a positive solution x^* defined on J. Furthermore, the sequence $\{x_n\}_{n=1}^\infty$ of successive approximations defined by (11) converges monotonically to x^*. This completes the proof.

References

1. H. Jiang, Existence results for fractional order functional differential equations with impulse. Comput. Math. Appl. **64**, 3477–3483 (2012)
2. N. Kosmatov, Integral equations and initial value problems for nonlinear differential equations of fractional order. Nonlinear Anal. **70**, 2521–2529 (2009)
3. V. Lakshmikantham, Theory of fractional functional differential equations. Nonlinear Anal. **69**(10), 3337–3343 (2008)
4. V. Lakshmikantham, A.S. Vatsala, Basic theory of fractional differential equations. Nonlinear Anal. **69**(8), 2677–2682 (2007)
5. I. Podlubny, *Fractional Differential Equations* (Academic Press, New York, 1993)
6. K.S. Miller, B. Ross, *An Introduction to the Fractional Calculus and Fractional Differential Equations* (Wiley, New York, 1993)
7. A.A. Kilbas, H.M. Srivastava, J.J. Trujillo, *Theory and Applications of Fractional Differential Equations*. North-Holland Mathematics Studies, vol. **204** (Elsevier, Amsterdam, 2006)
8. Y. Zhou, *Basic Theory of Fractional Differential Equations* (Xiangtan University, China, 2014)
9. B.C. Dhage, S.B. Dhage, Approximating positive solutions of nonlinear first order ordinary quadratic differential equations. Cogent Math. **2**, 1023671 (2015)

10. B.C. Dhage, V. Lakshmikantham, Basic results on hybrid differential equations. Nonlinear Anal. Hybrid Syst. **4**, 414–424 (2010)
11. B.C. Dhage, V. Lakshmikantham, Quadratic perturbations of periodic boundary value problems of second order ordinary differential equations. Differ. Equ. Appl. **2**, 465–486 (2010)
12. K. Hilal, A. Kajouni, Boundary value problems for hybrid differential equations. Math. Theory Model. 2224–5804 (2015)
13. K. Hilal, A. Kajouni, Boundary value problems for hybrid differential equations with fractional order. Adv. Differ. Equ. **183** (2015)
14. Y. Zhou, F. Jiao, J. Li, Existence and uniqueness for fractional neutral differential equations with infinite delay. Nonlinear Anal. **71**, 3249–3256 (2009)
15. R. Gorenflo, A.A. Kilbas, F. Mainardi, S.V. Rogosin, *Mittag-Leffler Functions, Related Topics and Applications* (Springer, Berlin, 2014)
16. X.L. Ding, Y.L. Jiang, waveform relaxation method for fractional differential-algebraic equations. Fract. Calc. Appl. Anal. **17**(3), 585–604 (2014)

Comments on Fuzzy Sets, Interval Type-2 Fuzzy Sets, General Type-2 Fuzzy Sets and Intuitionistic Fuzzy Sets

Oscar Castillo and Krassimir Atanassov

Abstract In this paper, we introduce specific types of intuitionistic fuzzy sets, inspired by the multi-dimensional intuitionistic fuzzy sets and general type-2 fuzzy sets. The newly proposed sets extend the opportunities of the general type-2 fuzzy sets when modeling particular types of uncertainty. Short comparison between concepts of interval type-2 fuzzy sets and intuitionistic fuzzy sets is presented. In addition, new future directions of research are outlined.

Keywords Interval type-2 fuzzy set · Intuitionistic fuzzy set

1 Introduction

In 1965, when Lofti A. Zadeh first proposed Fuzzy Sets (FSs) [1] his vision was set on giving more control over decision making, with his Fuzzy Logic an immeasurable amount of decision making situations could be easily modeled whereas hard logic, true or false, could not. This opened a new era in decision making with FSs that have been evolving since its initial days, first starting out with the concept of a Type-1 Fuzzy Sets, then coming into an Interval Type-2 Fuzzy Sets (IT2FSs) [2] and finally arriving at the current state of an advanced form of FS, which is a Generalized Type-2 Fuzzy Sets (GT2FSs) [3–5]. Another important extension of the FSs are Intuitionistic Fuzzy Sets (IFSs) [6, 7].

Here, as an extension of [8], we discuss the interpretability of the FSs, IT2FSs and GT2FSs by IFSs and one of their extensions—the m Multidimensional IFSs (MDIFSs) [9–12].

O. Castillo (✉)
Division of Graduate Studies, Tijuana Institute of Technology, Tijuana, Mexico
e-mail: ocastillo@tectijuana.mx

K. Atanassov
Bioinformatics and Mathematical Modelling Department, IBPhBME—Bulgarian Academy of Sciences, Sofia, Bulgaria
e-mail: krat@bas.bg

S. Melliani and O. Castillo (eds.), *Recent Advances in Intuitionistic Fuzzy Logic Systems*, Studies in Fuzziness and Soft Computing 372,
https://doi.org/10.1007/978-3-030-02155-9_3

2 Fuzzy Sets, Interval Type-2 Fuzzy Sets and General Type-2 Fuzzy Sets

A Type-1 Fuzzy Set A, denoted by $\mu_A(x)$ where $x \in X$, represented by

$$A = \{\langle x, \mu_A(x)\rangle | x \in X\},$$

which is a FS which takes on values between the interval [0, 1].

A type-2 fuzzy set A', is characterized by the membership function [13, 14]:

$$A' = \{((x, u), \mu_{A'}(x, u)) | x \in X, u \in J_x \subseteq [0, 1]\} \tag{1}$$

in which $0 \leq \mu_{A'}(x, u) \leq 1$. Another expression for A' is

$$A' = \int_{x \in X} \int_{u \in J_x} \mu_{A'}(x, u)/(x, u) \tag{2}$$

for $J_x \subseteq [0, 1]$, where $\iint$ denotes the union over all admissible input variables x and u. For discrete universes of discourse $\int$ is replaced by $\sum$, [15].

A General Type-2 Fuzzy Set A' is described by

$$A' = \int_X \mu_{A'}(x)/x = \int_X \left[\int_{J_x} f_x(u)/u \right] /x,$$

where $J_x \subseteq [0, 1]$, x is the partition of the primary membership function, and u is the partition of the secondary membership function. Applications of type-2 fuzzy have been made like in [16–20].

3 Intuitionistic Fuzzy Sets, Multidimensional Intuitionistic Fuzzy Sets and Their Interpretations of the Fuzzy Sets, Interval Type-2 Fuzzy Sets and General Type-2 Fuzzy Sets

Let X be a fixed universe. The intuitionistic fuzzy set over X has the form

$$A = \{\langle x, \mu_A(x), \nu_A(x)\rangle | x \in X\},$$

where $\mu_A(x)$, $\nu_A(x)$ are degrees of membership and non-membership of elements $x \in X$ to a fixed set $A \subset X$, $0 \leq \mu_A(x), \nu_A(x) \leq 1$ and

$$0 \le \mu_A(x) + \nu_A(x) \le 1.$$

Let the set X be fixed. The Intuitionistic Fuzzy Multi-Dimensional Set (IFMDS) A in X, $Z_1, Z_2, \ldots, Z_n$ is an object of the form

$$A(Z_1, Z_2, \ldots, Z_n) = \{\langle x, \mu_A(x, z_1, z_2, \ldots, z_n), \nu_A(x, z_1, z_2, \ldots, z_n)\rangle | \langle x, z_1, z_2, \ldots, z_n\rangle \in X \times Z_1 \times Z_2 \times \ldots \times Z_n\},$$

where: $0 \le \mu_A(x, z_1, z_2, \ldots, z_n), \nu_A(x, z_1, z_2, \ldots, z_n) \le 1$ are the degrees of membership and non-membership, respectively, of the elements $\langle x, z_1, z_2, \ldots, z_n\rangle \in X \times Z_1 \times Z_2 \times \ldots \times Z_n$ and

$$0 \le \mu_A(x, z_1, z_2, \ldots, z_n) + \nu_A(x, z_1, z_2, \ldots, z_n) \le 1.$$

Now, following the ideas for IFMDS, we can modify a given IF2-dimensional-S (IF2DS) to the next form:

$$A_1 = \{\langle x, \mu_{A_1}(x, u), \nu_{A_1}(x, u)\rangle | x \in X, u \in J_x \subseteq [0, 1]\}, \quad (3)$$

where $\mu_{A_1}(x, u)$ and $\nu_{A_1}(x, u)$ are the degrees of membership and non-membership of $x \in X$ and $u \in J_x \subseteq [0, 1]$ and

$$0 \le \mu_{A_1}(x, u) + \nu_{A_1}(x, u) \le 1.$$

For this set we construct the set

$$A_1^* = \{\langle z, \mu_{A_1^*}(z), \nu_{A_1^*}(z)\rangle | z \in E\},$$

that for the universe $E = \bigcup_{x \in X} (\{x\} \times J_x)$ and $J_x \in [0, 1]$ is an IFS. Obviously, for each fixed u, u being the second projection of $z \in E$, the new set is bijective one to set A_1, while for a fixed x, A_1 is a projection of the new set.

When $\nu_{A_1^*}(x, u) = \nu_x(u) = 1 - \mu_x(u) = 1 - \mu_{A_1^*}(x, u)$, we directly obtain an interpretation of the standard Type-1 Fuzzy Set for the case of universe E.

We must mention that A_1 can also be expressed as

$$A_1 = \int_{x \in X} \left(\int_{u \in J_x} \langle \mu_x(u), \nu_x(u)\rangle \right) / x \quad J_x \subseteq [0, 1] \quad (4)$$

or

$$A_1 = \left\{ \left\langle x, \bigcup_{u \in J_x} \langle \mu_x(u), \nu_x(u)\rangle \right\rangle | x \in X, J_x \subseteq [0, 1] \right\}$$

When X is a discrete set we have that

$$A_1 = \sum_{x \in X} \left(\sum_{u \in J_x} \langle \mu_x(u), \nu_x(u) \rangle \right) / x.$$

We can get different sets by expressions

$$A_1 = \left\{ \left\langle x, \underset{u \in J_x}{*} \langle \mu_x(u), \nu_x(u) \rangle \right\rangle | x \in X, J_x \subseteq [0, 1] \right\}$$

where * can denote the operation $\bigcup$, $\vee$, $\sum$ or another.

We can give the following definition as a generalization of IF2DS A_1. The set

$$A_2 = \{\langle x, \mu_{A_2}(x, u), \nu_{A_2}(x, w) \rangle | x \in X, u \in J_x \subseteq [0, 1], w \in H_x \subseteq [0, 1]\} \quad (5)$$

in which $\mu_{A_2}(x, u)$ and $\nu_{A_2}(x, w)$ are the degrees of membership and non-membership of $x \in X$ and $u \in J_x \subseteq [0, 1]$ and

$$0 \leq \mu_{A_2}(x, u) + \nu_{A_2}(x, w) \leq 1.$$

As above, A_1 can also be expressed in A_2-forms by

$$A_2 = \int_{x \in X} \left(\int_{u \in J_x} \left\langle \mu_x(u), \int_{w \in H_x} \nu_x(w) \right\rangle \right) / x \quad J_x \subseteq [0, 1], H_x \subseteq [0, 1] \quad (6)$$

and

$$A_2 = \left\{ \left\langle x, \bigcup_{u \in J_x} \left\langle \mu_x(u), \bigcup_{w \in H_x} \nu_x(w) \right\rangle \right\rangle | x \in X, J_x \subseteq [0, 1], \mathrm{H}_x \subseteq [0, 1] \right\}.$$

When X is a discrete set we have that

$$A_2 = \sum_{x \in X} \left(\sum_{u \in J_x} \left\langle \mu_x(u), \sum_{w \in H_x} \nu_x(w) \right\rangle \right) / x.$$

We can get different sets by expressions

$$A_2 = \left\{ \left\langle x, \underset{u \in J_x}{*} \left\langle \mu_x(u), \underset{w \in H_x}{*} \nu_x(w) \right\rangle \right\rangle | x \in X, J_x \subseteq [0, 1], \mathrm{H}_x \subseteq [0, 1] \right\}$$

where * can denote the operation $\bigcup$, $\vee$, $\sum$ or another.

At each value of x, say $x=x'$, the 2-D plane whose axes are u and $\mu_{\tilde{\tilde{A}}}(x', u)$ is called a *vertical slice* of $\mu_{\tilde{\tilde{A}}}(x, u)$. At each value of x, say $x=x'$, 2-D plane whose axes are w and $\nu_{\tilde{\tilde{A}}}(x', w)$ is called a *vertical slice* of $\nu_{\tilde{\tilde{A}}}(x, w)$. A *secondary membership function* is a vertical slice of $\mu_{\tilde{\tilde{A}}}(x, u)$. A *secondary non-membership function* is a vertical slice of $\nu_{\tilde{\tilde{A}}}(x, w)$. It is $\left\langle \mu_{\tilde{\tilde{A}}}(x', u), \nu_{\tilde{\tilde{A}}}(x', w)\right\rangle$ for every $u \in J_{x'} \subseteq [0, 1]$ and for every $w \in H_{x'} \subseteq [0, 1]$, i.e.,

$$\left\langle \mu_{\tilde{\tilde{A}}}(x', u), \nu_{\tilde{\tilde{A}}}(x', w)\right\rangle \equiv \left\langle \mu_{\tilde{\tilde{A}}}(x'), \nu_{\tilde{\tilde{A}}}(x')\right\rangle$$
$$= \int\limits_{u\in J_{x'}} \left\langle \mu_{x'}(u), \int\limits_{w\in H_{x'}} \nu_{x'}(w)\right\rangle / \langle u, w\rangle\, J_{x'} \subseteq [0, 1]\ H_{x'} \subseteq [0, 1], \tag{7}$$

in which $0 \leq \mu_{x'}(u) + \nu_{x'}(w) \leq 1$. Because for every $x' \in X$, we drop the prime notation on $\mu_{\tilde{\tilde{A}}}(x')$, and refer to $\mu_{\tilde{\tilde{A}}}(x)$ as a secondary membership function, we drop the prime notation on $\nu_{\tilde{\tilde{A}}}(x')$, and refer to $\nu_{\tilde{\tilde{A}}}(x)$ as a secondary non-membership function; it is an intuitionistic fuzzy set, which we also refer to as a *secondary set*.

Based on the concept of secondary sets, we can reinterpret a restricted 2-dimensional intuitionistic fuzzy set as the union of all secondary sets, i.e., using (7), *we can re-express in a vertical-slice manner*, as

$$\tilde{\tilde{A}} = \left\{\left\langle x, \mu_{\tilde{\tilde{A}}}(x), \nu_{\tilde{\tilde{A}}}(x)\right\rangle | x \in X\right\} \tag{8}$$

or, as

$$\tilde{\tilde{A}} = \int\limits_{x\in X} \left(\left\langle \mu_{\tilde{\tilde{A}}}(x), \nu_{\tilde{\tilde{A}}}(x)\right\rangle / \langle u, w\rangle\right)/x$$
$$= \int\limits_{x\in X} \left(\int\limits_{u\in J_x} \left\langle \mu_x(u), \int\limits_{w\in H_x} \nu_x(w)\right\rangle / \langle u, w\rangle\right)/x \tag{9}$$

When X is a discrete set we have that

$$\tilde{\tilde{A}} = \sum_{x\in X} \left(\sum_{u\in J_x} \left\langle \mu_x(u), \sum_{w\in H_x} \nu_x(w)\right\rangle / \langle u, w\rangle\right)/x.$$

We can get different sets by the expression

$$\tilde{\tilde{A}} = \left\{\left\langle x, \mathop{*}_{u\in J_x} \left\langle u, \mu_x(u), \mathop{*}_{w\in H_x} \langle w, \nu_x(w)\rangle\right\rangle\right\rangle | x \in X, J_x \subseteq [0, 1], \mathrm{H}_x \subseteq [0, 1]\right\},$$

where $*$ can denote the operation $\bigcup$, $\vee$, $\sum$ or another.

In the same case, the intuitionistic fuzzy set $\tilde{A}$ is characterized by the membership and non-membership functions, and has the form:

$$\tilde{A} = \{\langle (x, u), \mu_{\tilde{A}}(x, u), \nu_{\tilde{A}}(x, u)\rangle | x \in X, u \in J_x \subseteq [0, 1]\}, \tag{10}$$

in which $0 \leq \mu_{\tilde{A}}(x, u) \leq 1$, $0 \leq \nu_{\tilde{A}}(x, u) \leq 1$, $0 \leq \mu_{\tilde{A}}(x, u) + \nu_{\tilde{A}}(x, u) \leq 1$. The analogue of the second expression (2) now is:

$$\tilde{A} = \int_{x \in X} \int_{u \in J_x} \langle \mu_{\tilde{A}}(x, u), \nu_{\tilde{A}}(x, u)\rangle / (x, u) \tag{11}$$

for $J_x \subseteq [0, 1]$, where, as above, $\iint$ denotes the union over all admissible input variables x and u. For discrete universes of discourse $\int$ is replaced by $\sum$.

Let X be a universe, $g : X \to [0, 1]$ be an interval function and $X_g = \bigcup_{x \in X} x \times g(x)$. Let us consider the IFS

$$A_g^* = \{\langle z, \mu_{\tilde{A}}(z), \nu_{\tilde{A}}(z)\rangle | z \in X_g\}.$$

Then any type-2 fuzzy set $\tilde{A}$ can be represented by the IFS A_g^* with a suitable choice of g.

Below, we will express the same set $\tilde{A}$ by using the following definition.

At each value of x, say $x = x'$, the 2-D plane whose axes are u and $\mu_{\tilde{A}}(x', u)$ is called a *vertical slice* of $\mu_{\tilde{A}}(x, u)$. At each value of x, say $x = x'$, 2-D plane whose axes are w and $\nu_{\tilde{A}}(x', u)$ is called a *vertical slice* of $\nu_{\tilde{A}}(x, u)$. A *secondary membership function* is a vertical slice of $\mu_{\tilde{A}}(x, u)$. A *secondary non-membership function* is a vertical slice of $\nu_{\tilde{A}}(x, u)$. It is $\langle \mu_{\tilde{A}}(x', u), \nu_{\tilde{A}}(x', u)\rangle$ for every $u \in J_{x'} \subseteq [0, 1]$, i.e.,

$$\begin{aligned}\langle \mu_{\tilde{A}}(x', u), \nu_{\tilde{A}}(x', u)\rangle &\equiv \langle \mu_{\tilde{A}}(x'), \nu_{\tilde{A}}(x')\rangle \\ &= \int_{u \in J_{x'}} \langle \mu_{x'}(u), \nu_{x'}(u)\rangle / u \; J_{x'} \subseteq [0, 1],\end{aligned} \tag{12}$$

in which $0 \leq \mu_{x'}(u) + \nu_{x'}(u) \leq 1$. Because for every $x' \in X$, we drop the prime notation on $\mu_{\tilde{A}}(x')$, and refer to $\mu_{\tilde{A}}(x)$ as a secondary membership function, we drop the prime notation on $\nu_{\tilde{A}}(x')$, and refer to $\nu_{\tilde{A}}(x)$ as a secondary non-membership function; it is an intuitionistic fuzzy set, which we also refer to as a *secondary set*.

Based on the concept of secondary sets, we can reinterpret a restricted 2-dimensional intuitionistic fuzzy set as the union of all secondary sets, i.e., using (12), *we can re-express in a vertical-slice manner*, as

$$\tilde{A} = \{\langle x, \mu_{\tilde{A}}(x), \nu_{\tilde{A}}(x)\rangle | x \in X\} \tag{13}$$

or, as

$$\tilde{A} = \int_{x \in X} (\langle \mu_{\tilde{A}}(x), \nu_{\tilde{A}}(x) \rangle / u)/x = \int_{x \in X} \left(\int_{u \in J_x} \langle \mu_x(u), \nu_x(u) \rangle u \right) /x \tag{14}$$

When X is a discrete set we have that

$$\tilde{A} = \sum_{x \in X} \left(\sum_{u \in J_x} \langle \mu_x(u), \nu_x(u) \rangle / u \right) /x.$$

We can get different sets by expression

$$\tilde{A} = \left\{ \left\langle x, \underset{u \in J_x}{*} \langle u, \mu_x(u), \nu_x(u) \rangle \right\rangle | x \in X, J_x \subseteq [0, 1], \right\}$$

where $*$ can denote the operation $\bigcup, \vee, \sum$ or another.

The set $\tilde{A}$ can be extended to the form of $\tilde{\tilde{A}}$. The intuitionistic fuzzy set $\tilde{\tilde{A}}$ is characterized by a membership and a non-membership function and has the form:

$$\tilde{\tilde{A}} = \left\{ \left\langle (x, u, w), \mu_{\tilde{\tilde{A}}}(x, u), \nu_{\tilde{\tilde{A}}}(x, w) \right\rangle | x \in X, u \in J_x \subseteq [0, 1], w \in H_x \subseteq [0, 1] \right\}, \tag{15}$$

in which $0 \leq \mu_{\tilde{\tilde{A}}}(x, u) \leq 1, 0 \leq \nu_{\tilde{\tilde{A}}}(x, w) \leq 1, 0 \leq \mu_{\tilde{\tilde{A}}}(x, u) + \nu_{\tilde{\tilde{A}}}(x, w) \leq 1$. The analogue of the second expression (2), now is:

$$\tilde{\tilde{A}} = \int_{x \in X} \int_{u \in J_x} \int_{w \in H_x} \langle \mu_{\tilde{A}}(x, u), \nu_{\tilde{A}}(x, w) \rangle / (x, u, w) \tag{16}$$

for $J_x \subseteq [0, 1]$ and $H_x \subseteq [0, 1]$, where, as above, $\iint$ denotes the union over all admissible input variables x and u. For discrete universes of discourse $\int$ is replaced by $\sum$.

Let X be a universe, $g : X \to [0, 1]$ be an interval function. Let X be a universe, $f : X \to [0, 1]$ be an interval function and

$$X_{\langle f,g \rangle} = \bigcup_{x \in X} (\{x\} \times f(x) \times g(x)).$$

We can also note that another form of expressing the IFS $\tilde{\tilde{A}}$ is the following form:

$$A^*_{\langle f,g \rangle} = \left\{ \langle z, \mu_{\tilde{A}}(\mathrm{pr}_1 z, \mathrm{pr}_2 z), \nu_{\tilde{A}}(\mathrm{pr}_1 z, \mathrm{pr}_3 z) \rangle | z \in X_{\langle f,g \rangle} \right\}.$$

The implementation of this proposal is work is progress, but will enable the solution of problems with higher degrees of uncertainty.

4 Conclusion

In conclusion, we can mention that in future a detail comparison between both concepts will be prepared and relationships between the operations, relations and especially the operators will be compared. In addition, we envision that applications will be made of the type-2 intuitionistic fuzzy systems concepts presented here. The main reason for this is that some complex problems can possess different types of uncertainty, which could be captured using both the type-2 and intuitionistic concepts. We plan to prepare a series of papers along these lines of research.

References

1. L.A. Zadeh, Fuzzy sets. Inf. Control **8**, 338–353 (1965)
2. J.M. Mendel, X. Liu, Simplified interval Type-2 fuzzy logic systems. IEEE Trans. Fuzzy Syst. **21**(6), 1056–1069 (2013)
3. A. Rizzi, L. Livi, H. Tahayori, A. Sadeghian, Matching general type-2 fuzzy sets by comparing the vertical slices. in *2013 Joint IFSA World Congress and NAFIPS Annual Meeting (IFSA/NAFIPS)*, pp. 866–871 (2013)
4. M.A. Sanchez, O. Castillo, J.R. Castro, Generalized Type-2 Fuzzy Systems for controlling a mobile robot and a performance comparison with Interval Type-2 and Type-1 fuzzy systems. Expert Syst. Appl. **42**(14), 5904–5914 (2015)
5. L. Zhao, Y. Li, Y. Li, Computing with words for discrete general type-2 fuzzy sets based on α plane. in *Proceeding of 2013 IEEE International Conference on Vehicular Electronics and Safety*, pp. 268–272 (2013)
6. K. Atanassov, *Intuitionistic Fuzzy Sets* (Springer, Heidelberg, 1999)
7. K. Atanassov, *On Intuitionistic Fuzzy Sets Theory* (Springer, Berlin, 2012)
8. O. Castillo, P. Melin, R. Tsvetkov, K. Atanassov, Short remark on interval type-2 fuzzy sets and intuitionistic fuzzy sets. J. Notes on Intuitionistic Fuzzy Sets **20**(2), 1–5 (2014)
9. K. Atanassov, E. Szmidt, J. Kacprzyk, On intuitionistic fuzzy multi-dimensional sets. Issues Intuitionistic Fuzzy Sets Generalized Nets **7**, 1–6 (2008)
10. K. Atanassov, E. Szmidt, J. Kacprzyk, On intuitionistic fuzzy multi-dimensional sets. Part 3. in *Developments in Fuzzy Sets, Intuitionistic Fuzzy Sets, Generalized Nets and Related Topics,* vol. I*: Foundations*. (Warsaw, SRI Polish Academy of Sciences, 2010), pp. 19–26
11. K. Atanassov, E. Szmidt, J. Kacprzyk, On intuitionistic fuzzy multi-dimensional sets. Part 4. J. Notes on Intuitionistic Fuzzy Sets **17**(2), 1–7 (2011)
12. K. Atanassov, E. Szmidt, J. Kacprzyk, P. Rangasamy, On intuitionistic fuzzy multi-dimensional sets. Part 2. In *Advances in Fuzzy Sets, Intuitionistic Fuzzy Sets, Generalized Nets and Related Topics. Vol. I: Foundations* (Academic Publishing House EXIT, Warszawa, 2008), pp. 43–51
13. J.M. Mendel, *Uncertain Rule-Based Fuzzy Logic Systems: Introduction and new directions* (Prentice Hall, New Jersey, 2001)
14. J.M. Mendel, G.C. Mouzouris, Type-2 fuzzy logic systems. IEEE Trans. Fuzzy Syst. **7**, 643–658 (1999)
15. J.M. Mendel, R.I. Bob John, Type-2 fuzzy sets made simple. IEEE Trans. Fuzzy Syst. **10**, 117–127 (2002)
16. O. Castillo, P. Melin, Intelligent systems with interval type-2 fuzzy logic. Int. J. Innovative Comput. Inf. Control **4**(4), 771–783 (2008)
17. N.R. Cázarez-Castro, L.T. Aguilar, O. Castillo, Designing Type-1 and Type-2 fuzzy logic controllers via fuzzy lyapunov synthesis for nonsmooth mechanical systems. Eng. Appl. Artif. Intell. **25**(5), 971–979 (2012)

18. C. Leal-Ramírez, O. Castillo, P. Melin, A. Rodríguez-Díaz, Simulation of the bird age-structured population growth based on an interval type-2 fuzzy cellular structure. Inf. Sci. **181**(3), 519–535 (2011)
19. E. Ontiveros-Robles, P. Melin, O. Castillo, Comparative analysis of noise robustness of type 2 fuzzy logic controllers. Kybernetika **54**(1), 175–201 (2018)
20. E. Ontiveros-Robles, P. Melin, O. Castillo, New methodology to approximate type-reduction based on a continuous root-finding karnik mendel algorithm. Algorithms **10**(3), 77 (2017)

Quotient Rings Induced via Intuitionistic Fuzzy Ideals

Said Melliani, I. Bakhadach, H. Sadiki and L. S. Chadli

Abstract This work we give a construction of a quotient ring $R/<\mu, \nu>$ induced via an intuitionistic fuzzy ideal $<\mu, \nu>$ in a ring R. The Intuitionistic Fuzzy First, Second and Third Isomorphism Theorems are established. For some applications of this construction of quotient rings, we show that if $<\mu, \nu>$ is an intuitionistic fuzzy ideal of a commutative ring R, then $<\mu, \nu>$ is prime (resp, primary) if and only if $R/<\mu, \nu>$ is an integral domain (resp, every zero divisor in $R/<\mu, \nu>$ is nilpotent).

1 Introduction

The concept of fuzzy sets which was introduced by Zadeh [20] is applied to many mathematical branches. Atanassov [1, 2] introduced and developed the theory of intuitionistic fuzzy sets. Using the Atanassov sidea Biswas [4] established the intuitionistic fuzzification of the concept of sub-group of a group and introduced the notion of intuitionistic fuzzy subgroups. Hur et al. [6] introduced the concept of intuitionistic fuzzy ring. Sharma [17] introduced the notion of translates of intuitionistic fuzzy subrings. The notion of t-intuitionistic fuzzy quotient group has already been introduced by Sharma in [19].

The present note in Sect. 3 we give another way to construct quotient rings by intuitionistic fuzzy ideals. We also establish Intuitionistic Fuzzy First, Second and Third Isomorphism Theorems. As some applications we give the Intuitionistic fuzzy ideal characterizations of some classes of quotient rings in Sect. 4. In Sect. 2 we list some of necessary materials needed in sequel.

S. Melliani (✉) · I. Bakhadach · H. Sadiki · L. S. Chadli
Sultan Moulay Slimane University, BP 523 Beni Mellal, Morocco
e-mail: s.melliani@usms.ma

I. Bakhadach
e-mail: bakhadach@gmail.com

H. Sadiki
e-mail: razika.imi@gmail.com

L. S. Chadli
e-mail: sa.chadli@yahoo.fr

S. Melliani and O. Castillo (eds.), *Recent Advances in Intuitionistic Fuzzy Logic Systems*, Studies in Fuzziness and Soft Computing 372,
https://doi.org/10.1007/978-3-030-02155-9_4

2 Preliminaries

First we give the concept of intuitionistic fuzzy set defined by Atanassov as a generalization of the concept of fuzzy set given by Zadeh.

Definition 1 [1, 2] The intuitionistic fuzzy sets (in shorts IFS) defined on a non-empty set X as objects having the form

$$A = \{<x, \mu(x), \nu(x)> : x \in X\}$$

where the functions $\mu : X \to [0, 1]$ and $\nu : X \to [0, 1]$ denote the degree of membership and the degree of non-membership of each element $x \in X$ to the set A respectively, and $0 \leq \mu(x) + \nu(x) \leq 1$ for all $x \in X$.

For the sake of simplicity, we shall use the symbol $<\mu, \nu>$ for the intuitionistic fuzzy subset $A = \{<x, \mu(x), \nu(x)> : x \in X\}$.

Definition 2 [2] Let X be a nonempty set and let $A = <\mu_A, \nu_A>$ and $B = <\mu_B, \nu_B>$ IFSS of X. Then
$A \subset B$ iff $\mu_A \leq \mu_B$ and $\nu_A \geq \nu_B$
$A = B$ iff $A \subset B$ and $B \subset A$
$A^c = <\nu_A, \mu_A>$
$A \cap B = <\mu_A \wedge \mu_B, \nu_A \vee \nu_B>$
$A \cup B = <\mu_A \vee \mu_B, \nu_A \wedge \nu_B>$
$[\,]A = <\mu_A, 1 - \mu_A>, <> A = <1 - \nu_A, \nu_A>$

Definition 3 [13] Let R be a ring. An intuitionistic fuzzy set $A = \{<x, \mu(x), \nu(x)> : x \in R\}$ of R is said to be intuitionistic fuzzy subring of R (In short IFSR) of R if $\forall x, y \in R$

(i) $\mu(x - y) \geq \mu(x) \wedge \mu(y)$
(ii) $\nu(x - y) \leq \nu(x) \vee \nu(y)$
(iii) $\mu(xy) \geq \mu(x) \wedge \mu(y)$
(iv) $\nu(xy) \leq \nu(x) \vee \nu(y)$.

Definition 4 [17] Let R be a ring. An intuitionistic fuzzy set $A = \{<x, \mu(x), \nu(x)> : x \in R\}$ of R is said to be intuitionistic fuzzy ideal of R (In short IFI) of R if $\forall x, y \in R$

(i) $\mu(x - y) \geq \mu(x) \wedge \mu(y)$
(ii) $\nu(x - y) \leq \nu(x) \vee \nu(y)$
(iii) $\mu(xy) \geq \mu(x) \vee \mu(y)$
(iv) $\nu(xy) \leq \nu(x) \wedge \nu(y)$.

Definition 5 [5] Let X and Y be two non-empty sets and $f : X \longrightarrow Y$ be a mapping. Let A be an *IFS* in X and B be an *IFS* in Y: Then
(a) the image of A under f; denoted by $f(A)$; is the *IFS* in Y defined by
$f(A) = <f(\mu_A); f(\nu_A)>$; where for each $y \in Y$;

$$f(\mu_A)(y) = \begin{cases} \bigvee\limits_{x \in f^{-1}(y)} \mu_A(x) & \text{if } f^{-1}(y) \neq \emptyset \\ 0 & \text{if } f^{-1}(y) = \emptyset. \end{cases}$$

and

$$f(\nu_A)(y) = \begin{cases} \bigwedge\limits_{x \in f^{-1}(y)} \nu_A(x)\ , & \textit{if } f^{-1}(y) \neq \emptyset \\ 1 & \textit{if } f^{-1}(y) = \emptyset. \end{cases}$$

(b) the pre-image of B under f, denoted by $f^{-1}(B)$, is the *IFS* in X defined by $f^{-1}(B) = <f^{-1}(\mu_B), f^{-1}(\nu_B)>$ where $f^{-1}(\mu_B) = \mu_B \circ f$.

Definition 6 [18] Let A be Intuitionistic fuzzy set of a universe set X. Then $(\alpha, \beta)-$cut of A is a crisp subset

$$C_{(\alpha,\beta)}(A) = \{x : x \in X \text{ such that } \mu_A(x) \geqslant \alpha,\ \nu_A(x) \leqslant \beta\}$$

where $\alpha, \beta \in [0, 1]$ with $\alpha + \beta \leqslant 1$.

3 Quotient Rings Induced by Intuitionistic Fuzzy Ideals

Let $<\mu, \nu>$ be an intuitionistic fuzzy ideal of a ring R. For any $x, y \in R$, define a binary relation $\sim$ on R by $x \sim y$ if and only if $\mu(x - y) = \mu(0)$ and $\nu(x - y) = \nu(0)$.

Lemma 1 *$\sim$ is an equivalence relation of R.*

Proof (i) For any $x \in R$, we have $\mu(x - x) = \mu(0)$ and $\nu(x - x) = \nu(0)$. Hence $x \sim x$.

(ii) For any $x, y \in R$, if $x \sim y$, than $\mu(x - y) = \mu(0)$ and $\nu(x - x) = \nu(0)$.
Since $\mu(y - x) = \mu(-(x - y)) = \mu(x - y) = \mu(0)$,
and $\nu(y - x) = \nu(-(x - y)) = \nu(x - y) = \nu(0)$ hence $y \sim x$.

(iii) For any $x, y, z \in R$, if $x \sim y$ and $y \sim z$, then $\mu(x - y) = \mu(y - z) = \mu(0)$.
Since $\mu(x - z) = \mu((x - y) - (y - z)) \geq \mu(x - y) \wedge \mu(y - z) = \mu(0)$ and
$\nu(x - z) = \nu((x - y) - (y - z)) \leq \nu(x - y) \vee \nu(y - z) = \nu(0)$,
so $\mu(x - z) = \mu(0)$ and $\nu(x - z) = \nu(0)$.
Hence $x \sim z$. The proof is completed.

Lemma 2 *$x \sim y$ implies $x + z \sim y + z$ and so $z + x \sim z + y$ for all $x, y, z \in R$.*

Proof If $x \sim y$, then $\mu(x - y) = \mu(0)$.
Thus $\mu((x + z) - (y + z)) = \mu(x - y) = \mu(0)$, and $\nu((x + z) - (y + z)) = \nu(x - y) = \nu(0)$. That is $x + z \sim y + z$.
This completes the proof.

Lemma 3 *$x \sim y$ and $u \sim v$ imply $x + u \sim y + v$ and $xu \sim yv$ for all $x, y, u, v \in R$.*

Proof If $x \sim y$ and $u \sim v$, by Lemma 2 $x + u \sim y + u$ and $y + u \sim y + v$.
Using the transitivity of $\sim$, we get $x + u \sim y + v$.
Since $xu - yv =$
$xu - xv + xv - yv = x(u - v) + (x - y)v$, we have $\mu(xu - yv) = \mu(x(u - v) + (x - y)v) \geq \mu(x(u - v)) \wedge \mu((x - y)v) \geq \mu(u - v) \wedge \mu(x - y)$
If $x \sim y$ and $u \sim v$, then $\mu(x - y) = \mu(u - v) = \mu(0)$, thus $\mu(xu - yv) \geq \mu(0)$ and so $\mu(xu - yv) = \mu(0)$, Similarly we can show that $\nu(xu - yv) = \nu(0)$.
Hence $xu \sim yv$. The proof is completed.

Summarizing the above lemmas we have the following

Theorem 1 *$\sim$ is a congruence relation of R.*

Notation We denote $A_x = \{y \in R | y \sim x\}$ the equivalence class containing x and $R/{<}\mu_A, \nu_A{>} = \{A_x | x \in R\}$ the set of all equivalence classes of R.

Theorem 2 *If $A = {<}\mu_A, \nu_A{>}$ is an intuitionistic fuzzy ideal of a ring (resp. commutative ring, division ring, field) R, then $R/{<}\mu_A, \nu_A{>}$ is a ring (resp. commutative ring, division ring, field) under the binary operations $A_x + A_y = A_{x+y}$ and $A_x A_y = A_{xy}$, for any $x, y \in R$.*

Proof First we show that the above binary operations are well-defined.
In fact, if $A_x = A_u$ and $A_y = A_v$, then $x \sim u$ and $y \sim v$.
By Lemma 3 we have $x + y \sim u + v$ and $xy \sim uv$, and so $A_{x+y} = A_{u+v}$ and $A_{xy} = A_{uv}$.
Hence addition and multiplication are well-defined.
Clearly A_0 acts as the zero element in $R/{<}\mu_A, \nu_A{>}$, A_{-x} as additive inverse of A_x,A_1 as the unit element in $R/{<}\mu_A, \nu_A{>}$ and $A_{x^{-1}}$ as multiplicative inverse of A_x. The other conditions can be easily verified by routine algebraic methods.
The proof is completed.

Proposition 1 [14] *Let $f : R \to R'$ be a homomorphism of rings and B an intuitionistic fuzzy ideal of. Then $f^{-1}(B)$ is an intuitionistic fuzzy ideal of R.*

Theorem 3 (Intuitionistic Fuzzy First Isomorphism Theorem) *Let $f : R \to R'$ be an epimorphism of rings and $B = {<}\mu_B, \nu_B{>}$ an intuitionistic fuzzy ideal of R'. Then*

$$R/f^{-1}(B) \cong R'/B.$$

Proof It follows from Theorem 2 and Proposition 1, $R/f^{-1}(B)$ and R'/B are both rings.
Define $\xi : R/f^{-1}(B) \to R'/B$ by

$$\xi((f^{-1}(B))_x) = B_{f(x)}.$$

(i) ξ is well-defined: $(f^{-1}(B))_x = (f^{-1}(B))_y$ than we have $(f^{-1}(B))(x - y) = (f^{-1}(B))(0) \Rightarrow \mu_B(f(x) - f(y)) = \mu_B(f(0)) = \mu_B(0')$, and $\nu_B(f(x) - f(y)) = \nu_B(f(0)) = \nu_B(0')$, this implies that $B_f(x) = B_f(y)$.

(ii) ξ is a homomorphism: $\xi((f^{-1}(B))_x + (f^{-1}(B))_y) = \xi((f^{-1}(B))_{x+y}) = B_{f(x+y)}$ $= B_{f(x)+f(y)} = B_{f(x)} + B_{f(y)} = \xi((f^{-1}(B))_x) + \xi((f^{-1}(B))_y)$ and $\xi((f^{-1}(B))_x(f^{-1}(B))_y) = \xi((f^{-1}(B))_{xy}) = B_{f(xy)} = B_{f(x)f(y)} = B_{f(x)}B_{f(y)} = \xi$ $((f^{-1}(B))_x)\xi((f^{-1}(B))_y)$

(iii) ξ is an epimorphism: For any $B_y \in R'/B$, since f is epimorphic, there exists $x \in R$ such that $f(x) = y$. So $\xi((f^{-1}(B))_x) = B_{f(x)} = B_y$.

(iv) ξ is a monomorphism: $B_{f(x)} = B_{f(y)} \Rightarrow \mu_B(f(x) - f(y)) = \mu_B(0')$, and $\nu_B(f(x)$ $- f(y)) = \nu_B(0')$ So $\Rightarrow (f^{-1}(B))(x-y) = (f^{-1}(B))(0) \Rightarrow (f^{-1}(B))_x =$ $(f^{-1}(B))_y$.

Hence $R/f^{-1}(B) \cong R'/B$ and the proof is completed.

Lemma 4 *Let I be an ideal and* $A = <\mu_A, \nu_A>$ *an intuitionistic fuzzy ideal of a ring R. If A is restricted to I, then*

(i) $A = <\mu_A, \nu_A>$ *is an intuitionistic fuzzy ideal of I.*
(ii) I/A *is an ideal of* R/A.

Proof (i) Obviously.
(ii) We first show that $\{A_a | a \in I\}$ is an ideal of R/A. For any $A_a, A_b \in \{A_a | a \in I\}$, where $a, b \in I$.
Since I is an ideal, $a - b \in I$, hence $A_a - A_b = A_{a-b} \in \{\mu_a | a \in I\}$.
For any $A_a \in \{\mu_a | a \in I\}, A_x \in R/A$, where $a \in I$ and $x \in R$, then $ax, xa \in I$, hence $A_aA_x = A_{ax} \in \{A_a | a \in I\}$ and $A_xA_a = A_{xa} \in \{A_a | a \in I\}$.
Thus $\{A_a | a \in I\}$ is an ideal of R/A.
Next we define $\varphi : I/A \to R/A$ by $(A|_I)_a \mapsto A_a$ for all $(A|_I)_a \in I/A$. It is easy to verify that $I/A \cong \{A_a | a \in I\}$ under this mapping.
Hence in isomorphic sense we may regard I/μ as an ideal of R/μ, completing the proof.

Notation Let $A = <\mu_A, \nu_A>$ be an intuitionistic fuzzy ideal of a ring R. We denote $R_A = A_{A(0)} = \{x \in R | \mu_A(x) = \mu_A(0) \text{ and } \nu_A(x) = \nu_A(0)\}$. R_A is an ideal of R.

Theorem 4 *(Intuitionistic Fuzzy Second Isomorphism Theorem) Let* A, B *be two intuitionbistic fuzzy ideals of a ring R with* $A(0) = B(0)$. *Then*

$$R_A + R_B/B \cong R_A/A \cap B$$

Proof By Lemma 4, u is an intuitionistic fuzzy ideal of $R_A + R_B$ and $A \cap B$ is an intuitionisticfuzzy ideal of R_A.

For any $x \in R_A + R_B$, then $x = a + b$, where $a \in R_A, b \in R_B$. Define $f : R_A + R_\nu/B \to R_A/A \cap B$ by

$$f(B_x) = (A \cap B)_a.$$

If $B_x = B_{x'}$, where $x' = a' + b', a' \in R_A$ and $b' \in R_B$, then $B(x - x') = B((a + b) - (a' + b)) = B((a - a') - (b' - b)) = B(0)$, and so $B(a - a') = B(b' - b) = B(0)$.

Thus $(\mu_A \cap \mu_B)(a - a') = <\mu_A(a - a) \wedge \mu_B(a - a), \nu_A(a - a) \vee \nu_B(a - a)> = <\mu_A(0) \wedge \mu_B(0), \nu_A(0) \vee \nu_B(0)> = (\mu_A \cap \mu_B)(0)$.

That is $(A \cap B)_a = (A \cap B)_{a'}$.

Hence f is well-defined.

For any $B_x, B_y \in R_A + R_B/B$, where $x = a + b, y = a_1 + b_1, a, a_1 \in R_A$ and $b, b_1 \in R_B$, then $x + y = (a + a_1) + (b + b_1), xy = (a + b)(a_1 + b_1) = aa_1 + (ab_1 + ba_1 + bb_1) = aa_1 + b', b' = ab_1 + ba_1 + bb_1 \in R_B$.

We have

$$f(B_x + B_y) = f(B_{x+y}) = (A \cap B)_{a+a_1} = (A \cap B)_a + (A \cap B)_{a_1} = f(B_x) + f(B_y)$$

$$f(B_xB_y) = f(B_{xy}) = (A \cap B)_{aa_1} = (A \cap B)_a(A \cap B)_{a_1} = f(B_x)f(B_y)\,.$$

Hence f is a homomorphism.

For any $(A \cap B)_a \in R_A/A \cap B$, taking $b \in R_B$, then $x = a + b \in R_A + R_B$, and $f(B_x) = (A \cap B)_a$.

Hence f is an epimorphism.

For any $x, y \in R_A + R_B$, where $x = a + b, y = a_1 + b_1, a, a_1 \in R_A$ and $b, b_1 \in R_B$, if $(A \cap B)_a = (A \cap B)_{a_1}$, then $(A \cap B)(a - a_1) = (A \cap B)(0)$, i.e., $\min\{\mu_A(a - a_1), \mu_B(a - a_1)\} = \min\{\mu_A(0), \mu_B(0)\}$, and $\max\{\nu_A(a - a_1), \nu_B(a - a_1)\} = \max\{\nu_A(0), \nu_B(0)\}$.

Since $\mu_A(0) = \mu_B(0)$ and $\nu_A(0) = \nu_B(0)$, and $\mu_A(a - a_1) = \mu_A(0)$ and $\nu_A(a - a_1) = \nu_A(0)$, we have $\mu_B(a - a_1) = \mu_B(0)$, and $\nu_B(a - a_1) = \nu_B(0)$.

Hence $\mu_B(x - y) = \mu_B((a + b) - (a_1 + b_1)) = \mu_B((a - a_1) - (b_1 - b)) \geq \min\{\mu_B(a - a_1), \mu_B(b_1 - b)\} = \min\{\mu_B(0), \mu_B(0)\} = \mu_B(0)$, the seem for $\mu_B(x - y) = \mu_B(0)$ and so $B_x = B_y$.

Therefore f is a monomorphism.

Thus we have shown that $R_A + R_B/B \cong R_A/A \cap B$, completing the proof.

Theorem 5 (Intuitionistic Fuzzy Third Isomorphism Theorem) *Let A, B be two intuitionistic fuzzy ideals of a ring R with $A \leq B$ and $\mu_A(0) = \mu_B(0)$ and $\nu_A(0) = \nu_B(0)$. Then*

$$(R/B)/(R_A/B) \cong R/A$$

Proof By Lemma 4, R_A/B is an ideal of R/B.

Define $f : R/B \to R/A$ by

$$f(B_x) = A_x \text{ for all } x \in R.$$

If $B_x = B_y$, then $B(x - y) = B(0) = A(0)$.

Since $A \geq B$, we have $\mu_A(x - y) \geq \mu_B(x - y) = \mu_A(0)$, so $\mu_A(x - y) = \mu_A(0)$, and $\nu_A(x - y) \leq \nu_B(x - y) = \nu_A(0)$, so $\nu_A(x - y) = \nu_A(0)$, then $A_x = A_y$.

Hence f is well-defined.

$f(B_x + B_y) = f(B_{x+y}) = A_{x+y} = A_x + A_y = f(B_x) + f(B_y), f(B_xB_y) = f(B_{xy}) = A_{xy} = A_xA_y = f(B_x)f(B_y)$, they mean f is a homomorphism.

For any $A_x \in R/A$, there is $B_x \in R/B$ such that $f(B_x) = A_x$, so f is an epimorphism. Now we show $kerf = R_B/B$.

In fact, $kerf = \{B_x \in R/B | f(B_x) = A_0\} = \{B_x \in R/B | A_x = A_0\} = \{B_x \in R/B | A(x) = A(0)\} = \{B_x \in R/B | x \in R_A\} = R_A/B\Box$.

Therefore $(R/B)/(R_A/B) \cong R/A$. The proof is complete.

4 Intuitionistic Fuzzy Ideals Characterizations of Quotient Rings

Definition 7 [3] An intuitionistic fuzzy ideal $<\mu, \nu>$ of a ring R is called intuitionistic fuzzy prime if for all $x, y \in R$,
$\mu(xy) = \mu(0)$ implies $\mu(x) = \mu(0)$ or $\mu(y) = \mu(0)$ $\nu(xy) = \mu(0)$ implies $\nu(x) = \mu(0)$ or $\nu(y) = \mu(0)$.

Theorem 6 *Let R be a commutative ring and $A = <\mu, \nu>$ an intuitionistic fuzzy ideal of R. Then $A = <\mu, \nu>$ is prime if and only if R/A is an integral domain.*

Proof If $A = <\mu, \nu>$ is prime, then R/A is a commutative ring by Theorem 2. Assume that $A_x A_y = A_0$. Then $\mu(xy) = \mu(0)$ and $\nu(xy) = \nu(0)$.
By Definition 7 we have $\mu(x) = \mu(0)$ and $\nu(x) = \nu(0)$ or $\mu(y) = \mu(0)$ and $\nu(y) = \nu(0)$.
That is $A_x = A_0$ or $A_y = A_0$.
Hence R/A has no zero divisors and so R/A is an integral domain.
Conversely, if R/A is an integral domain, then R/A has no zero divisors.
Let $\mu(xy) = \mu(0)$ and $\nu(xy) = \nu(0)$.
That is $A_{xy} = A_x A_y = A_0$.
Then $A_x = A_0$ or $A_y = A_0$. Hence $\mu(x) = \mu(0)$ and $\nu(x) = \nu(0)$ or $\mu(y) = \mu(0)$ and $\nu(x) = \nu(0)$.
And so A is prime. The proof is completed.

Let I be an ideal of a ring R. Recall that a quotient ring R/I induced by an ideal I is determined by an equivalent relation $\sim$, where $x \sim y$ is defined by $x - y \in I$. For no confusion, we write $x \sim y(I)$ to show that x is equivalent to y with respect to the ideal I, and $x \sim y(\mu)$ to mean that x is equivalent to y with respect to the fuzzy ideal μ.

Lemma 5 *If I is an ideal in a ring R, then $x \sim y(I)$ if and only if $x \sim y(\chi_I)$.*

Proof $x \sim y(I)$ if and only if $x - y \in I$, if and only if $\chi_I(x - y) = 1$ and $1 - \chi_I(x - y) = 0$, if and only if $\chi_I(x - y) = \chi_I(0)$ and $1 - \chi_I(x - y) = 1 - \chi_I(0)$, if and only if $x \sim y(\chi_I)$. This proves the lemma.

Lemma 6 *Let I be a subset of a ring R. Then $<\chi_I, 1 - \chi_I>$ is an intuitionistic fuzzy prime ideal of R if and only if I is a prime ideal of R.*

Proof If $<\chi_I, 1-\chi_I>$ is an intuitionistic fuzzy prime ideal of R, then I is an ideal of R by Theorem 3 (i). For any $x, y \in R$, by Definition 7 we have $\chi_I(xy) = \chi_I(0) = 1$ implies $\chi_I(x) = \chi_I(0) = 1$ or $\chi_I(y) = \chi_I(0) = 1$.
It means $xy \in I$ implies $x \in I$ or $y \in I$.
Hence I is a prime ideal of R.
Conversely, if I is a prime ideal of R, then χ_I is an intuitionistic fuzzy ideal of R by Theorem 3 (i) again. For any $x, y \in R$, we have that $xy \in I$ implies $x \in I$ or $y \in I$.
It follows that $\chi_I(xy) = 1 = \chi_I(0)$ implies $\chi_I(x) = 1 = \chi_I(0)$ or $\chi_I(y) = 1 = \chi_I(0)$.
the seem for $(1-\chi_I)(xy) = 0 = (1-\chi_I)(0)$ implies $(1-\chi_I)(x) = 0 = (1-\chi_I)(0)$ or $(1-\chi_I)(y) = 0 = (1-\chi_I)(0)$
Thus $<\chi_I, 1-\chi_I>$ is an intuitionistic fuzzy prime ideal of R. The proof is completed.

By Lemma 6 we have $I_x = (\chi_I)_x$ and $R/I = R/\chi_I$. Combining Theorem 6 and Lemma 6 obtain the following corollary.

Corollary 1 *Let R be a commutative ring and I an ideal of R. Then I is prime if and only if R/I is an integral domain.*

Proof I is prime if and only if χ_I is prime, if and only if R/χ_I is an integral domain, if and only if R/I is an integral. The proof has completed.

Lemma 7 *If $A = <\mu, \nu>$ is an intuitionistic fuzzy ideal in a ring R, then $x \sim y(R_A)$ if and only if $x \sim y(A)$.*

Proof $x \sim y(R_A)$ if and only if $x - y \in R_A$, if and only if $\mu(x-y) = \mu(0)$ and $\nu(x-y) = \nu(0)$ if and only if $x \sim y(A)$.

Next we concern the quotient ring reduced by a fuzzy primary ideal.

Definition 8 An intuitionistic fuzzy ideal $<\mu, \nu>$ of a ring R is called intuitionistic fuzzy primary ideal if for all $x, y \in R$, either $\mu(xy) = \mu(0)$, $\mu(x) \neq \mu(0)$, $\nu(xy) = \nu(0)$ and $\nu(x) \neq \nu(0)$ imply $\mu(y^n) = \mu(0)$ and $\nu(y^n) = \nu(0)$ for some positive integer n.

Theorem 7 *Let R be a commutative ring and $A = <\mu, \nu>$ an intuitionistic fuzzy ideal of R. Then $<\mu, \nu>$ is primary if and only if every zero divisor in $R/<\mu, \nu>$ is nilpotent.*

Proof Assume that every zero divisor in $R/<\mu, \nu>$ is nilpotent. Let $\mu(xy) = \mu(0)$ and $\mu(x) \neq \mu(0)$ and $\nu(xy) = \nu(0)$ and $\nu(x) \neq \nu(0)$, i.e., $\mu_x\mu_y = \mu_{xy} = \mu_0$ and $\mu_x \neq \mu_0$.

Thus A_y is a zero divisor of $R/<\mu, \nu>$, and so there exists a positive integer n such that $(A_y)^n = A_{y^n} = A_0$.

Hence $A(y^n) = A(0)$ and $<\mu, \nu>$ is primary of R.

Conversely let $A = <\mu, \nu>$ be an intuitionistic fuzzy primary ideal, and A_y a zero divisor of $R/<\mu, \nu>$. Then there exists $A_x \in R/<\mu, \nu>$ and $A_x \neq A_0$ such that $A_xA_y = A_0$, i.e., $A(xy) = A(0)$ and $A(x) \neq A(0)$.

By the 8 we have $A(y^n) = A(0)$ for some positive integer n, i.e., $(A_y)^n = A_{y^n} = A_0$, ending the proof.

Lemma 8 *Let I be a subset of a ring R. Then* $<\chi_I, 1-\chi_I>$ *is an intuitionistic fuzzy primary ideal of R if and only if I is a primary ideal of R.*

Proof If $A = <\chi_I, 1-\chi_I>$ is an intuitionistic fuzzy primary ideal of R, then I is an ideal of R.

For any $x, y \in R$, by Definition 8 $\chi_I(xy) = \chi_I(0) = 1$ and $\chi_I(x) \neq \chi_I(0)$ and $(1-\chi_I)(xy) = (1-\chi_I)(0) = 0$ and $(1-\chi_I)(x) \neq (1-\chi_I)(0)$ imply $\chi_I(y^n) = \chi_I(0) = 1$ and $(1-\chi_I)(y^n) = (1-\chi_I)(0) = 0$ for some positive integer n.

That is $xy \in I$ and $x \notin I$ imply $y^n \in I$ for some positive integer n.

Therefore I is a primary ideal of R.

Conversely, if I is primary, then $A = <\chi_I, 1-\chi_I>$ is an intuitionistic fuzzy ideal of R.

For any $x, y \in R$, we have that $xy \in I$ and $x \notin I$ imply $y^n \in I$ for some positive integer n. Thus $\chi_I(xy) = 1 = \chi_I(0)$ and $\chi_I(x) \neq \chi_I(0)$ and $1-\chi_I(xy) = 0 = 1-\chi_I(0)$ and $1-\chi_I(x) \neq 1-\chi_I(0)$ imply $\chi_I(y^n) = \chi_I(0)$ and $(1-\chi_I)(y^n) = 1-\chi_I(0)$ for some positive integer n.

Hence $<\chi_I, 1-\chi_I>$ is an intuitionistic fuzzy primary ideal of R, completing the proof.

Combining Theorem 7, Lemmas 8 and 5, We have following

Corollary 2 *Let R be a commutative ring and I an ideal of R. Then I is primary if and only if every zero divisor in* R/I *is nilpotent.*

References

1. K. Atanassov, S. Stoeva, Intuitionistic fuzzy sets, in *Proceedings Polish Symposium on Interval and Fuzzy Mathematics* (Poznan, 1983), pp. 23–26
2. K. Atanassov, Intuitionistic fuzzy sets. Fuzzy Sets Syst. **20**, 87–96 (1986)
3. I. Bakhadach, S. Melliani, M. Oukessou, L.S. Chadli, Intuitionistic fuzzy ideal and intuitionistic fuzzy prime ideal in a ring. Notes Intuitionistic Fuzzy Sets **22**(2), 59–63 (2016)
4. R. Biswas, Intuitionistic fuzzy subgroup. Math. Forum **X**, 37–46 (1989)
5. D. Coker, An introduction to intuitionistic fuzzy topological spaces. Fuzzy Sets Syst. **88**, 81–89 (1997)
6. K. Hur, H.W. Kang, H.K. Song, Intuitionistic fuzzy subgroups and subrings. Honam Math. J. **25**(1), 19–41 (2003)
7. K. Hur, S.Y. Jang, H.W. Kang, Intuitionistic fuzzy ideal of a ring. J. Korea Soc. Math. Educ. Ser. B. Pure Appl. Math. **12**, 193–209 (2005)
8. N. Kuroki, Fuzzy bi-ideals in semigroups. Comment Math. Univ. St. Paul. **28**, 17–21 (1980)
9. N. Kuroki, Fuzzy semiprime ideals in semigroups. Fuzzy Sets Syst. **8**, 71–79 (1982)
10. D.S. Malik, J.N. Mordeson, Fuzzy prime ideals of a ring. Fuzzy Sets Syst. **37**, 93–98 (1990)
11. D.S. Malik, J.N. Mordeson, Fuzzy maximal, radical and primary ideals of a ring. Inf. Sci. **5**, 237–250 (1991)
12. M.F. Marashdeh, A.R. Salleh, Intuitionistic fuzzy rings. Int. J. Algebra **5**(1), 37–47 (2011)
13. K. Meena, K.V. Thomas, Intuitionistic L-fuzzy subrings. Int. Math. Forum **6**(52), 2561–2572 (2011)
14. S. Melliani, I. Bakhadach, L.S. Chadli, Intuitionistic Fuzzy Group With Extended Operations. Springer Proceedings in Mathematics and Statistics **228**, 55–65 (2016)

15. P.M. Pu, Y.M. Liu, Fuzzy topology. I. Neighborhood structure of a fuzzy point and Moore Smith convergence. J. Math. Anal. Appl. **76**, 571–599 (1980)
16. A. Rosenfeld, Fuzzy groups. J. Math. Anal. Appl **35**, 512–517 (1971)
17. P.K. Sharma, Translates of intuitionistic fuzzy subring. Int. Rev. Fuzzy Math. **6**(2), 77–84 (2011)
18. P.K. Sharma, (α, β) Cut of intuitionistic fuzzy groups. Int. Math. Forum **6**(53), 2605–2614 (2011)
19. P.K. Sharma, t-intuitionistic fuzzy quotient group. Adv. Fuzzy Math. **7**(1), 1–9 (2012)
20. L.A. Zadeh, Fuzzy sets. Inf. Control **8**, 338–353 (1965)

A Fourth Order Runge-Kutta Gill Method for the Numerical Solution of Intuitionistic Fuzzy Differential Equations

B. Ben Amma, Said Melliani and L. S. Chadli

Abstract In this paper we study the numerical methods for Intuitionistic Fuzzy Differential equations by an application of the Fourth Order Runge-Kutta Gill method for intuitionistic fuzzy differential equations. We give a numerical example to illustrate the theory.

1 Introduction

Generalizations of fuzzy sets theory [40] is considered to be one of intuitionistic fuzzy set (IFS). Later on Atanassov generalized the concept of fuzzy set and introduced the idea of intuitionistic fuzzy set [3, 7]. Atanassov [4] explored the concept of fuzzy set theory by intuitionistic fuzzy set (IFS) theory. Now-a-days, IFSs are being studied extensively and being used in different disciplines of Science and Technology. Amongst the all research works mainly on IFS we can include [2, 5, 6, 8–13, 19, 21, 22, 27, 28, 31, 34]. They are very necessary and powerful tool in modeling imprecision, valuable applications of IFSs have been flourished in many different fields, medical diagnosis [20], drug selection [23], along pattern recognition [24], microelectronic fault analysis [35], weight assessment [38], and decision-making problems [25, 39].

For intuitionistic fuzzy differential equations concepts, recently the authors [17, 18, 26, 29, 30, 36, 37] established, respectively, the cauchy problem for intuitionistic fuzzy differential equations, intuitionistic fuzzy functional differential equations, the intuitionistic fuzzy differential equations, intuitionistic fuzzy fractional equation, intuitionistic fuzzy differential equation with nonlocal condition, first order homoge-

B. Ben Amma · S. Melliani (✉) · L. S. Chadli
Sultan Moulay Slimane University, BP 523 Beni Mellal, Morocco
e-mail: s.melliani@usms.ma

B. Ben Amma
e-mail: bouchrabenamma@gmail.com

L. S. Chadli
e-mail: sa.chadli@yahoo.fr

S. Melliani and O. Castillo (eds.), *Recent Advances in Intuitionistic Fuzzy Logic Systems*, Studies in Fuzziness and Soft Computing 372,
https://doi.org/10.1007/978-3-030-02155-9_5

neous ordinary differential equation with initial value as triangular intuitionistic fuzzy number and system of differential equation with initial value as triangular intuitionistic fuzzy number and its application. They proved the existence and uniqueness of the intuitionistic fuzzy solution for these intuitionistic fuzzy differential equations using different concepts. There are only few applications of numerical methods such as the intuitionistic fuzzy Euler and Taylor methods, Runge-Kutta of order four, Adams-Bashforth, Adams-Moulton and Predictor-Corrector methods in intuitionistic fuzzy differential equations presented in [1, 14–16, 32, 33]. In this paper, intuitionistic fuzzy Cauchy problem is solved numerically by Runge-Kutta Gill method of order four.

The paper is organized as follows. In Sect. 2, some basic definitions and results are brought. Section 3 contains intuitionistic fuzzy differential equation whose numerical solution is the main interest of this paper. Solving numerically the intuitionistic fuzzy differential equation by using fourth order Runge-Kutta Gill method in Sect. 4. An example is presented in Sect. 5, and finally conclusion is drawn.

2 Preliminaries

2.1 Notations and Definitions

Consider the initial value problem:

$$\begin{cases} x'(t) = f(t, x(t)), & t \in [t_0, T] \\ x(t_0) = x_0 \end{cases} \tag{1}$$

The basis of all Runge-Kutta method is to express the difference between the value of y at t_{n+1} and t_n as

$$x_{n+1} - x_n = \sum_{i=1}^{i=m} w_i k_i, \tag{2}$$

where for $i = 1, 2, 3, \ldots, m$ the w_i's are constants and

$$k_i = hf\big(t_n + c_i h, x_n + \sum_{j=1}^{i-1} \beta_{ij} k_j\big) \tag{3}$$

The famous nonzero constants c_i, β_{ij} in the Runge-Kutta Gill method of order four are:

$$c_1 = 0, c_2 = \frac{1}{2}, c_3 = \frac{1}{2}, c_4 = 1,$$

$$\beta_{21} = \frac{1}{2}, \beta_{31} = \frac{1}{\sqrt{2}} - \frac{1}{2}, \beta_{32} = 1 - \frac{1}{\sqrt{2}},$$
$$\beta_{41} = 0, \beta_{42} = -\frac{1}{\sqrt{2}}, \beta_{43} = 1 + \frac{1}{\sqrt{2}}$$

where m = 4. Hence we have

$$k_1 = hf(t_i, x_i)$$
$$k_2 = hf(t_i + \frac{1}{2}h, x_i + \frac{1}{2}k_1)$$
$$k_3 = hf(t_i + \frac{1}{2}h, x_i + (\frac{1}{\sqrt{2}} - \frac{1}{2})k_1 + (1 - \frac{1}{\sqrt{2}})k_2)$$
$$k_4 = hf(t_i + h, x_i - \frac{1}{\sqrt{2}}k_2 + (1 + \frac{1}{\sqrt{2}})k_3)$$

$$x_{i+1} = x_i + \frac{1}{6}(k_1 + (2 - \sqrt{2})k_2 + (2 + \sqrt{2})k_3 + k_4)$$

where

$$t_0 < t_1 < t_2 < \cdots < t_N = T,\ h = \frac{T - t_0}{N},\quad t_i = t_0 + ih,\ \ i = 0, 1, \ldots, N \quad (4)$$

Throughout this paper, $(\mathbb{R}, B(\mathbb{R}), \mu)$ denotes a complete finite measure space.

Let us $P_k(\mathbb{R})$ the set of all nonempty compact convex subsets of $\mathbb{R}$.

we denote by

$$\mathbb{IF}_1 = \mathbb{IF}(\mathbb{R}) = \{\langle u, v\rangle : \mathbb{R} \to [0, 1]^2, |\forall\, x \in \mathbb{R}\ 0 \leq u(x) + v(x) \leq 1\}$$

An element $\langle u, v\rangle$ of $\mathbb{IF}_1$ is said an intuitionistic fuzzy number if it satisfies the following conditions

(i) $\langle u, v\rangle$ is normal i.e there exists $x_0, x_1 \in \mathbb{R}$ such that $u(x_0) = 1$ and $v(x_1) = 1$.
(ii) u is fuzzy convex and v is fuzzy concave.
(iii) u is upper semi-continuous and v is lower semi-continuous
(iv) $supp\langle u, v\rangle = \text{cl}\{x \in \mathbb{R} : | v(x) < 1\}$ is bounded.

So we denote the collection of all intuitionistic fuzzy number by $\mathbb{IF}_1$.

Definition 1 Let $\langle u, v\rangle$ an element of $\mathbb{IF}_1$ and $\alpha \in [0, 1]$, we define the following sets:

$$\Big[\langle u, v\rangle\Big]_l^+(\alpha) = \inf\{x \in \mathbb{R} \mid u(x) \geq \alpha\},\quad \Big[\langle u, v\rangle\Big]_r^+(\alpha) = \sup\{x \in \mathbb{R} \mid u(x) \geq \alpha\}$$
$$\Big[\langle u, v\rangle\Big]_l^-(\alpha) = \inf\{x \in \mathbb{R} \mid v(x) \leq 1 - \alpha\},\quad \Big[\langle u, v\rangle\Big]_r^-(\alpha) = \sup\{x \in \mathbb{R} \mid v(x) \leq 1 - \alpha\}$$

Remark 1

$$\Big[\langle u,v\rangle\Big]_\alpha = \Big[\Big[\langle u,v\rangle\Big]_l^+(\alpha), \Big[\langle u,v\rangle\Big]_r^+(\alpha)\Big]$$
$$\Big[\langle u,v\rangle\Big]^\alpha = \Big[\Big[\langle u,v\rangle\Big]_l^-(\alpha), \Big[\langle u,v\rangle\Big]_r^-(\alpha)\Big]$$

A Triangular Intuitionistic Fuzzy Number (TIFN) $\langle u, v\rangle$ is an intuitionistic fuzzy set in $\mathbb{R}$ with the following membership function u and non-membership function v:

$$u(x) = \begin{cases} \frac{x-a_1}{a_2-a_1} \text{ if } a_1 \le x \le a_2 \\ \frac{a_3-x}{a_3-a_2} \text{ if } a_2 \le x \le a_3 \\ 0 \quad \text{otherwise} \end{cases}$$

$$v(x) = \begin{cases} \frac{a_2-x}{a_2-a_1'} \text{ if } a_1' \le x \le a_2 \\ \frac{x-a_2}{a_3'-a_2} \text{ if } a_2 \le x \le a_3' \\ 1 \quad \text{otherwise} \end{cases}$$

where $a_1' \le a_1 \le a_2 \le a_3 \le a_3'$ and $u(x), v(x) \le 0.5$ for $u(x) = v(x)$, $\forall x \in \mathbb{R}$

This TIFN is denoted by $\langle u, v\rangle = \langle a_1, a_2, a_3; a_1', a_2, a_3'\rangle$ where,

$$[\langle u, v\rangle]_\alpha = [a_1 + \alpha(a_2 - a_1), a_3 - \alpha(a_3 - a_2)] \tag{5}$$

$$[\langle u, v\rangle]^\alpha = [a_1' + \alpha(a_2 - a_1'), a_3' - \alpha(a_3' - a_2)] \tag{6}$$

Theorem 1 ([28]) *d_∞ define a metric on IF_1.*

Theorem 2 ([28]) *The metric space (IF_1, d_∞) is complete.*

Remark 2 If $F : [a, b] \to \mathbb{F}_1$ is Hukuhara differentiable and its Hukuhara derivative F' is integrable over $[0, 1]$ then

$$F(t) = F(t_0) \oplus \int_{t_0}^{t} F'(s)ds$$

Definition 2 Let $\langle u, v\rangle$ and $\langle u', v'\rangle \in IF_1$, the H-difference is the IFN $\langle z, w\rangle \in IF_1$, if it exists, such that

$$\langle u, v\rangle \ominus \langle u', v'\rangle = \langle z, w\rangle \Longleftrightarrow \langle u, v\rangle = \langle u', v'\rangle \oplus \langle z, w\rangle$$

Definition 3 A mapping $F : [a, b] \to IF_1$ is said to be Hukuhara derivable at t_0 if there exist $F'(t_0) \in IF_1$ such that both limits:

$$\lim_{\Delta t \to 0^+} \frac{F(t_0 + \Delta t) \ominus F(t_0)}{\Delta t}$$

and

$$\lim_{\Delta t \to 0^+} \frac{F(t_0) \ominus F(t_0 - \Delta t)}{\Delta t}$$

exist and they are equal to $F'(t_0) = \langle u'(t_0), v'(t_0) \rangle$, which is called the Hukuhara derivative of F at t_0.

3 Intuitionistic Fuzzy Cauchy Problem

In this section we consider the initial value problem for the intuitionistic fuzzy differential equation

$$\begin{cases} x'(t) = f(t, x(t)), \quad t \in I \\ x(t_0) = \langle u_{t_0}, v_{t_0} \rangle \in IF_1 \end{cases} \tag{7}$$

where $x \in IF_1$ is unknown $I = [t_0, T]$ and $f : I \times IF_1 \to IF_1$.

$x(t_0)$ is intuitionistic fuzzy number.

Denote the α-level set

$$[x(t)]_\alpha = \left[[x(t)]_l^+(\alpha), [x(t)]_r^+(\alpha) \right]$$
$$[x(t)]^\alpha = \left[x(t)]_l^-(\alpha), [x(t)]_r^-(\alpha) \right]$$

and

$$[x(t_0)]_\alpha = \left[[x(t_0)]_l^+(\alpha), [x(t_0)]_r^+(\alpha) \right]$$
$$[x(t_0)]^\alpha = \left[x(t_0)]_l^-(\alpha), [x(t_0)]_r^-(\alpha) \right]$$

$$[f(t, x(t))]_\alpha = \left[f_l^+(t, x(t); \alpha), [f_r^+(t, x(t); \alpha) \right]$$
$$[f(t, x(t))]^\alpha = \left[f_l^-(t, x(t); \alpha), [f_r^-(t, x(t); \alpha) \right]$$

where

$$
\begin{aligned}
f_1^+(t,x(t);\alpha) &= \min\left\{f(t,u)|u\in\left[[x(t)]_l^+(\alpha),[x(t)]_r^+(\alpha)\right]\right\}\\
f_2^+(t,x(t);\alpha) &= \max\left\{f(t,u)|u\in\left[[x(t)]_l^+(\alpha),[x(t)]_r^+(\alpha)\right]\right\}\\
f_3^-(t,x(t);\alpha) &= \min\left\{f(t,u)|u\in\left[x(t)\right]_l^-(\alpha),[x(t)]_r^-(\alpha)\right]\right\}\\
f_4^-(t,x(t);\alpha) &= \max\left\{f(t,u)|u\in\left[x(t)\right]_l^-(\alpha),[x(t)]_r^-(\alpha)\right]\right\} \qquad (8)
\end{aligned}
$$

Denote

$$
\begin{aligned}
f_1^+(t,x(t);\alpha) &= G\big(t,[x(t)]_l^+(\alpha),[x(t)]_r^+(\alpha)\big)\\
f_2^+(t,x(t);\alpha) &= H\big(t,[x(t)]_l^+(\alpha),[x(t)]_r^+(\alpha)\big)\\
f_3^-(t,x(t);\alpha) &= L\big(t,[x(t)]_l^-(\alpha),[x(t)]_r^-(\alpha)\big)\\
f_4^-(t,x(t);\alpha) &= K\big(t,[x(t)]_l^-(\alpha),[x(t)]_r^-(\alpha)\big) \qquad (9)
\end{aligned}
$$

Sufficient conditions for the existence of an unique solution to Eq. (7) are:

1. Continuity of f
2. Lipschitz condition: for any pair $\big(t,\langle u,v\rangle\big),\big(t,\langle u',v'\rangle\big)\in I\times \mathbb{F}_1$, we have

$$
d_\infty\Big(f\big(t,\langle u,v\rangle\big),f\big(t,\langle u',v'\rangle\big)\Big)\le K d_\infty\Big(\langle u,v\rangle,\langle u',v'\rangle\Big) \qquad (10)
$$

where $K>0$ is a given constant.

Theorem 3 [17] *Let us suppose that the following conditions hold:*

(a) Let $R_0=[t_0,t_0+p]\times\overline{B}(\langle u,v\rangle_{t_0},q)$, $p,q\ge 0,\langle u,v\rangle_{t_0}\in IF_1$ *where* $\overline{B}(\langle u,v\rangle_{t_0},q)=\{\langle u,v\rangle\in IF_1: d_\infty(\langle u,v\rangle,\langle u,v\rangle_{t_0})\le q\}$ *denote a closed ball in* IF_1 *and let* $f:R_0\longrightarrow IF_1$ *be a continuous function such that* $d_\infty\big(f(t,\langle u,v\rangle),0_{(1,0)}\big)\le M$ *for all* $(t,\langle u,v\rangle)\in R_0$.
(b) Let $g:[t_0,t_0+p]\times[0,q]\longrightarrow\mathbb{R}$ *such that* $g(t,0)\equiv 0$ *and* $0\le g(t,x)\le M_1$, $\forall t\in[t_0,t_0+p], 0\le x\le q$ *such that* $g(t,x)$ *is non-decreasing in* u *and* g *is such that the initial value problem*

$$
x'(t)=g(t,x(t)),x(t_0)=x_0. \qquad (11)
$$

has only the solution $x(t)\equiv 0$ *on* $[t_0,t_0+p]$
(c) We have $d_\infty\Big(f\big(t,\langle u,v\rangle\big),f\big(t,\langle z,w\rangle\big)\Big)\le g(t,d_\infty(\langle u,v\rangle,\langle z,w\rangle)),\forall(t,\langle u,v\rangle)$, $(t,\langle z,w\rangle)\in R_0$ *and* $d_\infty(\langle u,v\rangle,\langle z,w\rangle)\le q$. *Then the intuitionistic fuzzy initial value problem*

$$
\begin{cases}\langle u,v\rangle' &= f(t,\langle u,v\rangle),\\ \langle u,v\rangle(t_0) &= \langle u,v\rangle_{t_0}\end{cases} \qquad (12)
$$

has an unique solution $\langle u,v\rangle\in C^1[[t_0,t_0+r],B(x_0,q)]$ *on* $[t_0,t_0+r]$ *where* $r=\min\{p,\frac{q}{M},\frac{q}{M_1},d\}$ *and the successive iterations*

$$\langle u, v\rangle_0(t) = \langle u, v\rangle_{t_0}, \quad \langle u, v\rangle_{n+1}(t) = \langle u, v\rangle_{t_0} + \int_{t_0}^{t} f(s, \langle u, v\rangle_n(s))ds \quad (13)$$

converge to $\langle u, v\rangle(t)$ *on* $[t_0, t_0 + r]$.

4 Fourth Order Runge-Kutta Gill Method

Let the exact solutions

$$[X(t_n)]_\alpha = \Big[[X(t_n)]_l^+(\alpha), [X(t_n)]_r^+(\alpha)\Big], \quad [X(t_n)]^\alpha = \Big[[X(t_n)]_l^-(\alpha), [X(t_n)]_r^-(\alpha)\Big]$$

be approximated by

$$[x(t_n)]_\alpha = \Big[[x(t_n)]_l^+(\alpha), [x(t_n)]_r^+(\alpha)\Big], \quad [x(t_n)]^\alpha = \Big[[x(t_n)]_l^-(\alpha), [x(t_n)]_r^-(\alpha)\Big]$$

at t_n, $0 \le n \le N$

be approximated solutions at t_n, $0 \le n \le N$

The solutions are calculated by grid points at

$$t_0 < t_1 < t_2 < \cdots < t_N = T, \; h = \frac{T - t_0}{N}, \quad t_n = t_0 + nh, \quad n = 0, 1, \ldots, N \quad (14)$$

From (2) and (3) we define

$$[x(t_{n+1})]_l^+(\alpha) - [x(t_n)]_l^+(\alpha) = \sum_{i=1}^{i=4} w_i [k_i]_l^+(\alpha) \quad (15)$$

$$[x(t_{n+1})]_r^+(\alpha) - [x(t_n)]_r^+(\alpha) = \sum_{i=1}^{i=4} w_i [k_i]_r^+(\alpha) \quad (16)$$

$$[x(t_{n+1})]_l^-(\alpha) - [x(t_n)]_l^-(\alpha) = \sum_{i=1}^{i=4} w_i [k_i]_l^-(\alpha) \quad (17)$$

$$[x(t_{n+1})]_r^-(\alpha) - [x(t_n)]_r^-(\alpha) = \sum_{i=1}^{i=4} w_i [k_i]_r^-(\alpha) \quad (18)$$

where the $w_i's$ are constants and

$$
\begin{cases}
[k_i]_\alpha = \Big[[k_i]_l^+(\alpha), [k_i]_r^+(\alpha)\Big], \quad i = 1, 2, 3, 4 \\
[k_i]^\alpha = \Big[[k_i]_l^-(\alpha), [k_i]_r^-(\alpha)\Big] \\
[k_i]_l^+(\alpha) = hG\Big(t_n + c_i h, [x(t_n)]_l^+ + \sum_{j=1}^{i-1} \beta_{ij}[k_j]_l^+(\alpha), \\
\qquad [x(t_n)]_r^+ + \sum_{j=1}^{i-1} \beta_{ij}[k_j]_r^+(\alpha)\Big) \\
[k_i]_r^+(\alpha) = hH\Big(t_n + c_i h, [x(t_n)]_l^+ + \sum_{j=1}^{i-1} \beta_{ij}[k_j]_l^+(\alpha), \\
\qquad [x(t_n)]_r^+ + \sum_{j=1}^{i-1} \beta_{ij}[k_j]_r^+(\alpha)\Big) \\
[k_i]_l^-(\alpha) = hL\Big(t_n + c_i h, [x(t_n)]_l^- + \sum_{j=1}^{i-1} \beta_{ij}[k_j]_l^-(\alpha), \\
\qquad [x(t_n)]_r^- + \sum_{j=1}^{i-1} \beta_{ij}[k_j]_r^-(\alpha)\Big) \\
[k_i]_r^-(\alpha) = hK\Big(t_n + c_i h, [x(t_n)]_l^- + \sum_{j=1}^{i-1} \beta_{ij}[k_j]_l^-(\alpha), \\
\qquad [x(t_n)]_r^- + \sum_{j=1}^{i-1} \beta_{ij}[k_j]_r^-(\alpha)\Big)
\end{cases}
$$

Assuming the following Runge-Kutta Gill method with four slopes:

$$
\begin{aligned}
[x(t_{n+1})]_l^+(\alpha) = [x(t_n)]_l^+(\alpha) + \frac{1}{6}([k_1]_l^+(\alpha) + (2 - \sqrt{2})[k_2]_l^+(\alpha) \\
+ (2 + \sqrt{2})[k_3]_l^+(\alpha) + [k_4]_l^+(\alpha))
\end{aligned} \tag{19}
$$

where

$$
\begin{cases}
[k_1]_l^+(\alpha) = hG\big(t_n, [x(t_n)]_l^+(\alpha), [x(t_n)]_r^+(\alpha)\big) \\
[k_2]_l^+(\alpha) = hG\big(t_n + \frac{1}{2}h, [x(t_n)]_l^+(\alpha) + \frac{1}{2}[k_1]_l^+(\alpha), [x(t_n)]_r^+(\alpha) + \frac{1}{2}[k_1]_r^+(\alpha)\big) \\
[k_3]_l^+(\alpha) = hG\big(t_n + \frac{1}{2}h, [x(t_n)]_l^+(\alpha) + (\frac{\sqrt{2}-1}{2})[k_1]_l^+(\alpha) + (\frac{2-\sqrt{2}}{2})[k_2]_l^+(\alpha), \\
\qquad [x(t_n)]_r^+(\alpha) + (\frac{\sqrt{2}-1}{2})[k_1]_r^+(\alpha) + (\frac{2-\sqrt{2}}{2})[k_2]_r^+(\alpha)\big) \\
[k_4]_l^+(\alpha) = hG\big(t_n + h, [x(t_n)]_l^+(\alpha) - \frac{1}{\sqrt{2}}[k_2]_l^+(\alpha) + (1 + \frac{1}{\sqrt{2}})[k_3]_l^+(\alpha), \\
\qquad [x(t_n)]_r^+(\alpha) - \frac{1}{\sqrt{2}}[k_2]_r^+(\alpha) + (1 + \frac{1}{\sqrt{2}})[k_3]_r^+(\alpha)\big)
\end{cases}
$$

and

$$
\begin{aligned}
[x(t_{n+1})]_r^+(\alpha) = [x(t_n)]_r^+(\alpha) + \frac{1}{6}([k_1]_r^+(\alpha) + (2 - \sqrt{2})[k_2]_r^+(\alpha) \\
+ (2 + \sqrt{2})[k_3]_r^+(\alpha) + [k_4]_r^+(\alpha))
\end{aligned} \tag{20}
$$

where

$$\begin{cases} [k_1]_r^+(\alpha) = hH\big(t_n, [x(t_n)]_l^+(\alpha), [x(t_n)]_r^+(\alpha)\big) \\ [k_2]_r^+(\alpha) = hH\big(t_n + \frac{1}{2}h, [x(t_n)]_l^+(\alpha) + \frac{1}{2}[k_1]_l^+(\alpha), [x(t_n)]_r^+(\alpha) + \frac{1}{2}[k_1]_r^+(\alpha)\big) \\ [k_3]_r^+(\alpha) = hH\big(t_n + \frac{1}{2}h, [x(t_n)]_l^+(\alpha) + (\frac{\sqrt{2}-1}{2})[k_1]_l^+(\alpha) + (\frac{2-\sqrt{2}}{2})[k_2]_l^+(\alpha), \\ \qquad [x(t_n)]_r^+(\alpha) + (\frac{\sqrt{2}-1}{2})[k_1]_r^+(\alpha) + (\frac{2-\sqrt{2}}{2})[k_2]_r^+(\alpha)\big) \\ [k_4]_r^+(\alpha) = hH\big(t_n + h, [x(t_n)]_l^+(\alpha) - \frac{1}{\sqrt{2}}[k_2]_l^+(\alpha) + (1 + \frac{1}{\sqrt{2}})[k_3]_l^+(\alpha), \\ \qquad [x(t_n)]_r^+(\alpha) - \frac{1}{\sqrt{2}}[k_2]_r^+(\alpha) + (1 + \frac{1}{\sqrt{2}})[k_3]_r^+(\alpha)\big) \end{cases}$$

and

$$[x(t_{n+1})]_l^-(\alpha) = [x(t_n)]_l^-(\alpha) + \frac{1}{6}([k_1]_l^-(\alpha) + (2 - \sqrt{2})[k_2]_l^-(\alpha) + (2 + \sqrt{2})[k_3]_l^-(\alpha)) + [k_4]_l^-(\alpha)) \tag{21}$$

where

$$\begin{cases} [k_1]_l^-(\alpha) = hL\big(t_n, [x(t_n)]_l^-(\alpha), [x(t_n)]_r^-(\alpha)\big) \\ [k_2]_l^-(\alpha) = hL\big(t_n + \frac{1}{2}h, [x(t_n)]_l^-(\alpha) + \frac{1}{2}[k_1]_l^-(\alpha), [x(t_n)]_r^-(\alpha) + \frac{1}{2}[k_1]_r^-(\alpha)\big) \\ [k_3]_l^-(\alpha) = hL\big(t_n + \frac{1}{2}h, [x(t_n)]_l^-(\alpha) + (\frac{\sqrt{2}-1}{2})[k_1]_l^-(\alpha) + (\frac{2-\sqrt{2}}{2})[k_2]_l^-(\alpha), \\ \qquad [x(t_n)]_r^-(\alpha) + (\frac{\sqrt{2}-1}{2})[k_1]_r^-(\alpha) + (\frac{2-\sqrt{2}}{2})[k_2]_r^-(\alpha)\big) \\ [k_4]_l^-(\alpha) = hL\big(t_n + h, [x(t_n)]_l^-(\alpha) - \frac{1}{\sqrt{2}}[k_2]_l^-(\alpha) + (1 + \frac{1}{\sqrt{2}})[k_3]_l^-(\alpha), \\ \qquad [x(t_n)]_r^-(\alpha) - \frac{1}{\sqrt{2}}[k_2]_r^-(\alpha) + (1 + \frac{1}{\sqrt{2}})[k_3]_r^-(\alpha)\big) \end{cases}$$

and

$$[x(t_{n+1})]_r^-(\alpha) = [x(t_n)]_r^-(\alpha) + \frac{1}{6}([k_1]_r^-(\alpha) + (2 - \sqrt{2})[k_2]_r^-(\alpha) + (2 + \sqrt{2})[k_3]_r^-(\alpha) + [k_4]_r^-(\alpha)) \tag{22}$$

where

$$\begin{cases} [k_1]_r^-(\alpha) = hK\big(t_n, [x(t_n)]_l^-(\alpha), [x(t_n)]_r^-(\alpha)\big) \\ [k_2]_r^-(\alpha) = hK\big(t_n + \frac{1}{2}h, [x(t_n)]_l^-(\alpha) + \frac{1}{2}[k_1]_l^-(\alpha), [x(t_n)]_r^-(\alpha) + \frac{1}{2}[k_1]_r^-(\alpha)\big) \\ [k_3]_r^-(\alpha) = hK\big(t_n + \frac{1}{2}h, [x(t_n)]_l^-(\alpha) + (\frac{\sqrt{2}-1}{2})[k_1]_l^-(\alpha) + (\frac{2-\sqrt{2}}{2})[k_2]_l^-(\alpha), \\ \qquad [x(t_n)]_r^-(\alpha) + (\frac{\sqrt{2}-1}{2})[k_1]_r^-(\alpha) + (\frac{2-\sqrt{2}}{2})[k_2]_r^-(\alpha)\big) \\ [k_4]_r^-(\alpha) = hK\big(t_n + h, [x(t_n)]_l^-(\alpha) - \frac{1}{\sqrt{2}}[k_2]_l^-(\alpha) + (1 + \frac{1}{\sqrt{2}})[k_3]_l^-(\alpha), \\ \qquad [x(t_n)]_r^-(\alpha) - \frac{1}{\sqrt{2}}[k_2]_r^-(\alpha) + (1 + \frac{1}{\sqrt{2}})[k_3]_r^-(\alpha)\big) \end{cases}$$

Let $G(t, u^+, v^+)$, $H(t, u^+, v)^+$, $L(t, u^-, v)^-$ and $K(t, u^-, v^-)$ be the functions of (9), where u^+, v^+, u^- and v^- are the constants and $u^+ \leq v^+$ and $u^- \leq v^-$.

The domain of G and H is

$$M_1 = \{(t, u^+, v^+) \backslash\ t_0 \leq t \leq T,\ \infty < u^+ \leq v^+,\ \ -\infty < v^+ < +\infty\}$$

and the domain of L and K is

$$M_2 = \{(t, u^-, v^-) \backslash \ t_0 \leq t \leq T, \ \infty < u^- \leq v^-, \ -\infty < v^- < +\infty\}$$

where $M_1 \subseteq M_2$

Theorem 4 *Let $G(t, u^+, v^+)$, $H(t, u^+, v^+)$ belong to $C^4(M_1)$ and $L(t, u^-, v^-)$, $K(t, u^-, v^-)$ belong to $C^4(M_2)$ and the partial derivatives of G, H and L, K be bounded over M_1 and M_2 respectively. Then, for arbitrarily fixed $0 \leq \alpha \leq 1$, the numerical solutions of (19), (20), (21) and (22) converge to the exact solutions $[X(t)]_l^+(\alpha)$, $[X(t)]_r^+(\alpha)$, $[X(t)]_l^-(\alpha)$ and $[X(t)]_r^-(\alpha)$ uniformly in t.*

Proof See [14].

5 Example

Example Consider the intuitionistic fuzzy initial value problem

$$\begin{cases} x'(t) = x(t) \text{ for all } t \in [0, T] \\ x_0 \quad = \Big((\alpha - 1.0013, 1.0013 - \alpha), (-1.58\alpha, 1.58\alpha)\Big) \end{cases} \tag{23}$$

Then, we have the following parametrized differential system:

$$\begin{cases} [x(t)]_l^+(\alpha) = (\alpha - 1.0013)\exp(t) \\ [x(t)]_r^+(\alpha) = (1.0013 - \alpha)\exp(t) \\ \\ [x(t)]_l^-(\alpha) = 1.58(\alpha - 1.0013)\exp(t) \\ \\ [x(t)]_r^-(\alpha) = 1.58(1.0013 - \alpha)\exp(t) \end{cases}$$

Therefore the exact solutions is given by

$$[X(t)]_\alpha = \Big[(\alpha - 1.0013)\exp(t), (1.0013 - \alpha)\exp(t)\Big]$$

$$[X(t)]^\alpha = \Big[(1.58(\alpha - 1.0013)\exp(t), 1.58(1.0013 - \alpha)\exp(t)\Big]$$

which at $t = 2$ are

$$[X(2)]_\alpha = \Big[(\alpha - 1.0013)\exp(2), (1.0013 - \alpha)\exp(2)\Big]$$

$$[X(2)]^\alpha = \Big[(1.58(\alpha - 1.0013)\exp(2), 1.58(1.0013 - \alpha)\exp(2)\Big]$$

Comparison of results of the fourth order Runge-Kutta Gill method and the Euler and 2nd-order Taylor method in [14] for $N = 25$ and $t = 2$:

	RK-Gill		Exact	
α	$([x]_l^+, [x]_r^+)$	$([x]_l^-, [x]_r^-)$	$([X]_l^+, [X]_r^+)$	$([X]_l^-, [X]_r^-)$
0	(−7.39865714, 7.39865714)	(0,0)	(−7.39866187, 7.39866187)	(0,0)
0.2	(−5.92084687, 5.92084687)	(−2.33494023, 2.33494023)	(−5.92085065, 5.92085065)	(−2.33494172, −2.33494172)
0.4	(−4.44303659, 4.44303659)	(−4.66988047, 4.66988047)	(−4.44303943, 4.44303943)	(−4.66988345, 4.66988345)
0.6	(−2.96522631, 2.96522631)	(−7.00482070, 7.00482070)	(−2.96522821, 2.96522821)	(−7.00482518, 7.00482518)
0.8	(−1.48741604, 1.48741604)	(−9.33976094, 9.33976094)	(−1.48741699, 1.48741699)	(−9.33976690, 9.33976690)
1	(−0.00960576, 0.00960576)	(−11.67470117, 11.67470117)	(−0.00960577, 0.00960577)	(−11.67470863, 11.67470863)

	Euler		2nd-Order Taylor	
α	$([x]_l^+, [x]_r^+)$	$([x]_l^-, [x]_r^-)$	$([x]_l^+, [x]_r^+)$	$([x]_l^-, [x]_r^-)$
0	(−6.85737821, 6.85737821)	(0,0)	(−7.38380977, 7.38380977)	(0,0)
0.2	(−5.48768317, 5.48768317)	(−2.16411816, 2.16411816)	(−5.90896511, 5.90896511)	(−2.33025455, 2.33025455)
0.4	(−4.11798813, 4.11798813)	(−4.32823632, 4.32823632)	(−4.43412046, 4.43412046)	(−4.66050911, 4.66050911)
0.6	(−2.74829309, 2.74829309)	(−6.49235448, 6.49235448)	(−2.95927580, 2.95927580)	(−6.99076367, 6.99076367)
0.8	(−1.37859805, 1.37859805)	(−8.65647264, 8.65647264)	(−1.48443114, 1.48443114)	(−9.32101823, 9.32101823)
1	(−0.00890301, 0.00890301)	(−10.82059081, 10.82059081)	(−0.00958649, 0.00958649)	(−11.65127279, 11.65127279)

α	Error in Euler	Error in 2nd-order Taylor	Error in RK4-Gill
0	2.7064×10^{-1}	7.4260×10^{-3}	2.3026×10^{-6}
0.2	3.0199×10^{-1}	8.2863×10^{-3}	2.6364×10^{-6}
0.4	3.3334×10^{-1}	9.1466×10^{-3}	2.9101×10^{-6}
0.6	3.6470×10^{-1}	1.0006×10^{-2}	3.1838×10^{-6}
0.8	3.9605×10^{-1}	1.0867×10^{-2}	3.4575×10^{-6}
1	4.2741×10^{-1}	1.1727×10^{-2}	3.7312×10^{-6}

The exact and approximate solutions obtained by the Euler method and the fourth order Runge-Kutta Gill method are compared and plotted at $t = 2$ and $N = 25$ in Fig. 1.

The error between the Euler and 2nd-order Taylor method and 4th-order Runge-Kutta Gill method is plotted in Fig. 2.

6 Conclusion

In this paper we have applied iterative solution of fourth order Runge-Kutta Gill method for numerical solution of intuitionistic fuzzy differential equations. The efficiency and the accuracy of the proposed method have been illustrated by a suitable example. Comparison of the solutions of the example shows that the proposed method gives a better solution than the Euler method and 2nd-order Taylor method.

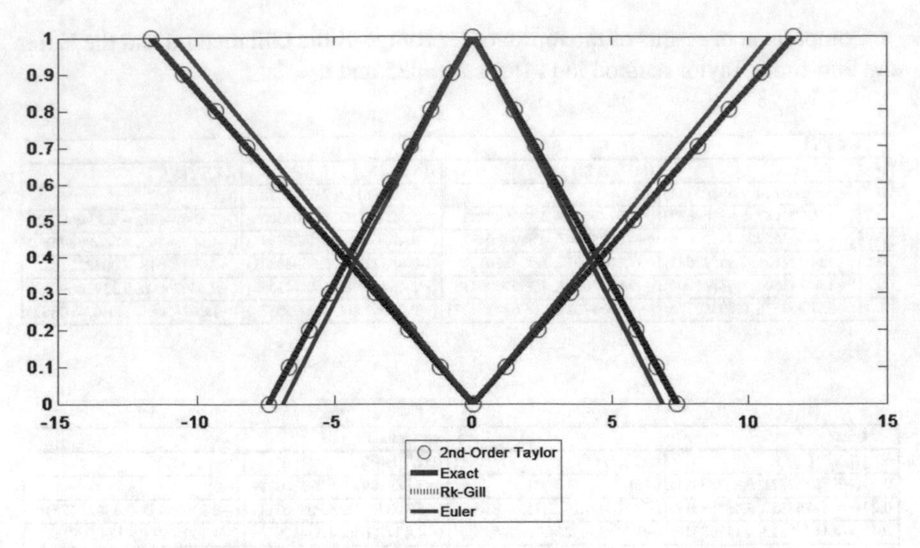

Fig. 1 h = 0.08

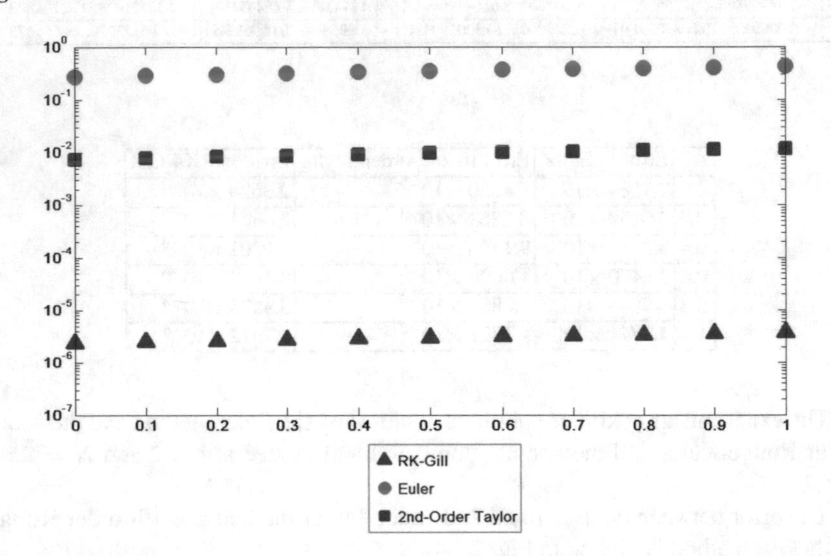

Fig. 2 h = 0.08

References

1. S. Abbasbandy, T. Allahviranloo, Numerical solution of fuzzy differential equation by Runge-Kutta method and the intutionistic treatment. Notes IFS **8**(3), 45–53 (2002)
2. A.K. Adak, M. Bhowmik, M. Pal, Intuitionistic fuzzy block matrix and its some properties. Ann. Pure Appl. Math. **1**(1), 13–31 (2012)
3. K.T. Atanassov, *Intuitionistic fuzzy sets. VII ITKRs session*, Sofia (deposited in Central Science and Technical Library of the Bulgarian Academy of Sciences 1697/84) (1983)

4. K.T. Atanassov, Intuitionistic fuzzy sets. Fuzzy Sets Syst. **20**, 87–96 (1986)
5. K.T. Atanassov, G. Gargov, Interval-valued intuitionistic fuzzy sets. Fuzzy Sets Syst. **31**(3), 43–49 (1989)
6. K.T. Atanassov, More on intuitionistic fuzzy sets. Fuzzy Sets Syst. **33**(1), 37–45 (1989)
7. K.T. Atanassov, Operators over interval valued intuitionistic fuzzy sets. Fuzzy Sets Syst. **64**(2), 159–174 (1994)
8. K.T. Atanassov, G. Gargov, Elements of intuitionistic fuzzy logic, part I. Fuzzy Sets Syst. **95**(1), 39–52 (1998)
9. K.T. Atanassov, *Intuitionistic Fuzzy Sets* (Physica-Verlag, Heidelberg, 1999)
10. K.T. Atanassov, Two theorems for Intuitionistic fuzzy sets. Fuzzy Sets Syst. **110**, 267–269 (2000)
11. T. Buhaesku, On the convexity of intuitionistic fuzzy sets, *Itinerant Seminar on Functional Equations*, Approximation and Convexity, Cluj-Napoca (1988), pp. 137–144
12. T. Buhaesku, Some observations on intuitionistic fuzzy relations, *Itinerant Seminar of Functional Equations*, Approximation and Convexity (1989), pp. 111–118
13. A.I. Ban, Nearest interval approximation of an intuitionistic fuzzy number, in *Computational Intelligence, Theory and Applications* (Springer, Berlin, 2006), pp. 229–240
14. B. Ben Amma, S. Melliani, L.S. Chadli, Numerical solution of intuitionistic fuzzy differential equations by Euler and Taylor methods. Notes Intuitionistic Fuzzy Sets **22**(2), 71–86 (2016)
15. B. Ben Amma, S. Melliani, L.S. Chadli, Numerical solution of intuitionistic fuzzy differential equations by Adams three order predictor-corrector method. Notes Intuitionistic Fuzzy Sets **22**(3), 47–69 (2016)
16. B. Ben Amma, L.S. Chadli, Numerical solution of intuitionistic fuzzy differential equations by Runge-Kutta Method of order four. Notes Intuitionistic Fuzzy Sets **22**(4), 42–52 (2016)
17. B. Ben Amma, S. Melliani, L.S. Chadli, The Cauchy problem of intuitionistic fuzzy differential equations. Notes Intuitionistic Fuzzy Sets **24**(1), 37–47 (2018)
18. B. Ben Amma, S. Melliani, L.S. Chadli, Intuitionistic fuzzy functional differential equations, in *Fuzzy Logic in Intelligent System Design: Theory and Applications*, ed. by P. Melin, O. Castillo, J. Kacprzyk, M. Reformat, W. Melek (Springer International Publishing, Cham, 2018), pp. 335–357
19. C. Cornelis, G. Deschrijver, E.E. Kerre, Implication in intuitionistic fuzzy and interval-valued fuzzy set theory: construction, application. Int. J. Approximate Reasoning **35**, 55–95 (2004)
20. S.K. De, R. Biswas, A.R. Roy, An application of intuitionistic fuzzy sets in medical diagnosis. Fuzzy Sets Syst. **117**, 209–213 (2001)
21. G. Deschrijver, E.E. Kerre, On the relationship between intuitionistic fuzzy sets and some other extensions of fuzzy set theory. J. Fuzzy Math. **10**(3), 711–724 (2002)
22. T. Gerstenkorn, J. Manko, Correlation of intuitionistic fuzzy sets. Fuzzy Sets Syst. **44**, 39–43 (1991)
23. A. Kharal, Homeopathic drug selection using intuitionistic fuzzy sets. Homeopathy **98**, 35–39 (2009)
24. D.F. Li, C.T. Cheng, New similarity measures of intuitionistic fuzzy sets and application to pattern recognitions. Pattern Recognit Lett. **23**, 221–225 (2002)
25. D.F. Li, Multiattribute decision making models and methods using intuitionistic fuzzy sets. J. Comput. Syst. Sci. **70**, 73–85 (2005)
26. S. Melliani, L.S. Chadli, Intuitionistic fuzzy differential equation. Notes Intuitionistic Fuzzy Sets **6**, 37–41 (2000)
27. G.S. Mahapatra, T.K. Roy, Reliability evaluation using triangular intuitionistic fuzzy numbers arithmetic operations. Proc. World Acad. Sci. Eng. Technol. **38**, 587–595 (2009)
28. S. Melliani, M. Elomari, L.S. Chadli, R. Ettoussi, Intuitionistic fuzzy metric space. Notes Intuitionistic Fuzzy sets **21**(1), 43–53 (2015)
29. S. Melliani, M. Elomari, L.S. Chadli, R. Ettoussi, Intuitionistic fuzzy fractional equation. Notes Intuitionistic Fuzzy sets **21**(4), 76–89 (2015)
30. S. Melliani, M. Atraoui Elomari, L.S. Chadli, Intuitionistic fuzzy differential equation with nonlocal condition. Notes Intuitionistic Fuzzy sets **21**(4), 58–68 (2015)

31. M. Nikolova, N. Nikolov, C. Cornelis, G. Deschrijver, Survey of the research on intuitionistic fuzzy sets. Adv. Stud. Contemp. Math. **4**(2), 127–157 (2002)
32. V. Nirmala, Numerical approach for solving intuitionistic fuzzy differential equation under generalised differentiability concept. Appl. Math. Sci. **9**(67), 3337–3346 (2015)
33. V. Parimala, P. Rajarajeswari, V. Nirmala, Numerical solution of intuitionistic fuzzy differential equation by Milne's Predictor-Corrector Method under generalised differentiability. Int. J. Math. Appl. **5**, 45–54 (2017)
34. E. Szmidt, J. Kacprzyk, Distances between intuitionistic fuzzy sets. Fuzzy Sets Syst. **114**(3), 505–518 (2000)
35. M.H. Shu, C.H. Cheng, J.R. Chang, Using intuitionistic fuzzy sets for fault-tree analysis on printed circuit board assembly. Microelectron. Reliab. **46**(12), 2139–2148 (2006)
36. P.M. Sankar, T.K. Roy, First order homogeneous ordinary differential equation with initial value as triangular intuitionistic fuzzy number. J. Uncertainty Math. Sci. **2014**, 1–17 (2014)
37. P.M. Sankar, T.K. Roy, System of differential equation with initial value as triangular intuitionistic fuzzy number and its application. Int. J. Appl. Comput. Math **1**(3), 449–474 (2015)
38. Z. Wang, K.W. Li, W. Wang, An approach to multiattribute decision making with interval-valued intuitionistic fuzzy assessments and incomplete weights. Inform. Sci. **179**(17), 3026–3040 (2009)
39. J. Ye, Multicriteria fuzzy decision-making method based on a novel accuracy function under interval valued intuitionistic fuzzy environment. Expert Syst. Appl. **36**, 6899–6902 (2009)
40. L.A. Zadeh, Fuzzy sets. Inf. Control **8**(3), 338–353 (1965)

Task Parameter Impacts in Fuzzy Real Time Scheduling

Mohammed Blej and Mostafa Azizi

Abstract The classical analysis of real-time systems tries to ensure that the instance of every task finishes before its absolute deadline (strict guarantee). The probabilistic approach tends to estimate the probability that it will happen. The deterministic timed behaviour is an important parameter for analysing the robustness of the system. Most of related works are mainly based on the determinism of time constraints. However, in most cases, these parameters are non-precise. The vagueness of parameters suggests the use of fuzzy logic to decide in what order the requests should be executed to reduce the chance of a request being missed. The choice of task parameters and numbers of rules in fuzzy inference engine influences directly generated outputs. Our main contribution is proposing a fuzzy approach to perform real-time scheduling in which the scheduling parameters are treated as fuzzy variables. A comparison of the results of the use of each parameter as linguistic variable is also given.

1 Introduction

Real time systems are present and vital in very diverse fields such as avionics, process control, air traffic control systems, and mission critical computations [1]. In the literature, these systems have been defined as: systems in which the correctness depends not only on the logical results of computation, but also on the time at which the results are delivered [2]. Real-time activities are traditionally characterized by temporal constraints, called deadlines [3]. Each task is a stream of jobs; each one has a time of release, it is characterized by a computation time, and must finish before its absolute deadline. Real-time task is said to be hard, soft, or firm depending on the criticity of the consequence of a deadline miss. For a hard task, no deadline miss

M. Blej (✉)
CRMEFO, MATSI Lab, Mohammed 1st University, Oujda, Morocco
e-mail: moblej@gmail.com

M. Azizi
ESTO, MATSI Lab, Mohammed 1st University, Oujda, Morocco
e-mail: azizi.mos@ump.ac.ma

S. Melliani and O. Castillo (eds.), *Recent Advances in Intuitionistic Fuzzy Logic Systems*, Studies in Fuzziness and Soft Computing 372,
https://doi.org/10.1007/978-3-030-02155-9_6

accepted, since a single job of task finishes after its deadline could jeopardize the entire system [4]. A soft task tolerates jobs that finish after their deadlines, whereas a firm task can only tolerate some job failures. More precisely, a firm job must be firm job that misses its deadline is useless, though it does not jeopardize the system [5]. Because of their critical nature, hard real-time tasks must be guaranteed off-line; a hard task is accepted only if it is guaranteed that every job is executed before its deadline. For firm tasks, admission tests are usually online, a job is activated only if it is certain that it will end its execution before its deadline, otherwise it will be rejected. A schedule is said to be feasible if all the tasks matched both their deadlines and any additional specified constraints [6]. Real-time scheduling can be classified in two categories: static [3] and dynamic [7]. A static real-time scheduling algorithm such as Rate Monotonic schedules all real-time tasks offline; this requires complete knowledge about tasks and system settings [8]. For dynamic scheduling, the feasibility of the algorithm is calculated online and tasks are invoked dynamically. These algorithms use dynamic parameters like deadline, latency and laxity [9–11]. However, the use of fuzzy logic to facilitate searching for a feasible schedule is motivated by several reasons. First, in a dynamic hard real-time system, not all the characteristics of tasks (e.g., precedence constraints, resource allocations) are known a priori. For example, for aperiodic tasks, the arrival time for the next task is unknown. In particular, there is a wide uncertainty in hard real-time environment which will worsen scheduling problems (e.g. arbitrary arrival time, system load, and uncertain computation time). Furthermore, in overload case, we must slow down the execution of tasks while ensuring that the most important tasks are run first, thus allowing a ratio of flexibility in the scheduler under adverse conditions to determine which tasks are run and which are not [12]. The remaining of this paper is organized as follows. Section 2 describes scheduling algorithms, classic and fuzzy, and model tasks. Section 3 discusses the fuzzy inference engine. In Sect. 4, we present our model and the concept of Fuzzy Inference Engine with the difference between Mamdani and Sugeno types. Section V provides some concluding remarks.

2 Scheduling Model

2.1 Classical Case

The scheduling in real time systems based on the use of the CPU time and other resources to execute all the tasks in question in order to meet the time constraints [2]. However, scheduling in real-time systems is more important than in classic systems [13, 14]. The real-time tasks must be performed correctly while respecting the deadline [15]. There are brief descriptions of the main scheduling algorithms: FCFS algorithm (First-Come-First-Served) selects the task with the earliest arrival time [16]. The release time of periodic tasks in the system will be considered, but this algorithm makes no efforts to consider task deadline, so it is not feasible for hard real time task. Rate Monotonic (RM) algorithm assigns priority according to period; A task with a shorter period has a higher priority [3]. This algorithm is an optimal static-priority scheduling [17]. Earliest Deadline First (EDF) algorithm always chooses the task with the earliest deadline [18], a job with the earliest deadline is executed. Since it cannot consider priority and therefore cannot analyse it. This algorithm is optimal dynamic priority scheduling. Least Laxity First (LLF scheduling algorithm assigns higher priority to a task with the least laxity and has been proved to be optimal for uniprocessor systems [2].

2.2 *Fuzzy Logic*

After at least two decades since its elaboration, fuzzy logic was finally accepted as the basis for an emerging technology, ranging from consumer products, to industrial process control, to automotive applications [19]. It is so closer to human brain thinking compared to conventional logical systems. Fuzzy logic is a multi-valued logic. It deals with approximation rather than exactness. In contrast to classical sets (A classical set takes true or false values), fuzzy logic variables (also known as linguistic variables) can have a truth value that ranges into an floating interval between 0 (completely true) and 1 (completely false) [20]. The linguistic variable degree may be determined with a specific method [21]. Nowadays, fuzzy logic is largely used in real world issues, and it is also a research interest of a great number of researchers. It has been shown to be a strong methodology of design and analysis in control theory, enabling the implementation of advanced knowledge-based control methods for complex dynamic systems like those rising applications for systems and artificial biology [20]. Hiwarkar et al. gave a large list of use cases of fuzzy logic in [21]. Verma et al. discussed Type 1 fuzzy systems and the origin of type 2 fuzzy logic systems [22] and its application in the field of engineering, finance and medicine. Xia Feng et al. designed a schedule for control of embedded systems based on fuzzy logic [23]. A priority scheduler has been developed for mobile by Gomathy et al. in [24].

2.3 *Fuzzy Algorithms*

The main conventional algorithms are based on binary logic; the new versions of these algorithms are introduced based on fuzzy logic. This logic is used in different conventional algorithms in the case of the highest-priority task, giving birth to fuzzy versions of these algorithms. In the literature, most of the work is about soft real-time systems. In what follows, we quote the most famous results: M. Sabeghi et al. introduced a fuzzy algorithm for scheduling periodic tasks on multiprocessor soft realtime systems [25]. P. Vijayakumar et al. presented a fuzzy EDF algorithm for soft real-time systems [26], they used laxity and deadline as fuzzy parameters.

Fuzzy inference rules include 15 fuzzy inference rules. The miss deadline ratio of the tasks will be reduced compared to the traditional EDF algorithm. V. Salmani et al. proposed a new fuzzy-based algorithm for scheduling real-time tasks on uniform parallel machines in [27]. It is shown that the proposed approach ensured a performance close to that of EDF in non-overloaded conditions and it has supremacy over EDF in overloaded situations in many aspects. In [28], Hamzeh et al. proposed a new fuzzy scheduler. They use laxity, CPU time and deadline as a fuzzy parameter. This scheduler has low complexity due to the simplicity of fuzzy inference engine. Sheo Das et al. have described the use of fuzzy logic to multiprocessor realtime scheduling [29]. They have showed that using deadline as a fuzzy parameter in multiprocessor real-time scheduling is more promising than laxity, and that this model is efficient when the system has heterogeneous tasks with different constraints. For embedded systems, Springer et al. [30] presented a new scheduling approach for real-time tasks in an embedded system. The results are a demonstrated reduction in deadline misses for all tasks during periods of overload as compared to traditional fixed priority based scheduling mechanisms. In the case of hard real-time systems, publications on fuzzy logic are scarce. Very early John Yen et al. introduced a Designing of a Fuzzy Scheduler for Hard Real-time Systems [12]. They worked on the scheduling problem as a search problem, using a set of fuzzy rules to guide the search for a feasible schedule, and the scheduler is triggered by a newly arrival task. In conclusion, most papers dealt with soft real-time systems, but not enough for hard real-time systems. Fuzzy versions of different conventional algorithms are given and showed that they are more promising than conventional ones. The choice of fuzzy parameters and the comparison of different results are studied in special cases, and research over the generalization is an open work.

3 Fuzzy Inference Engine

Fuzzy Inference Systems are conceptually very simple. They consist of an input, a processing, and an output stage [31]. The input stage receives inputs like deadline, execution time, laxity and so on, and maps these to appropriate membership functions and truth values. In the processing stage, each specific rule is invoked and the corresponding result is generated. Then results are combined so that it will be given as an input to the output stage. In output stage, the combined result is converted back into a specific value [32]. The membership function of a fuzzy set is a generalization of the indicator function in classical sets. In fuzzy logic, it represents the degree of truth as an extension of valuation. It can be expressed in the form of a curve that defines how each point in the input space is mapped to a membership value (or degree of membership) between 0 and 1. It can also have many forms (triangular, trapezoidal and Bell curves) [32]. The processing stage also called inference engine is based on a set of logical rules in the form of IF-THEN statements. An example of fuzzy IF-THEN rules is: IF Speed is Low AND Race is Dry THEN Braking is Soft, Where the IF part is called the "antecedent" and the THEN part is called the "consequent". The terms Speed, Race and Braking are linguistic variables, and Low, Dry and Soft are linguistic terms. Each linguistic term corresponds to a value of the membership function. Typically, Fuzzy Inference Systems have dozens of rules [31]. The inference engine processes the inputs and generates outputs based on the rules already defined. There are five steps in the fuzzy inference:

- Fuzzify inputs,
- Apply the fuzzy operator,
- Apply the implication method,
- Aggregate all outputs,
- Defuzzify outputs.

Below is a brief overview of these five steps. The first step is to take the inputs and determine the degree to which they belong to each of the appropriate fuzzy sets via membership functions. Fuzzification of the input amounts to either a table lookup or a function evaluation. After fuzzifying the inputs, we know then the degree to which each part of the antecedent has been fullfilled for each rule. So if the antecedent of a given rule has more than one part, the fuzzy operator is applied to obtain one number that represents the result of the antecedent for that rule. This number will then be applied to the output function. The number of inputs to the fuzzy operator is two or more membership values from fuzzified input variables. Whereas the output is a single truth value. The output fuzzy set has been modified by the implication function to the degree specified by the antecedent. Since decisions are based on the testing of all of the rules in the Fuzzy Inference Subsystem (FIS), the results from each rule must be combined in order to generate the final decision. Aggregation is the process by which the fuzzy sets that represent the outputs of each rule are combined into a single fuzzy set. It occurs only once for each output variable, just prior to the fifth and final step, defuzzification. The input for the defuzzification process is an aggregate output fuzzy set, and the output is a single number. At run-time, based on the parameter of tasks, the fuzzy scheduler selects the highest priority task that is ready for execution. Several parameters determine the priority of tasks: task deadline, task criticality, task execution time, laxity. The task deadline is the time before the task should be completed. The task criticality relates to the consequences of missing a deadline. The worst case execution time of task is his execution time. Laxity is the time that separates the task deadline and the worst case execution time of task. These parameters constitute the linguistics variables and then fuzzified. Fuzzy rules are then applied to the linguistic variables to compute the service value. The linguistic values for the chosen parameters are defined. The fuzzification is applied to the conclusion; the way in which this happens depends on the inference model. There are two common inference methods [32, 33]: Mamuzzy dani's finference method proposed in 1975 by Mamdani [34] and Takagi-Sugeno-Kang, method of fuzzy inference introduced in 1985 [35]. Mamdani's efforts are based on the work of Lotfi A. Zadeh on fuzzy algorithms for complex systems and decision processes [36]. The main difference between Mamdani-type FIS and Sugeno-type FIS resides in the way the crisp output is generated from the fuzzy inputs [37]. While Mamdani-type FIS uses the technique of defuzzification of a fuzzy output, Sugeno-type FIS uses weighted average to compute the crisp output [38], so the Sugenos output membership functions are either linear or constant but Mamdanis inference expects the output membership functions to be fuzzy sets. Furthermore, Sugeno method has better processing time since the weighted average replace the time consuming defuzzification process [38].

4 The Proposed Model

In the proposed model, the input stage consists of two linguistic variables. The first one is a priority; each task arrives in the system with its priority. This priority is assigned to the task from the outside world and its static. The other input variable can be one of other task parameters. This input can be tasks interval, laxity, wait time, or so on, for different scheduling algorithms. Each parameter may cause the system to react in a different way, since changing the input variables the corresponding membership functions may be changed accordingly. We will compare the impact of the change of the second parameter, while keeping the same task priority (Fig. 1).

4.1 Laxity and Priority

Laxity of a task at any time is defined as remaining time to deadline minus the amount of remaining execution. We choose this parameter as the second linguistic variable. A tasks priority shows the importance of the task, the notion of laxity is used in the proposed approach to facilitate the computation. The inputs of these parameters are justified. In our example, we choose task priority and laxity as inputs parameters. We considered 4 triangular membership functions for tasks priority. High, Normal, Low and Very low are these membership functions. Laxity membership function considered 3 and also triangular. Critical, Medium and sufficient are the name of

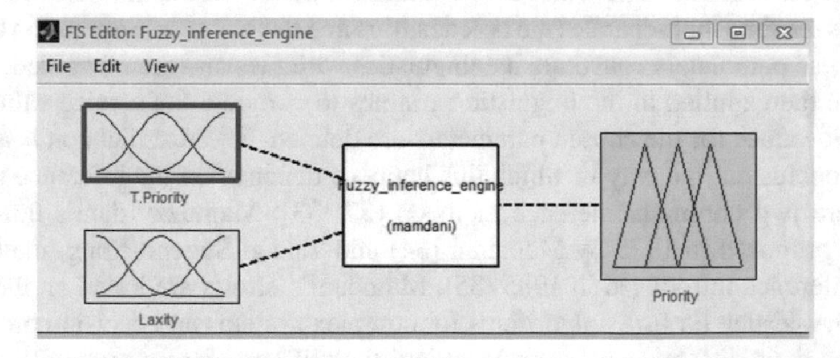

Fig. 1 Proposed fuzzy inference engine

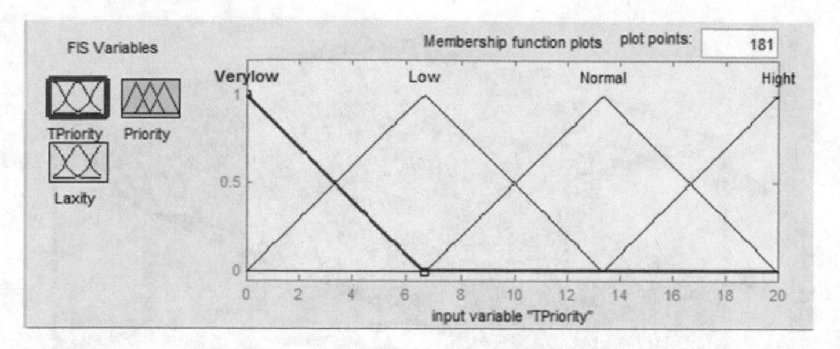

Fig. 2 Input task priority membership function

these functions. Fuzzy rules try to combine these parameters as they are connected in real worlds. Some of these rules are mentioned here:

- If (T.Priority is High) and (Laxity is Critical), then (Priority is Very high).
- If (T.Priority is High) and (Laxity is Medium), then (Priority is high).
- If (T.Priority is High) and (Laxity is sufficient), then (Priority is Medium) (Fig. 2).

4.2 Deadline and Priority

To see the impact of changing the parameters of the tasks in the input stage on the final output priority, we choose the task deadline as second paramcters in the input system. It is very important to remember that assigning priority to tasks according to their deadlines is a simple yet successful strategy for uniprocessor real-time scheduling. EDF (Earliest Deadline First) and DM (Deadline Monotonic) are shown as optimal dynamic- and static-priority policies for preemptive uniprocessor scheduling, although they are not in the case multiprocessor. We considered 3 triangular membership functions for tasks deadline. Low, medium and sufficient are these membership functions (Fig. 3).

5 Results Analysis

To compare the external priority which is the priority assigned to the task from the outside world and a new priority based on fuzzy logic, we use this external priority as an input linguistic variable and we use separately two parameters, namely laxity and deadline, as second linguistic variable. Fuzzy rules compute the crisp value using centroid Defuzzification method of Mamdani inference, and an output priority of task is calculated. The output of the system is priority that determines which is used as a parameter for making a decision. Third step, we use both parameters at the same time in an input stage then we look and compare different results. The following

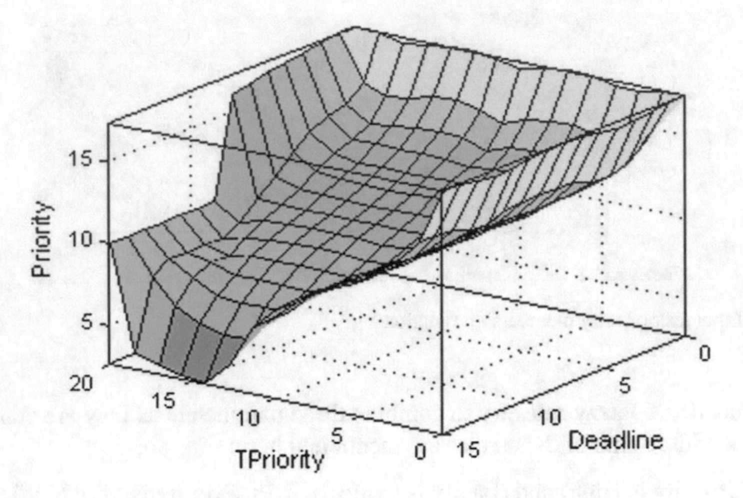

Fig. 3 The decision surface corresponding to inference rules

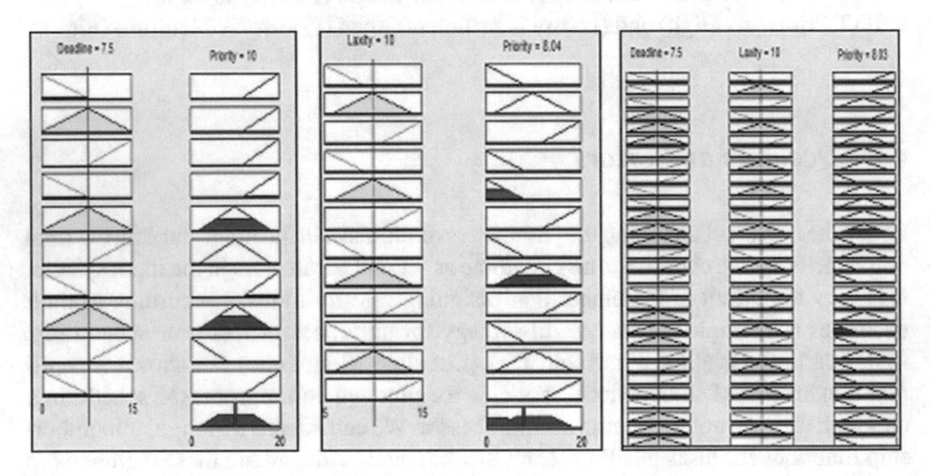

Fig. 4 Computation of new priority for the three cases

figure shows the different results. The priority calculed in the third case is the least best. This can be explained by the fact that in this case the number of rules is very large, since we have three input linguistic variables. When we have more rules and member functions, it will directly affect the overall system accuracy, even though performance of the system can be made better when number of rules are reduce. For the other two cases, we have the same number of rules for the two linguistics variables, namely deadline and laxity. As it is depicted in the Fig. 4, the simulation results show that the output priority based on deadline is much better than the output priority based on laxity, knowing that the initial priority is the same for the two cases.

6 Conclusion

For scheduling, fuzzy logic is used in the different conventional algorithms with the choice of the highest-priority task, giving thus birth to fuzzy versions of these algorithms. This paper presents a comparison of fuzzy priorities based on deadline and priorities based on laxity. As results, we conclude that using deadline as a fuzzy parameter is more promising than laxity. The choice of parameters and numbers of rules in fuzzy inference engine influences directly generated outputs. Being based on Mamdani or Sugeno, Fuzzy Inference Systems (FIS) are still on-going research areas. As a future work, we plan to redo this work but based this time on Sugeno-type FIS and to compare the results of the two models.

References

1. L.A. Sha, Real-time scheduling theory and ada. IEEE Comput. **23**(4), 53–62 (1990)
2. K. Ramamnitham, J.A. Stankovic, Scheduling algorithms and operating systems. Proc. IEEE **82**(1), 55–67 (1994)
3. C.L. Layland, Scheduling alghorithms for multiprogramming systems. J. ACM **20**(1), (1973)
4. J. Zhu, T.G. Scheduling, in hard real-time applications. IEEE Softw. **12**, 54–63 (1994)
5. Haibin, W.L. Research on a soft real-time scheduling algorithm based on hybrid adaptive control architecture, in *Proceedings of the American Control Conference*, vol. 5 (Lisbon, Portugal, 2003), pp. 4022–4027
6. F. Abdelzaher Tarek, K.G. Shin, Comment on a pre-runtime scheduling algorithm for hard realtime systems. IEEE Trans. Softw. Eng. **23**, 599–600 (1997)
7. K. Ramamritham, J.A. Stankovic, Dynamic task scheduling in hard real-time distributed systems. IEEE Softw. **1**, 65–75 (1984)
8. P.A. Laplante, The certainty of uncertainty in real-time systems. IEEE Instrum. Meas. Mag. **7**, 44–50 (2004)
9. A.S. Kreuzinger, Real-time scheduling on multithreaded, in *Proceedings of the 7th International Conference on Real-Time Computing Systems* (Cheju Island, South Korea, 2000), pp. 155–159
10. Z. Deng, J.W. Liu, Dynamic scheduling of hard realtime applications in open system environment. Technical Report (University of Illinois at Urbana-Champaign, 1996)
11. G. Buttazzo, J.A. Stankovic, RED: robust earliest deadline scheduling, in *Proceedings of the 3rd International Workshop Responsive Computing* (Lincoln, NH, 1993) pp. 100–111
12. J.L. John Yen, *Designing a Fuzzy Scheduler for Hard Real-Time Systems* (Department of Computer Science Texas A.M University, College Station, TX, 1993), p. 77843
13. F. Gruian, Energy-centric scheduling for real-time systems. Ph.D. Dissertation (Department of Computer Science, Lund University, 2002) p. 164
14. W. Lifeng, Y. Haibin Research on a soft real-time scheduling algorithm based on hybrid adaptive control architecture, in *Proceedings of the American Control Conference*, vol.5 (Lisbon, Portugal, 2003), pp. 4022–4027
15. M. Silly-Chetto, Dynamic acceptance of aperiodic tasks with periodic tasks under resource sharing constraints. IEEE Proc. Softw. **146**, 120–127 (1999)
16. A.S. Tanenbaum, *Distributed Operating Systems* (Prentice Hall, Upper Saddle River, 1994)
17. S.A. Yoshifumi Manabe, *A Feasibility Decision Algorithm for Rate Monotonic Scheduling of Periodic Real-Time Tasks* (NTT Basic Research Laboratories, Atsugi-shi, Kanagawa 243-01 Japan, 1995)

18. N.D. Thai, Real-time scheduling in distributed systems, in *Proceedings of the International Conference on Parallel Computing in Electrical Engineering* (Warsaw, Poland, 2002), pp. 165–170
19. J. Yen, R. Langari, *Fuzzy Logic* (Pearson Education, Upper Saddle River, 2004)
20. V.K. Rajani Kumari, Design and implementation of modified fuzzy based CPU scheduling algorithm. Int. J. Comput. Appl. (0975 8887) **77**(17) (2013)
21. A. Dr. Tryambak, R. Hiwarkar, T.A. Sridhar Iyer, New applications of soft computing, artificial intelligence, fuzzy logic, genetic algorithm in bioinformatics. Int. J. Comput. Sci. Mobile Comput. **2**(5), 202–207 (2013)
22. K.A. Varma, Applications of type-2 fuzzy logic in power systems: a literature survey, in *12th International Conference on Environment and Electrical Engineering (EEEIC)* (IEEE 2013)
23. F. Xia, *Fuzzy logic-based feedback scheduler for embedded control systems, in Advances in Intelligent Computing* (Springer, Berlin, Heidelberg, 2005), pp. 453–462
24. C. Gomathy, An efficient fuzzy based priority scheduler for mobile and hoc networks and performance analysis for various mobility models, in *Wireless Communications and Networking Conference, WCNC* vol. 2 (IEEE, 2004)
25. M. Sabeghi, A fuzzy algorithm for real-time scheduling of soft periodic tasks. IJCSNS Int. J. Comput. Sci. Netw. Secur. **6**(2A), 227–235 (2006)
26. P. Vija yakumar, Fuzzy EDF algorithm for soft real time aystems. Int. J. Comput. Commun. Inf. Syst. 2(1) (2010). ISSN 09761349
27. V.R.N. Salmani, A Fuzzy-based multi-criteria scheduler for uniform multiprocessor real-time systems, in *10th International Conference on Information Technology* (IEEE, 2007) ISBN 0-7695-3068-0
28. M.S.M. Hamzeh, Soft real-time fuzzy task scheduling for multiprocessor systems. Int. J. Intell. Technol. **2**(4) (2007). ISSN 1305-6417
29. P.G. Sheo, Das, A fuzzy approach scheduling on more than one processor system in real time environment. Int. J. Sci. Res. Eng. Technol. **1**(5), 289–293 (2012)
30. S.P. Tom Springer, *Fuzzy Logic Based Adaptive Hierarchical Scheduling for Periodic Real-Time Tasks* (Springer, EWiLi15, Amsterdam, The Netherlands, 2015)
31. H. Deldari, M. Sabeghi, A fuzzy algorithm for scheduling periodic tasks on multiprocessor. IJCSN Int. J. Comput. Sci. Netw. Secur. **6**(3A), 88 (2006)
32. L.-X. Wang, *A Course in Fuzzy Systems and Control* (Prentice Hall, Paperback, 1996)
33. M. Blej, M. Azizi, Comparison of Mamdani-type and Sugeno-type fuzzy inference systems for fuzzy real time scheduling Int. J. Appl. Eng. Res. **11**(22), 11071-11075 (2016). ISSN 0973-4562
34. E.H. Mamdani, S. Assilian, An experiment in linguistic synthesis with a fuzzy logic controller. Int. J. Man-Mach. Stud. **7**(1), 1–13 (1975)
35. M. Sugeno, *Industrial Applications of Fuzzy Control* (Elsevier Science Inc., New York, NY, 1985)
36. L. Zadeh, Outline of a new approach to the analysis of complex systems and decision processes. IEEE Trans. Syst. Man Cybern. **3**(1), 28–44 (1973)
37. A. Hamam, N.D. Georganas, *A comparison of Mamdani and Sugeno fuzzy inference systems, in IEEE International Workshop on Haptic Audio-Visual Environments and their Applications* (Ottawa, Canada, 2008), pp. 18–19
38. M. Blej, M. Azizi, Survey on fuzzy logic in real-time system. Int. J. Adv. Comput. Technol. (IJACT) (2016). ISSN 2319-7900

Framework for Optimization of Intuitionistic and Type-2 Fuzzy Systems in Control Applications

Oscar Castillo

Abstract In this paper a framework for finding the optimal design of intuitionistic fuzzy systems in control applications is presented. Traditional models deal with type-0 values, which mean using precise numbers in the models, but since the seminal work of Prof. Zadeh in 1965, type-1 fuzzy models emerged as a powerful way to represent human knowledge and natural phenomena. Later type-2 fuzzy models were also proposed by Prof. Zadeh in 1975 and more recently have been studied and applied in real world problems by many researchers. In addition, as another extension of type-1 fuzzy logic, Prof. Atanassov proposed Intuitionistic Fuzzy Logic, which is a very powerful theory in its own right. Previous works of the author and other researchers have shown that certain problems can be appropriately solved by using type-1, and others by interval type-2, while others by using intuitionistic fuzzy logic. Bio-inspired and meta-heuristic optimization algorithms have been commonly used to find optimal designs of type-1, type-2 or intuitionistic fuzzy models for applications in control, robotics, pattern recognition, time series prediction, just to mention a few. However, the question still remains about if even more complex problems (meaning non-linearity, noisy, dynamic environments, etc.) may require even higher types, orders or extensions of type-1 fuzzy models to obtain better solutions to real world problems. In this paper a framework for solving this problem of finding the optimal fuzzy model for a particular problem is presented. To the knowledge of the author, this is the first work to propose a systematic approach to solve this problem, and we envision that in the future this approach will serve as a basis for developing more efficient algorithms for the same task of finding the optimal fuzzy system.

Keywords Intuitionistic fuzzy systems · Type-2 fuzzy systems
Type-1 fuzzy systems

O. Castillo (✉)
Division of Graduate Studies and Research, Tijuana Institute of Technology,
Calzada Tecnologico s/n, Fracc. Tomas Aquino, Tijuana 22379, Mexico
e-mail: ocastillo@tectijuana.mx

S. Melliani and O. Castillo (eds.), *Recent Advances in Intuitionistic Fuzzy Logic Systems*, Studies in Fuzziness and Soft Computing 372,
https://doi.org/10.1007/978-3-030-02155-9_7

1 Introduction

Nowadays it is well accepted that either type-1, type-2 or intuitionistic fuzzy systems can solve many real world problems. Initially, type-1 fuzzy logic was proposed and applied in a plethora of real world problems ranging from control, to pattern recognition and time series prediction [1–3]. Although, more recently it has been recognized that type-1 fuzzy sets do not really handle uncertainty, because they only use precise values for the membership values. For this reason, type-2 fuzzy logic emerged as an extension of its type-1 counterpart. Type-2 fuzzy systems use membership functions whose values are fuzzy sets [4–6]. Of course, as a special in between case, we have interval type-2 fuzzy systems that use membership functions whose values are intervals. In the literature, we can find many cases in which type-1 fuzzy systems have adequately solve real world problems. However, for more complex situations, meaning dynamic or noisy environments or highly nonlinear problems, type-2 fuzzy systems have shown that can outperform type-1 fuzzy systems. In addition, as another extension of type-1, intuitionistic fuzzy systems (IFSs) emerged as another way to model uncertainty in real world problems. Even with all of these previous works, an open question still remains, which is, if type-1 or type-2 fuzzy systems are sufficient to model all existing problems in the real world.

The main contribution of this paper is a proposed mathematical framework to find the optimal fuzzy system (type-1, type-2 or intuitionistic) for a particular problem or class of problems, in the sense that the fuzzy model should approximate in the best way possible the real dynamical system. Of course, the proposed approach is posed as an optimization problem because we need to minimize the error in some particular metric measuring the difference between the outputs of the fuzzy model and the real dynamical system under consideration.

The remaining of the paper is organized as follows. Section 2 outlines some related works on optimization of fuzzy systems and granularity. Section 3 outlines the proposed method for finding the optimal fuzzy system for a particular application. Section 4 describes an implementation approach of the proposed method using genetic algorithms. Finally, Sect. 5 offers some conclusions and possible future works in this area.

2 Related Works

As related works we can discuss papers that are related to type-*n* and intuitionistic fuzzy systems, but from the granularity point of view, like in the works of Witold Pedrycz that we briefly describe below. First, there is the work by Pedrycz [7] on Algorithmic Developments of Information Granules of Higher Type and Higher Order and Their Applications in which it is described very clearly how type-n information granules are used in the augmentation of numeric models. Another work by Pedrycz [8] is on the development of granular meta-structures and their use in a multifaceted representation of data and models presents a constructive way of forming type-2 fuzzy sets via the principle of justifiable granularity exhibits a significant level of originality and offers a general way of designing information granules. In addition, the work by Pedrycz [9] on Hierarchical Architectures of Fuzzy Models: From Type-1 fuzzy sets to Information Granules of Higher Type, which describes the enhanced interpretability of fuzzy sets by elaborating on the role of type-2 fuzzy sets (which offers an effective vehicle of linguistic quantification of numeric membership degrees) and shadowed sets (with their ability to express uncertainty). Finally, also in the work by Pedrycz [10] on Concepts and Design Aspects of Granular Models of Type-1 and Type-2 some interesting ideas on forming type 2 fuzzy models are presented.

There also several recent works on developing type-2 fuzzy models in diverse areas of application that can be viewed as related work. In the work by Melin et al. [5] on Edge-detection method for image processing based on generalized type-2 fuzzy logic, an approach using type-2 fuzzy for edge detection that outperforms other methods is presented. In the work of Rubio et al. [11] an Extension of the Fuzzy Possibilistic Clustering Algorithm using Type-2 Fuzzy Logic Techniques is presented. In the work of Olivas et al. [12] a Comparative Study of Type-2 Fuzzy Particle Swarm, Bee Colony and Bat Algorithms in Optimization of Fuzzy Controllers is outlined. In the work of Castillo et al. [13] a Review of Recent Type-2 Fuzzy Image Processing Applications is presented. In the work of Gonzalez et al. [14] an optimization method of interval type-2 fuzzy systems for image edge detection is described. In the work of Tai et al. [15] a Review of Recent Type-2 Fuzzy Controller Applications is presented. In the work of Sepulveda et al. [16] an Experimental study of intelligent controllers under uncertainty using type-1 and type-2 fuzzy logic is described. In the work of Castillo and Melin [4] the Design of Intelligent Systems with Interval Type-2 Fuzzy Logic is presented, and this work includes the Theory and Applications of the design process. Finally, in Sanchez et al. [17] Generalized Type-2 Fuzzy Systems for controlling a mobile robot and a performance comparison with Interval Type-2 and Type-1 Fuzzy Systems are presented.

3 Proposed Method

Initially, we can define fuzzy models of type-1 that can be represented as follows:

$$\begin{aligned} &A\colon X \to [0, 1] \\ &X = \{x_1, x_2, \ldots x_n\} \end{aligned} \tag{1}$$

where A is the so called membership function, and X is the domain of interest.

Then these models can be extended by considering uncertainty in the membership functions, in this way obtaining a definition of type-2 fuzzy models as follows:

$$A: X \rightarrow F([0,1])$$
$$X = \{x_1, x_2, \ldots x_n\} \tag{2}$$

where F represents families of type-1 fuzzy sets, and in this case the interval type-2 fuzzy models are also included.

In the same way, we can extend the fuzzy models once more by considering uncertainty in the type-2 fuzzy models, in this way obtaining a definition of type-3 fuzzy models as follows:

$$A: X \rightarrow F^2([0,1])$$
$$X = \{x_1, x_2, \ldots x_n\} \tag{3}$$

where F^2 stands in this case for families of type-2 fuzzy sets.

For the case of intuitionistic fuzzy models, the uncertainty is represented by a membership and non-membership function, so the representation is as follows:

$$A: X \rightarrow F([0,1],[0,1]) \tag{4}$$

It is also possible that intuitionistic type-2 fuzzy models can be represented as follows:

$$A: X \rightarrow F^2([0,1],[0,1]) \tag{5}$$

Now we can pose the problem of designing a fuzzy system that models the uncertainty represented by ε as the maximization of data coverage as follows:

$$\text{Max}_{\varepsilon} \sum_{k=1}^{N} \text{coverage (target data)}$$

Such that

$$\sum_{i=1}^{p} \varepsilon_i = \varepsilon \text{ and } \varepsilon_i \geq 0$$

where "target data" means available input-output training data to construct the model, N means the number of data points, p is the number of fuzzy sets used (depending on the performed granulation). The meaning of "coverage" is defined by the ratio of data points covered by the fuzzy model out of the total of points, which is at most one for a perfect model.

The concept of coverage of a fuzzy set, cov(.) is discussed with regards to some experimental data existing in R^n, that is $\{X_1, X_2, \ldots, X_N\}$. As the name itself stipulates, coverage is concerned with an ability of a fuzzy set to represent (cover) these data. In general, the larger number of data is being "covered", the higher the coverage of the fuzzy set. Formally, the coverage can be sought as a non-decreasing function of the number of data that are represented by the given fuzzy set A. The monotonicity property of the coverage measure is obvious: the higher the values of ε, the higher the resulting coverage. Hence the coverage is a non-decreasing function of ε.

The problem can be re-structured in the following form in which the objective function is a product of the coverage and specificity-determine optimal allocation of information granularity $[\varepsilon_1, \varepsilon_2, \ldots, \varepsilon_p]$ so that the coverage and specificity criteria become maximized.

Finally, we can pose the general optimization of the fuzzy model, meaning finding the appropriate value of n for a fuzzy model according to the target data of the problem. In this case, it is a minimization problem that can be stated in general as follows:

$$\mathrm{Min}_n \left\| F^n(M) - RS \right\| < \tau$$

where τ is accuracy threshold, which is application related. In this case, the Euclidean distance can be used as the norm, but others could be used depending on the application area. The minimization problem can be solved by any optimization method in the literature, although due to the complexity issue we prefer to use meta-heuristic algorithms that can provide a sufficiently good approximation to the optimal solution of the problem.

We envision that different n values would be the optimal ones for certain classes of problems depending on the complexities and nonlinearities. For the moment, we can solve this problem by using bio-inspired or meta-heuristic optimization algorithms due to the high computational overhead required in this hierarchical optimization problem.

4 Intuitionistic Fuzzy Logic Systems

According to Atanassov [18–22], an IFS on the universum $X \neq \varnothing$ is an expression A given by:

$$A = \{\langle x, \mu_A(x), \nu_A(x)\rangle | x \in X\}, \tag{6}$$

where the functions

$$\mu_A, \nu_A : X \to [0, 1] \tag{7}$$

satisfy the condition

$$0 \leq \mu_A(x) + \nu_A(x) \leq 1 \tag{8}$$

and describe, respectively, the degree of the membership $\mu_A(x)$ and the non-membership $\nu_A(x)$ of an element x to A. Let

$$\pi_A(x) = 1 - \mu_A(x) - \nu_A(x), \tag{9}$$

therefore, function π_A determines the degree of uncertainty.

According to [22] the geometrical forms of the intuitionistic fuzzy numbers can be generalized as follows: For the first case functions μ_A and ν_A satisfied the conditions [22]:

$$\sup_{y \in E} \mu_A(y) = \mu_A(x) = a, \quad \inf_{y \in E} \nu_A(y) = \nu_A(x) = b.$$

for each $x \in [x_1, x_2]$, and for the second case [22]:

$$\sup_{y \in E} \mu_A(y) = \mu_A(x_0) = a, \quad \inf_{y \in E} \nu_A(y) = \nu_A(x_0) = b.$$

For the first case we have:	For the second case we have:
μ_A is increasing function from $-\infty$ to x_1	μ_A is increasing function from $-\infty$ to x_0
μ_A is decreasing function from x_2 to $+\infty$	μ_A is decreasing function from x_0 to $+\infty$
ν_A is decreasing function from $-\infty$ to x_1	ν_A is decreasing function from $-\infty$ to x_0
ν_A is increasing function from x_2 to $+\infty$.	ν_A is increasing function from x_0 to $+\infty$.

Obviously, in both cases the functions μ_A and ν_A can be represented in the form

$$\mu_A = \mu_A^{\text{left}} \cup \mu_A^{\text{right}}, \; \nu_A = \nu_A^{\text{left}} \cup \nu_A^{\text{right}},$$

where μ_A^{left} and ν_A^{left} are the left, while μ_A^{right} and ν_A^{right} are the right sides of these functions.

Therefore, the above conditions can be re-written in the (joint) form [22]:

$$\sup_{y \in E} \mu_A(y) = \mu_A(x) = a, \quad \inf_{y \in E} \nu_A(y) = \nu_A(x) = b,$$

for each $x \in [x_1, x_2]$ and in the particular case, when $x_1 = x_0 = x_2$, μ_A^{left} is increasing function; μ_A^{right} is decreasing function; ν_A^{left} is decreasing function and ν_A^{right} is increasing function. Following [22], we will consider, ordered by generality, the definitions:

1. In the graphical representation in both cases above $a = 1, b = 0$.
2. $\sup_{y \in E} \mu_A(y) = \mu_A(x_0) > 0.5 > \nu_A(x_0) = \inf_{y \in E} \nu_A(y)$.
3. $\sup_{y \in E} \mu_A(y) = \mu_A(x_0) \geq 0.5 \geq \nu_A(x_0) = \inf_{y \in E} \nu_A(y)$.
4. $\sup_{y \in E} \mu_A(y) = \mu_A(x_0) > \nu_A(x_0) = \inf_{y \in E} \nu_A(y)$.

5. $\sup_{y \in E} \mu_A(y) = \mu_A(x_0) \geq \nu_A(x_0) = \inf_{y \in E} \nu_A(y)$.
6. $\sup_{y \in E} \mu_A(y) = \mu_A(x_0) > 0$.
7. $\inf_{y \in E} = \nu_A(x_0) < 1$.

5 Conclusions

In this paper a framework for finding the optimal fuzzy systems was presented. In addition, a framework for solving this problem of finding the optimal type-1, type-2 or intuitionistic fuzzy model was presented. We envision that in the future even more general intuitionistic type-*n* fuzzy models will be used more frequently for solving complex problems, as real world situations are becoming more complicated by dynamic and non-linear environments, as well as huge amounts of data being available for processing very quickly in real time decision making. In particular, we expect to test the proposed approach with problems of nonlinear control, time series prediction, and pattern recognition. As future work, theoretical as well as applied works are envisioned that will be done. For example, in regards to the applications of this framework we can consider the following cases are a good choice to test the proposed approach [23–25].

References

1. O. Castillo, P. Melin, E. Ramírez, J. Soria, Hybrid intelligent system for cardiac arrhythmia classification with fuzzy K-nearest neighbors and neural networks combined with a fuzzy system. Expert Syst. Appl. **39**(3), 2947–2955 (2012)
2. P. Melin, O. Castillo, *Modelling, Simulation and Control of Non-linear Dynamical Systems: An Intelligent Approach Using Soft Computing and Fractal Theory* (CRC Press, 2001)
3. P. Melin, O. Castillo, Intelligent control of complex electrochemical systems with a neuro-fuzzy-genetic approach. IEEE Trans. Ind. Electron. **48**(5), 951–955 (2001)
4. O. Castillo, P. Melin, Design of intelligent systems with interval type-2 fuzzy logic, in *Type-2 Fuzzy Logic: Theory and Applications* (2008), pp. 53–76
5. P. Melin, C.I. Gonzalez, J.R. Castro, O. Mendoza, O. Castillo, Edge-detection method for image processing based on generalized type-2 fuzzy logic. IEEE Trans. Fuzzy Syst. **22**(6), 1515–1525 (2014)
6. G.M. Mendez, O. Castillo, Interval type-2 TSK fuzzy logic systems using hybrid learning algorithm, in *The 14th IEEE International Conference on Fuzzy Systems, FUZZ'05* (2005), pp. 230–235
7. W. Pedrycz, Algorithmic developments of information granules of higher type and higher order and their applications, *WILF* (2016), pp. 27–41
8. W. Pedrycz, the development of granular metastructures and their use in a multifaceted representation of data and models. Kybernetes **39**(7), 1184–1200 (2010)
9. W. Pedrycz, Hierarchical architectures of fuzzy models: from type-1 fuzzy sets to information granules of higher type. Int. J. Comput. Intell. Syst. **3**(2), 202–214 (2010)
10. W. Pedrycz, Concepts and design aspects of granular models of type-1 and type-2. Int. J. Fuzzy Logic Intell. Syst. **15**(2), 87–95 (2015)

11. E. Rubio, O. Castillo, F. Valdez, P. Melin, C.I. González, G. Martinez, An extension of the fuzzy possibilistic clustering algorithm using type-2 fuzzy logic techniques. Adv. Fuzzy Syst. 7094046:1–7094046:23 (2017)
12. F. Olivas, L. Amador-Angulo, J. Pérez, C. Caraveo, F. Valdez, O. Castillo, Comparative study of type-2 fuzzy particle swarm, bee colony and bat algorithms in optimization of fuzzy controllers. Algorithms **10**(3), 101 (2017)
13. O. Castillo, M.A. Sanchez, C.I. González, G.E. Martinez, Review of recent type-2 fuzzy image processing applications. Information **8**(3), 97 (2017)
14. C.I. González, P. Melin, J.R. Castro, O. Castillo, O. Mendoza, Optimization of interval type-2 fuzzy systems for image edge detection. Appl. Soft Comput. **47**, 631–643 (2016)
15. K. Tai, A.-R. El-Sayed, M. Biglarbegian, C.I. González, O. Castillo, S. Mahmud, Review of recent type-2 fuzzy controller applications. Algorithms **9**(2), 39 (2016)
16. R. Sepulveda, O. Castillo, P. Melin, A. Rodríguez Díaz, O. Montiel, Experimental study of intelligent controllers under uncertainty using type-1 and type-2 fuzzy logic. Inf. Sci. **177**(10), 2023–2048 (2007)
17. M.A. Sanchez, Oscar Castillo, Juan R. Castro, Generalized type-2 fuzzy systems for controlling a mobile robot and a performance comparison with interval type-2 and type-1 fuzzy systems. Expert Syst. Appl. **42**(14), 5904–5914 (2015)
18. K. Atanassov, Intuitionistic fuzzy sets. Fuzzy Set Syst. **20**(1), 87–96 (1986)
19. K. Atanassov, Intuitionistic fuzzy sets, in *VII ITKR Session* (Sofia, 20–23 June 1983), Reprinted: Int. J. Bioautom. **20**(S1), S1–S6 (2016)
20. K. Atanassov, *Intuitionistic Fuzzy Sets: Theory and Applications* (Springer, Heidelberg, 1999)
21. K. Atanassov, *On Intuitionistic Fuzzy Sets Theory* (Springer, Berlin, 2012)
22. K. Atanassov, P. Vassilev, R. Tsvetkov, *Intuitionistic Fuzzy Sets, Measures and Integrals* (Academic Publishing House "Prof. Marin Drinov", Sofia, 2013)
23. C. Leal Ramírez, O. Castillo, P. Melin, A. Rodríguez Díaz, Simulation of the bird age-structured population growth based on an interval type-2 fuzzy cellular structure. Inf. Sci. **181**(3), 519–535 (2011)
24. N.R. Cázarez-Castro, L.T. Aguilar, O. Castillo, Designing type-1 and type-2 fuzzy logic controllers via fuzzy lyapunov synthesis for nonsmooth mechanical systems. Eng. Appl. AI **25**(5), 971–979 (2012)
25. O. Castillo, P. Melin, Intelligent systems with interval type-2 fuzzy logic. Int. J. Innov. Comput. Inf. Control **4**(4), 771–783 (2008)

Approximation of Intuitionistic Fuzzy Systems for Time Series Analysis in Plant Monitoring and Diagnosis

Oscar Castillo

Abstract We describe in this paper a proposed approach for approximating the fuzzy inference process in intuitionistic fuzzy systems. The new approach combines the outputs of two traditional type-1 fuzzy systems to obtain the final output of the intuitionistic fuzzy system. The new method provides an efficient way of calculating the output of an intuitionistic fuzzy system and as consequence can be applied to real-world problems in many areas of application. We illustrate the new approach with a simple example to motivate the ideas behind this work. We also illustrate the new approach for fuzzy inference with a more complicated example of monitoring a non-linear dynamic plant.

Keywords Intuitionistic fuzzy logic · Time series analysis · Monitoring

1 Introduction to Intuitionistic Fuzzy Logic

The intuitionistic fuzzy sets where defined as an extension of the ordinary fuzzy sets [1]. As opposed to a fuzzy set in X [15, 16], given by

$$B = \{(x, \mu_B(x)) | x \in X\} \tag{1}$$

where $\mu_B: X \to [0, 1]$ is the membership function of the fuzzy set B, an intuitionistic fuzzy set A is given by

$$A = \{(x, \mu_A(x), \nu_A(x)) | x \in X\} \tag{2}$$

where $\mu_A: X \to [0, 1]$ and $\nu_A: X \to [0, 1]$ are such that

O. Castillo (✉)
Division of Graduate Studies and Research, Tijuana Institute of Technology, Tijuana, Mexico
e-mail: ocastillo@tectijuana.mx

S. Melliani and O. Castillo (eds.), *Recent Advances in Intuitionistic Fuzzy Logic Systems*, Studies in Fuzziness and Soft Computing 372,
https://doi.org/10.1007/978-3-030-02155-9_8

$$0 \leq \mu_A + \nu_A \leq 1 \tag{3}$$

and $\mu_A(x)$; $\nu_A(x) \in [0, 1]$ denote a degree of membership and a degree of non-membership of $x \in A$, respectively.

For each intuitionistic fuzzy set in X, we will call

$$\pi_A(x) = 1 - \mu_A(x) - \nu_A(x) \tag{4}$$

a "hesitation margin" (or an "intuitionistic fuzzy index") of $x \in A$ and, it expresses a hesitation degree of whether x belongs to A or not. It is obvious that $0 \leq \pi_A(x) \leq 1$, for each $x \in X$.

On the other hand, for each fuzzy set B in X, we evidently have that

$$\pi_B(x) = 1 - \mu_B(x) - [1 - \mu_B(x)] = 0 \text{ for each } x \in X. \tag{5}$$

Therefore, if we want to fully describe an intuitionistic fuzzy set, we must use any two functions from the triplet [14]:

- Membership function,
- Non-membership function,
- Hesitation margin.

In other words, the application of intuitionistic fuzzy sets instead of fuzzy sets means the introduction of another degree of freedom into a set description (i.e. in addition to μ_A we also have ν_A or π_A).

Since the intuitionistic fuzzy sets being a generalization of fuzzy sets give us an additional possibility to represent imperfect knowledge, they can make it possible to describe many real problems in a more adequate way.

Basically, intuitionistic fuzzy sets based models maybe adequate in situations when we face human testimonies, opinions, etc. involving two (or more) answers of the type [14]:

- Yes,
- No,
- I do not know, I am not sure, etc.

Voting can be a good example of such a situation, as human voters may be divided into three groups of those who:

- Vote for,
- Vote against,
- Abstain or give invalid votes.

This third group is of great interest from the point of view of, say, customer behavior analysis, voter behavior analysis, etc., because people from this third undecided group after proper enhancement (e.g., different marketing activities) can finally become sure, i.e. become persons voting for (or against), customers wanting to buy advertised products, etc.

2 Intuitionistic Fuzzy Inference Systems

Fuzzy inference in intuitionistic systems has to consider the fact that we have the membership μ functions as well as the non-membership ν functions. In this case, we propose that the conclusion of an intuitionistic fuzzy system can be a linear combination of the results of two classical type-1 fuzzy systems, one corresponding to the membership functions and the other for the non-membership functions.

Assume that IFS stands for the output of an intuitionistic fuzzy system, and then with the following equation we can calculate the total output as a linear combination:

$$\mathrm{IFS} = (1 - \pi)\mathrm{FS}_{\mu} + \pi\mathrm{FS}_{\nu} \tag{6}$$

where FS_{μ} is the traditional output of a fuzzy system using only the membership function μ, and FS_{ν} is the output of a fuzzy system using the non-membership function ν. Of course, Eq. (6) for $\pi = 0$ will reduce to the output of a traditional fuzzy system, but for other values of π the result of IFS will be different as we are now taking into account the hesitation margin π.

The advantage of this method for computing the output IFS of an intuitionistic fuzzy system is that we can use our previous machinery of traditional fuzzy systems for computing both FS_{μ} and FS_{ν}. Then, we only perform a weighted average of both results to obtain the final output IFS of the intuitionistic fuzzy inference system. We consider below a simple example to illustrate these ideas.

Example Let us assume that we have a simple intuitionistic fuzzy system of only two rules:

IF x is small THEN y is big
IF x is big THEN y is small

We will consider for simplicity uniform rectangular membership functions for both linguistic variables. We show in Fig. 1 the membership functions for the linguistic values "small" and "big" of the input linguistic variable. We also show in Fig. 2 the non-membership functions for the linguistic values of the output variable. It is clear from Fig. 2 that in this case the membership and non-membership functions are not complementary, which is due to the fact that we have an intuitionistic fuzzy system.

From Fig. 2 we can clearly see that the hesitation margin π is 0.05 for both cases. As a consequence Eq. (6) can be written for our example as follows:

$$\mathrm{IFS} = 0.95\mathrm{FS}_{\mu} + 0.05\mathrm{FS}_{\nu} \tag{7}$$

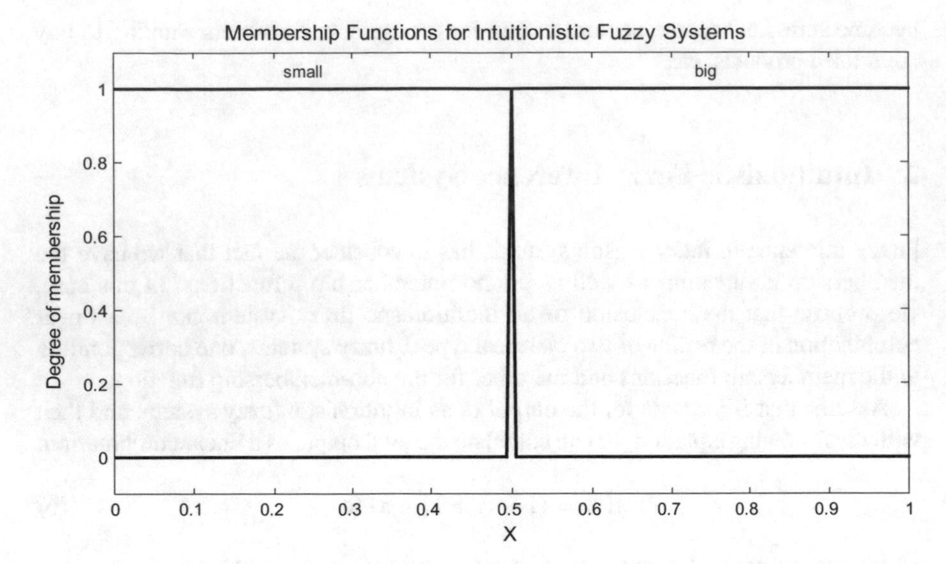

Fig. 1 Membership functions for the "small" and "big" linguistic values of the input variable

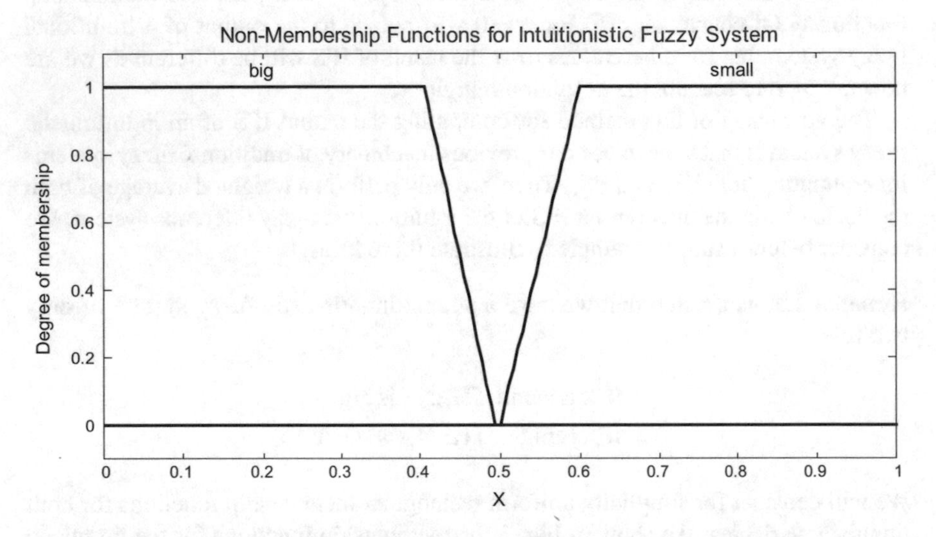

Fig. 2 Non-membership functions for the "small" and "big" linguistic values of the output variable

Now, let us assume that we want to calculate the output of the fuzzy system for a given input value of $x = 0.45$. In this case, we have that x is small with $\mu = 1$ and x is not small with $\nu = 0$, and x is big with $\mu = 0$ and x is not big with $\nu = 0.5$. As a consequence of these facts we have that,

$$\mathrm{IFS} = \mathrm{IFS}_{\mathrm{small}} + \mathrm{IFS}_{\mathrm{big}}$$

Table 1 Sample results of the intuitionistic fuzzy system for several input values

Input values, x	Membership result	Non-membership result	Intuitionistic result
0.2500	0.7500	0.7766	0.741330
0.3500	0.7500	0.7766	0.741330
0.4500	0.7500	0.7650	0.740750
0.5500	0.2500	0.2359	0.249295
0.6500	0.2500	0.2250	0.248750
0.7500	0.2500	0.2250	0.248750

$$\begin{aligned}
\mathrm{IFS} &= 0.95\mathrm{FS}_{\mu\,\mathrm{small}} + 0.05\mathrm{FS}_{\nu\mathrm{small}} + 0.95\mathrm{FS}_{\mu\mathrm{big}} + 0.05\mathrm{FS}_{\nu\mathrm{big}} \\
\mathrm{IFS} &= 0.95\mathrm{FS}_{\mu\,\mathrm{small}} + 0.05\mathrm{FS}_{\nu\mathrm{big}} \\
\mathrm{IFS} &= 0.95(0.75) + 0.05(0.765) \\
\mathrm{IFS} &= 0.74075
\end{aligned}$$

Of course, we can compare this intuitionistic fuzzy output with the traditional one (of 0.75), the difference between these two output values is due to the hesitation margin. We have to mention that in this example the difference is small because the hesitation margin is also small. We show in Table 1 the results of the intuitionistic fuzzy system for several input values.

We can appreciate from Table 1 the difference between the outputs of the intuitionistic fuzzy system and the output of the classical one.

Finally, we show in Figs. 3 and 4 the non-linear surfaces for the fuzzy systems of the membership and non-membership functions, respectively. We can appreciate from these figures that the surfaces are similar, but they differ slightly because of the hesitation margin.

3 Monitoring and Diagnosis Using Fuzzy Logic

Monitoring means checking or regulating the performance of a machine, a process, or a system [6]. Diagnosis, on the other hand, means deciding the nature and the cause of a diseased condition of a machine, a process, or a system by examining the symptoms. In other words, monitoring is detecting suspect symptoms, whereas diagnosis is determining the cause of the symptoms. There are several words and phrases that have similar or slightly different meanings such as fault detection, fault prediction, in-process evaluation, on-line inspection, identification, and estimation [12, 13]. The problems of engineering monitoring and diagnosis vary. The following list mentions a few examples:

(1) monitoring the deterioration of a high-pressure vessel based on acoustic emission signals;

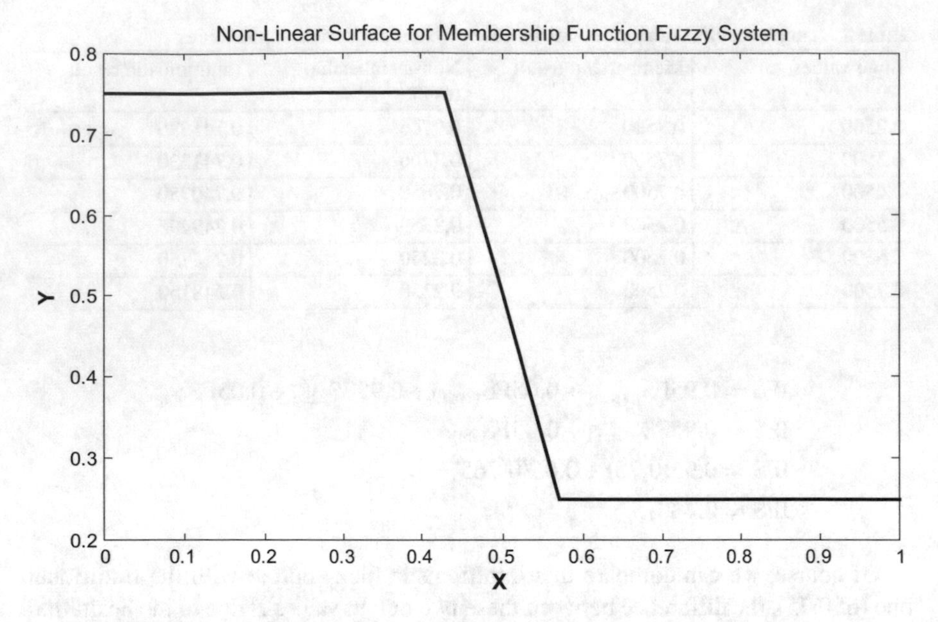

Fig. 3 Non-linear surface of the membership function fuzzy system

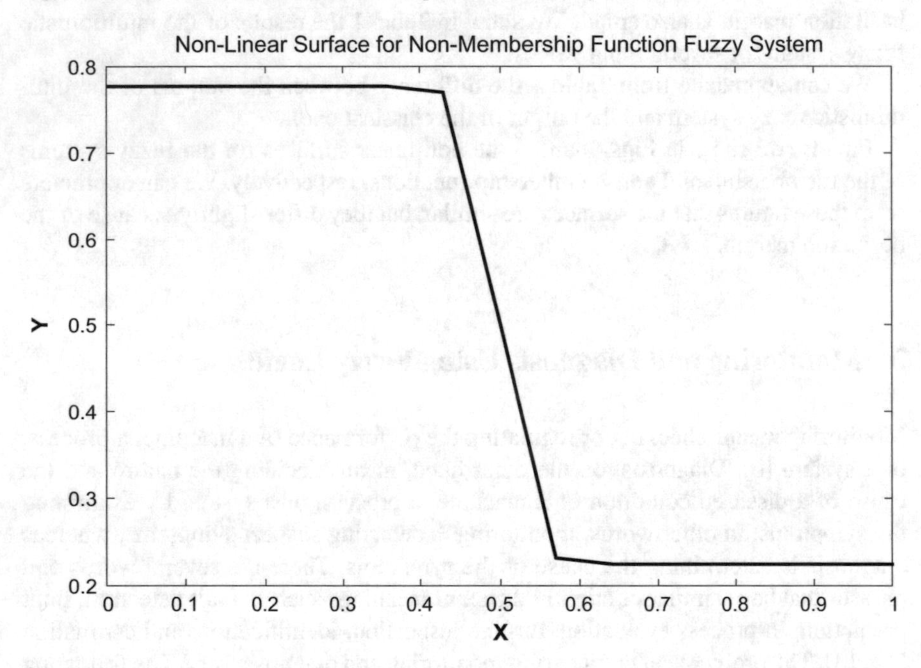

Fig. 4 Non-linear surface of non-membership function fuzzy system

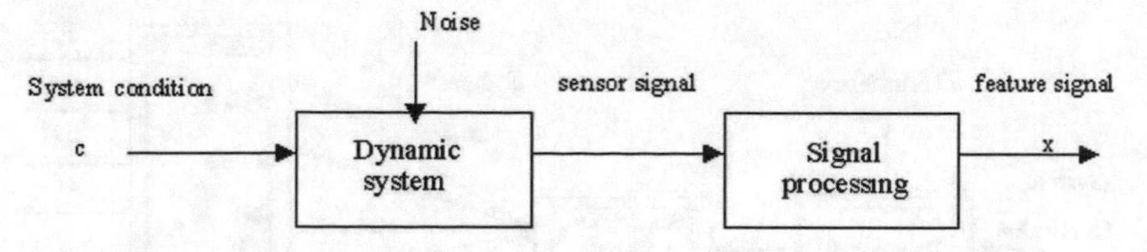

Fig. 5 Plant monitoring and diagnosis illustration

(2) monitoring the tool condition in a transfer machining station based on motor current signals;
(3) diagnosis of the condition of a turbine machinery set based on vibration signals.

The importance of monitoring and diagnosis of plant processes now is widely recognized because it results in increased productivity, improved product quality and decreased production cost [4, 5]. As a result, in the past decade, a large number of research and development projects have been carried and many monitoring and diagnosis methods have been developed [10, 11]. The commonly used monitoring and diagnosis methods include modeling-based methods [13], pattern recognition methods [12], fuzzy systems methods [2, 3], knowledge-based systems methods [11], and artificial neural networks [7]. It is interesting to note that even though these methods are rather different, they share a very similar structure as shown in Fig. 5.

In Fig. 5 the engineering system is considered to be dynamic, i.e. changing its condition with time. For this reason, there is a block in Fig. 5 called "dynamic system". Also, in Fig. 5 there is block indicating that "signal processing" has to be performed on the sensor signals to extract the feature signals, which in turn are used to be able to monitor the dynamic process. Of course, Fig. 5 only shows the general structure for engineering monitoring and diagnosis.

The "condition" of a machine, a process, or an engineering system (which will be referred to as system condition and denoted by $c \in \{c_1, c_2, \ldots, c_m\}$) can be considered as the "input", the system working conditions and noises (including system noise and sampling noise) can be considered as the "noise", and the sensor signals are the "outputs" from the system. Typically, the sensor signals are processed by a computer, after which the signals are transformed into a set of features called feature signals, denoted as $\mathbf{x} = \{x_1, x_2, \ldots, x_n\}$. In general, the systems conditions are predefined, such as normal, critical, etc. On the other hand, the features may be the mean of a temperature signal, the variance of a displacement signal, etc. Sensing and signal processing are very important to the success of plant monitoring and diagnosis [5, 6].

More formally, the goal of monitoring is to use the feature signals, x, to determine whether the plant is in an acceptable condition(s) (a subset of $\{c_1, c_2, \ldots, c_m\}$). On the other hand, the objective of diagnosis is to use the feature signals, x, to determine the system condition, $c \in \{c_1, c_{2,\ldots,} c_m\}$. No matter how monitoring and diagnosis methods may differ, monitoring and diagnosis always consist of two phases: training and decision making. Training is to establish a relationship between the feature

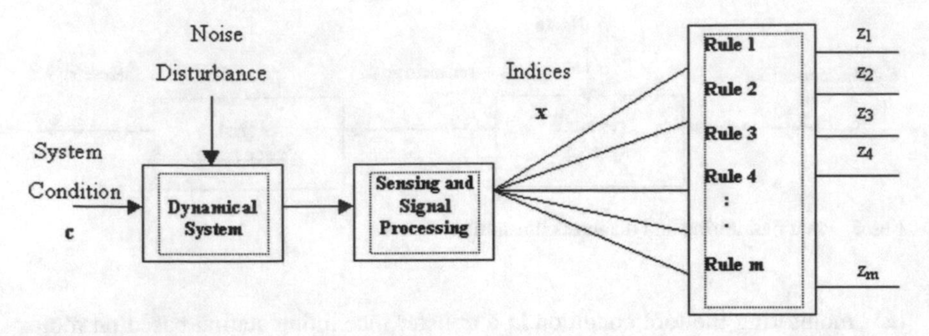

Fig. 6 Fuzzy system for plant monitoring and diagnosis

signals and the systems conditions. Without losing generality, this relationship can be represented as

$$\mathbf{x} = \mathrm{F}(\mathrm{c}). \tag{8}$$

It should be pointed out that F(c) represents a fuzzy system, a neural network or another method that could be used to obtain this relationship [11]. In fact, it is the form of the relationship that determines the methods of monitoring and diagnosis, as well as the performance of the methods. The relationship F(c) is established based on training samples, denoted by $\mathbf{x}_1, \mathbf{x}_2, \ldots, \mathbf{x}_k, \ldots, \mathbf{x}_N$, where the system condition for each training sample is known [and denoted as $c(\mathbf{x}_k)$].

After the relationship is established, when a new sample is given (from an unknown system condition), its corresponding condition is estimated based on the inverse relationship

$$\mathrm{c} = \mathrm{F}^{-1}(\mathbf{x}). \tag{9}$$

This is called decision-making, or classifying. Whereas it is not likely that the training samples will cover all possible cases, decision making often involves reasoning or inferencing. Of course, if the inverse relationship doesn't exist then a "pseudo-inverse" approach can be used instead [11].

In particular, when a fuzzy system is used, the relationship is given by a set of rules as shown in Fig. 6. The input to the fuzzy system is the feature signal and the output of the fuzzy system is the estimated plant condition(s) [i.e., $\mathbf{z} = (z_1, z_2, \ldots, z_m)$ is an estimate of $\mathbf{c} = (c_1, c_2, \ldots, c_m)$]. In other words, the fuzzy system models the inverse relationship between the system conditions and the feature signals.

4 Plant Monitoring Using the Intuitionistic Fuzzy Logic Approach

In this section, we outline a method to implement a fuzzy rule-based expert monitoring system with two basic sensors: temperature, and pressure. We also use as input the fuzzy fractal dimension of the time series of the measured variables [8, 9]. Individual sensors can identify two linguistic values (high and low). The three inputs can be combined to give 8 different scenarios.

Let x_1 be the temperature, x_2 the pressure, x_3 the fuzzy fractal dimension, and y the diagnostic statement. Let L_i and H_i, represent the two sets of low range and high range for input data x_i, where $i = 1, 2$, or 3. Furthermore, let $C_1, C_2, \ldots, C_8$ be the individual scenarios that could happen for each combination of the different data sets. The fuzzy rules have the general form:

$$
\begin{aligned}
&R^{(0)}: \text{IF } x_1 \text{ is } L_1 \text{ AND } x_2 \text{ is } L_2 \text{ AND } x_3 \text{ is } L_3 \text{ THEN y is } C_1 \\
&\quad \ldots \qquad\qquad\qquad\qquad \ldots \\
&R^{(i)}: \text{IF } x_1 \text{ is } V_1 \text{ AND } x_2 \text{ is } V_2 \text{ AND } x_3 \text{ is } V_3 \text{ THEN y is } C_i \\
&\quad \ldots \qquad\qquad\qquad\qquad \ldots \\
&R^{(26)}: \text{IF } x_1 \text{ is } H_1 \text{ AND } x_2 \text{ is } H_2 \text{ AND } x_3 \text{ is } H_3 \text{ THEN y is } C_8
\end{aligned}
\tag{10}
$$

In this case, V_i represents L_i or H_i, depending on the condition for the plant. Experts have to provide their knowledge in plant monitoring to label the individual cases C_i for $i = 1, 2, \ldots, 8$. Also, the membership functions for the linguistic values of variables have to be defined according to historical data of the problem and expert knowledge.

We can use the Fuzzy Logic Toolbox of the MATLAB programming language to implement the intuitionistic fuzzy monitoring system described above. In this case, we need to specify the particular fuzzy rules, and the corresponding membership functions and non-membership functions for the problem [17, 18]. We show below a sample implementation of a health monitoring system using the MATLAB language. First, we show in Fig. 7 the general architecture of the fuzzy monitoring system. In this figure, we can see the input linguistic variables (temperature, pressure, and fractal dimension) and the output linguistic variable (condition of the plant) of the fuzzy monitoring system. Of course, in this case the fractal dimension is estimated using the box counting algorithm, which was implemented also as a computer program in MATLAB The actual 8 rules were defined according to expert knowledge on the process and are shown below.

1. If (temp is low) and (pressure is low) and (fracdim is low) then (condition is C1)
2. If (temp is low) and (pressure is high) and (fracdim is low) then (condition is C2)
3. If (temp is low) and (pressure is low) and (fracdim is high) then (condition is C3)
4. If (temp is high) and (pressure is low) and (fracdim is low) then (condition is C4)
5. If (temp is low) and (pressure is high) and (fracdim is high) then (condition is C5)

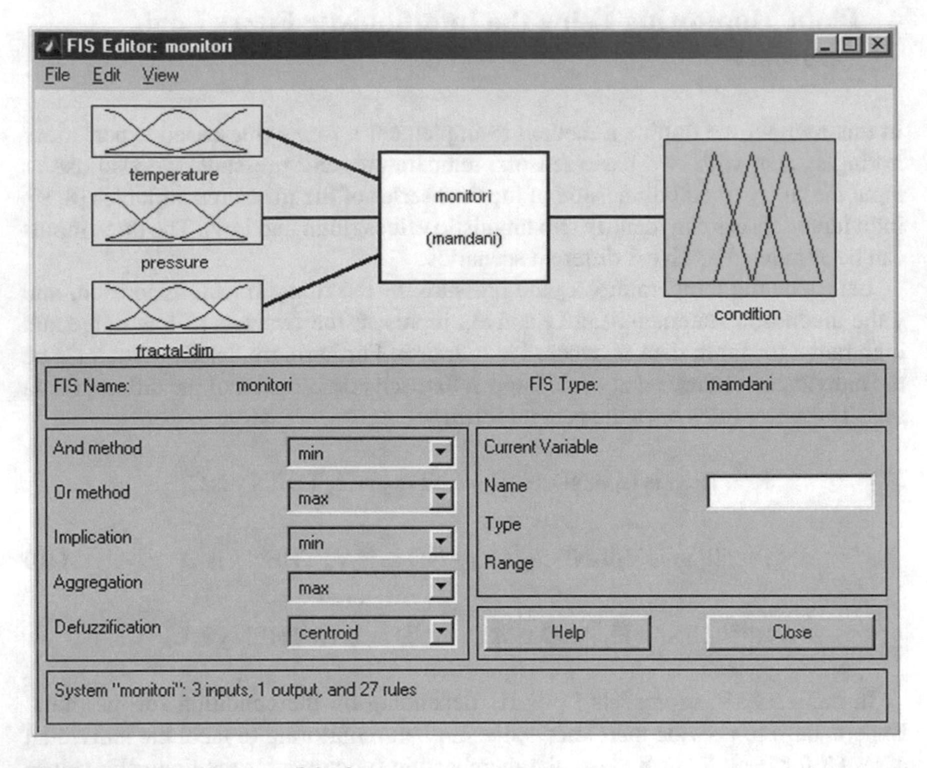

Fig. 7 General architecture of the fuzzy system

6. If (temp is high) and (pressure is low) and (fracdim is high) then (condition is C6)
7. If (temp is high) and (pressure is high) and (fracdim is low) then (condition is C7)
8. If (temp is high) and (pressure is high) and (fracdim is high) then (condition is C8)

In Fig. 8 the trapezoidal non-membership functions for the temperature variable are shown. On the other hand, Fig. 9 shows the membership functions of the temperature variable. We can appreciate from these figures that the membership and non-membership functions are not complementary, which is expected in an intuitionistic fuzzy system. In Fig. 10 we show the non-linear surface for the fuzzy system of plant monitoring with membership functions. In Fig. 11 we show the non-linear surface of the fuzzy system for non-membership functions. We have to point out that a Mamdani fuzzy inference system with max-min operators was used in the implementation of this fuzzy system.

Finally, in Table 2 we summarize the results of 14 different cases (conditions of the plant). Again, we can notice that in some cases the intuitionistic output is very close to the traditional fuzzy output. However, in other cases the difference between them is significant. Of course, we consider that the intuitionistic fuzzy result is closer to the real one because it models human experts more closely.

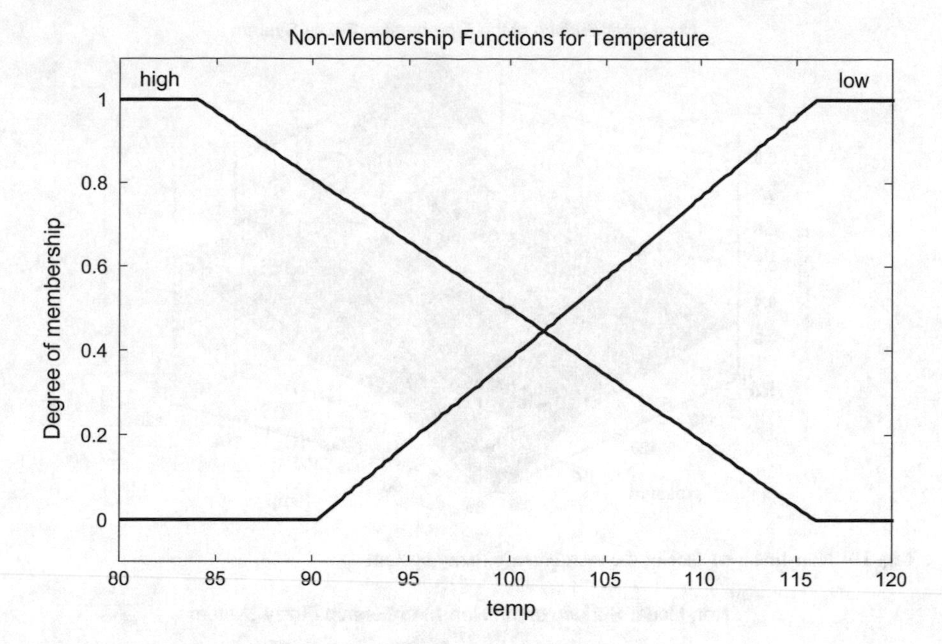

Fig. 8 Trapezoidal non-membership functions of the temperature variable

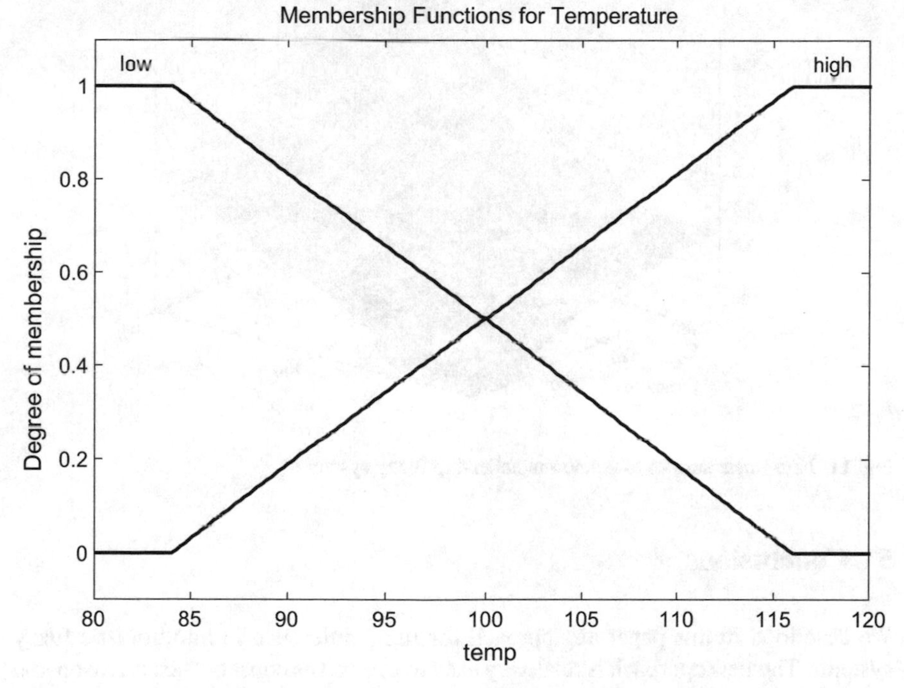

Fig. 9 Trapezoidal membership functions of the temperature variable

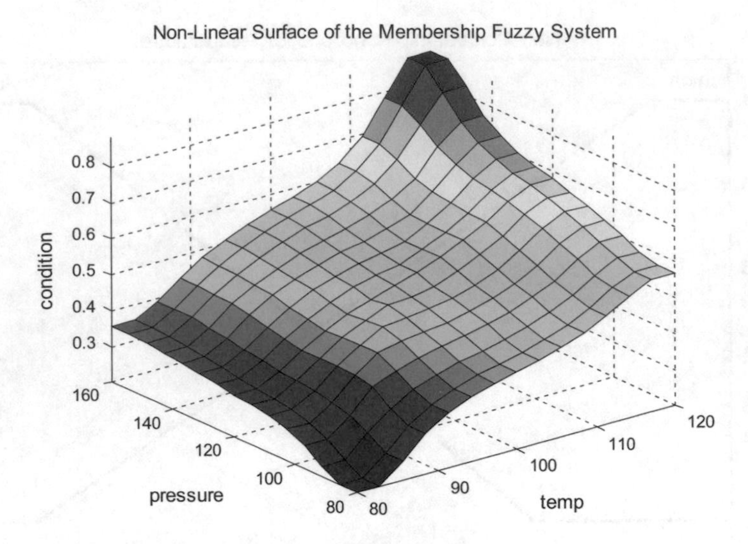

Fig. 10 Non-linear surface of the membership fuzzy system

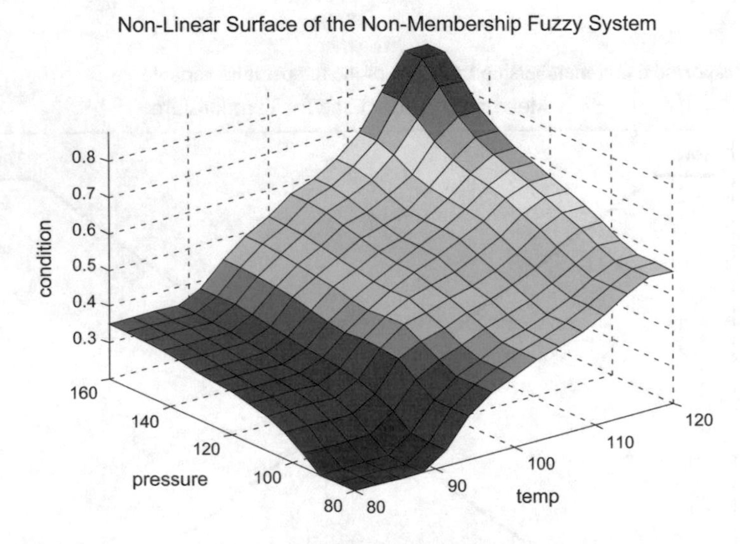

Fig. 11 Non-linear surface of the non-membership fuzzy system

5 Conclusions

We described in this paper an approach for fuzzy inference in intuitionistic fuzzy systems. The new approach is relatively easy to use and as consequence is reasonable to apply it in an efficient way in real-world problems. We have described the approach and have shown a simple example to explain the ideas behind the method. We have

Table 2 Summary of the results of fourteen different cases for the non-linear plant

Inputs			Output		
Temperature	Pressure	Fractal dim.	Member	Non-member	Intuitionistic
90	125	1.5	0.4181	0.3221	0.4094
95	120	1.3	0.4491	0.3959	0.4443
110	140	1.7	0.5974	0.5961	0.5972
115	150	1.9	0.8622	0.8619	0.8622
85	90	1.2	0.2030	0.0869	0.1925
90	95	1.1	0.3036	0.0522	0.2808
100	105	1.3	0.4590	0.4420	0.4575
105	115	1.4	0.5096	0.4967	0.5085
120	130	1.8	0.7542	0.7519	0.7540
120	130	1.1	0.6911	0.6882	0.6908
80	85	1.5	0.2041	0.2013	0.2038
100	100	1.7	0.4887	0.4575	0.4858
125	95	1.2	0.5218	0.4724	0.5174
130	115	1.6	0.6814	0.6763	0.6809

also shown a more complete example of plant monitoring in which intuitionistic fuzzy logic has the potential of modeling the uncertainty in the identification of problems in a dynamic process. We think that this new approach for fuzzy inference can also be applied to other problems like in intelligent control, pattern recognition and time series prediction.

References

1. K. Atanassov, *Intuitionistic Fuzzy Sets: Theory and Applications* (Springer, Heidelberg, Germany, 1999)
2. O. Castillo, P. Melin, *Automated Mathematical Modelling for Financial Time Series Prediction Combining Fuzzy Logic and Fractal Theory", Edited Book " Soft Computing for Financial Engineering* (Springer, Heidelberg, Germany, 1999), pp. 93–106
3. O. Castillo, P. Melin, *Soft Computing and Fractal Theory for Intelligent Manufacturing* (Springer, Heidelberg, Germany, 2003)
4. O. Castillo, P. Melin, E. Ramírez, J. Soria, Hybrid intelligent system for cardiac arrhythmia classification with Fuzzy K-nearest neighbors and neural networks combined with a fuzzy system. Expert Syst. Appl. **39**(3), 2947–2955 (2012)
5. G. Chen, T.T. Pham, *Introduction to Fuzzy Sets, Fuzzy Logic, and Fuzzy Control Systems* (CRC Press, Boca Raton, Florida, USA, 2001)
6. L.H. Chiang, E.L. Russell, R.D. Braatz, *Fault Detection and Diagnosis in Industrial Systems* (Springer, Heidelberg, Germany, 2000)
7. J.R. Jang, C.T. Sun, E. Mizutani, *Neuro-Fuzzy and Soft Computing* (Prentice Hall, 1997)
8. B. Mandelbrot, *The Fractal Geometry of Nature* (W.H. Freeman and Company, New York, USA 1987)
9. B. Mandelbrot, *Fractals and Scaling in Finance* (Springer, Heidelberg, Germany, 1997)

10. P. Melin, O. Castillo, A new method for adaptive model-based control of non-linear dynamic plants using a neuro-fuzzy-fractal approach. J. Soft Comput. **5**(2), 171–177 (2001). Springer
11. P. Melin, O. Castillo, Intelligent control of complex electrochemical systems with a neuro-fuzzy-genetic approach. IEEE Trans. Ind. Electron **48**(5), 951–955 (2001)
12. R.J. Patton, P.M. Frank, R.N. Clark, *Issues of Fault Diagnosis for Dynamic Systems* (Springer, Heidelberg, Germany, 2000)
13. E.L. Russell, L.H. Chiang, R.D. Braatz, *Data-driven Methods for Fault Detection and Diagnosis in Chemical Processes* (Springer, Heidelberg, Germany, 2000)
14. E. Szmidt, J. Kacprzyk, Analysis of agreement in a group of experts via distances between intuitionistic preferences. in *Proceedings of IPMU'2002* (Universite Savoie, France, 2002), pp 1859–1865
15. L.A. Zadeh, Fuzzy sets. Inf. Control **8**, 338–353 (1965)
16. L.A. Zadeh, Quantitative fuzzy semantics. Inf. Sci. **3**, 159–176 (1971)
17. L.A. Zadeh, Outline of a new approach to the analysis of complex systems and decision processes. IEEE Trans. Syst. Man Cybern. **3**(1), 28–44 (1973)
18. L.A. Zadeh, The concept of a linguistic variable and its application to approximate reasoning. Inf. Sci. **8**, 43–80 (1975)

New Fractional Derivative in Colombeau Algebra

Said Melliani, A. Chafiki and L. S. Chadli

Abstract In this paper we introduce an approach to fractional derivatives involving singularities based on the theory of algebras of generalized functions in the Colombeau algebra $\mathscr{G}$, using new definition of fractional derivative called conformable fractional derivative introduced by the authors Khalil et al. in [1].

1 Introduction

This paper is extension of fractional derivatives in Colombeau algebra of generalized functions in order to solve problems involving a multiplication of distributions and other nonlinear operations with singularities, provided by Colombeau theory, but including non-integer derivatives and operations among them. There are many fractional order equations with a lack of the solution in classical spaces, especially in the space of distributions involving nonlinear operations and singularities. In this way, many problems with fractional derivatives involving such kind of operation, would have been solved. Another reason for introducing fractional derivatives into Colombeau theory is an extension of the Colombeau theory to derivatives of arbitrary order, i.e. to non-integer ones [2].

In the last decades, fractional, or non-integer, differentiation has played a very important role in various fields such as mechanics, electricity, chemistry, biology, economics, modeling problems, anomalous diffusion and notably control theory and signal and image processing. It has been found that fractional differential equations play a crucial role in modeling anomalous diffusion, time-dependent materials and

S. Melliani (✉) · A. Chafiki · L. S. Chadli
Sultan Moulay Slimane University, BP 523 Beni Mellal, Morocco
e-mail: s.melliani@usms.ma

A. Chafiki
e-mail: ahmedchafiki815@gmail.com

L. S. Chadli
e-mail: sa.chadli@yahoo.fr

S. Melliani and O. Castillo (eds.), *Recent Advances in Intuitionistic Fuzzy Logic Systems*, Studies in Fuzziness and Soft Computing 372,
https://doi.org/10.1007/978-3-030-02155-9_9

processes with long-range dependence, allometric scaling laws, as well as power law in complex systems.

The theory of algebra of generalized functions provides extension to derivatives of arbitrary order [2, 3]. It allows us to solve nonlinear partial differential problems with fractional order of temporal or spatial derivatives. These problems sometimes better describe the structure of the problems in nature than ODEs or PDEs do. The paper is organized as follows, in the first section we give some basic preliminaries such as notations and definitions of the objects we shall work with. We also introduce different spaces of Colombeau algebra of generalized functions and results concerning conformable fractional derivative. In the second section we prove the Fractional derivatives of Colombeau generalized. In the third section we prove the fractional integral of Colombeau generalized.

2 Preliminaries

2.1 *Definition of the Colombeau Algebra*

We use the following notations

$$\mathscr{A}_q = \{\varphi \in \mathscr{D}(\mathbb{R}^+)/ \int_{\mathbb{R}^+} \varphi(x)dx = 1, \quad \int_{\mathbb{R}^+} x^\alpha \varphi(x)dx = 0 \quad \text{for} \quad 1 \le |\alpha| \le q\}$$

$q = 1, 2, \dots$

$$\varphi_\varepsilon(x) = \frac{1}{\varepsilon}\varphi(\frac{x}{\varepsilon}) \quad \text{for} \quad \varphi \in \mathscr{D}(\mathbb{R}^+)$$

We denote by

$$\mathscr{E}(\mathbb{R}^+) = \{u : \mathscr{A}_1 \times \mathbb{R}^+ \to \mathbb{C}/ \quad \text{with} \quad u(\varphi, x) \quad \text{is} \quad \mathscr{C}^\infty \quad \text{to the second variable} \quad x\}$$

$$u(x, \varphi_\varepsilon) = u_\varepsilon(x) \ \forall \varphi \in \mathscr{A}_1$$

$$\begin{aligned}\mathscr{E}_M(\mathbb{R}^+) = \{&(u_\varepsilon)_{\varepsilon>0} \subset \mathscr{E}(\mathbb{R}^+)/\forall K \subset\subset \mathbb{R}^+, \forall m \in \mathbb{N}, \exists N \in \mathbb{N} \quad \text{such that} \\ &\sup_{x\in K} |D^m u_\varepsilon(x)| = \mathscr{O}(\varepsilon^{-N}) \quad \text{as} \quad \varepsilon \to 0\}\end{aligned}$$

$$\begin{aligned}\mathscr{N}(\mathbb{R}^+) = \{&(u_\varepsilon)_{\varepsilon>0} \subset \mathscr{E}(\mathbb{R}^+)/\forall K \subset\subset \mathbb{R}^+, \forall m \in \mathbb{N}, \forall p \in \mathbb{N} \quad \text{such that} \\ &\sup_{x\in K} |D^m u_\varepsilon(x)| = \mathscr{O}(\varepsilon^{p}) \quad \text{as} \quad \varepsilon \to 0\}\end{aligned}$$

The Colombeau algebra is defined as a factor set $\mathcal{G}(\mathbb{R}^+) = \mathcal{E}_M(\mathbb{R}^+)/\mathcal{N}(\mathbb{R}^+)$, where the elements of the set $\mathcal{E}_M(\mathbb{R}^+)$ are moderate while the elements of the set $\mathcal{N}(\mathbb{R}^+)$ are negligible.

Finally, we introduce C^k-Colombeau generalized in the following way.

Denote by

$$\mathcal{E}_M^k(\mathbb{R}^+) = \{(u_\varepsilon)_{\varepsilon>0} \subset \mathcal{E}(\mathbb{R}^+)/\forall K \subset\subset \mathbb{R}^+, \forall m \in \mathbb{N}, \exists N \in \mathbb{N} \text{ such that } \sup_{x\in K} |D^m u_\varepsilon(x)| = \mathcal{O}(\varepsilon^{-N}), m \in \{0, \dots, k\}, \text{ as } \varepsilon \to 0\}$$

$$\mathcal{N}^k(\mathbb{R}^+) = \{(u_\varepsilon)_{\varepsilon>0} \subset \mathcal{E}(\mathbb{R}^+)/\forall K \subset\subset \mathbb{R}^+, \forall m \in \mathbb{N}, \forall p \in \mathbb{N} \text{ such that } \sup_{x\in K} |D^m u_\varepsilon(x)| = \mathcal{O}(\varepsilon^{p}), m \in \{0, \dots, k\} \text{ as } \varepsilon \to 0\}$$

The C^k - Colombeau algebra is defined as a factor set $\mathcal{G}^k(\mathbb{R}^+) = \mathcal{E}_M^k(\mathbb{R}^+)/\mathcal{N}^k(\mathbb{R}^+)$.

2.2 New Definition of Fractional Derivative

We recall some notations, definitions, and results concerning conformable fractional derivative which are used throughout this paper. By $C(I, R)$ we denote the Banach space of all continuous functions from I into R with the norm $\|f\|_\infty = \sup_{t\in I} |f(t)|$

Definition 1 [1] Given a function $f : [0, \infty) \to R$, then the "conformable fractional derivative" of f of order α is defined by

$$T_\alpha(f)(t) = \lim_{\varepsilon\to 0} \frac{f(t + \varepsilon t^{1-\alpha}) - f(t)}{\varepsilon}$$

for all $t > 0$, $\alpha \in (0, 1)$. If f is α-differentiable in some $(0, a)$, $a > 0$ and $\lim_{t\to 0^+} T_\alpha(f)(t)$ exists, then define $f^{(\alpha)}(0) = \lim_{t\to 0^+} T_\alpha(f)(t)$.

Definition 2 [1] The $\alpha-$fractional integral of a continuous function f starting from $a \geq 0$ of order $\alpha \in (0, 1)$ is defined by

$$I_\alpha^a(f)(t) = I_1^a(f)(t) = \int_a^t s^{\alpha-1} f(s) ds$$

Lemma 1 [1] *Assume that* $f[a, \infty) \to R$, *such that is continuous and* $0 < \alpha \leq 1$. *Then, for all* $t > a$ *we have*

$$T_\alpha^a I_\alpha^a f(t) = f(t)$$

In the right case we have:

Lemma 2 [4] *Assume that* $f(-\infty, b] \to R$*, such that is continuous and* $0 < \alpha \leq 1$*. Then, for all* $t < b$ *we have*

$$T_\alpha^b I_\alpha^b f(t) = f(t)$$

Lemma 3 [4] *Let* $f, h : [a, \infty) \to R$ *be functions such that* $T_\alpha^a f(t)$ *exists for all* $t > a$*,* f *is differentiable on* $(0, a)$ *and*

$$T_\alpha^a f(t) = (t - a)^{1-\alpha} h(t) \tag{1}$$

Then $h(t) = f'(t)$ *for all* $t > a$*.*

Lemma 4 [4] *Let* $f : (a, b) \to R$ *be differentiable and* $0 < \alpha \leq 1$*. Then for all* $t > a$ *we have*

$$I_\alpha^a T_\alpha^a f(t) = f(t) - f(a)$$

3 Fractional Derivatives of Colombeau Generalized

Let $(u_\varepsilon)_{\varepsilon>0}$ be a representative of a Colombeau generalized $u \in \mathscr{G}([0, \infty))$. By (1), the fractional derivative of $(u_\varepsilon)_{\varepsilon>0}$, is defined by

$$D^\alpha u_\varepsilon(x) = x^{1-\alpha} \frac{d}{dx} u_\varepsilon(x) \tag{2}$$

Lemma 5 *Let* $(u_\varepsilon)_{\varepsilon>0}$ *be a representative of* $u \in \mathscr{G}([0, \infty))$*. Then, for every* $\alpha > 0$*,* $\sup_{x\in[0,T]} \left| D^\alpha u_\varepsilon(x) \right|$ *has a moderate bound.*

Proof

$$\begin{aligned} \sup_{x\in[0,T]} \left| D^\alpha u_\varepsilon(x) \right| = \sup_{x\in[0,T]} \left| x^{1-\alpha} \frac{d}{dx} u_\varepsilon(x) \right| &\leq T^{1-\alpha} \sup_{x\in[0,T]} \left| \frac{d}{dx} u_\varepsilon(x) \right| \\ &\leq T^{1-\alpha} C \varepsilon^{-N} \\ &\leq C_{\alpha,T} \varepsilon^{-N} \end{aligned}$$

Lemma 6 *Let* $(u_{1,\varepsilon})_{\varepsilon>0}$ *and* $(u_{2,\varepsilon})_{\varepsilon>0}$ *be two different representative of* $u \in \mathscr{G}([0, \infty))$*.*

Then, for every $\alpha > 0$*,* $\sup_{x\in[0,T]} \left| D^\alpha u_{1,\varepsilon}(x) - D^\alpha u_{2,\varepsilon}(x) \right|$ *is negligible.*

Proof

$$\begin{aligned}\sup_{x\in[0,T]}\left|D^{\alpha}u_{1,\varepsilon}(x)-D^{\alpha}u_{2,\varepsilon}(x)\right| &= \sup_{x\in[0,T]}\left|x^{1-\alpha}\frac{d}{dx}u_{1,\varepsilon}(x)-x^{1-\alpha}\frac{d}{dx}u_{2,\varepsilon}(x)\right| \\ &= \sup_{x\in[0,T]}\left|x^{1-\alpha}(\frac{d}{dx}u_{1,\varepsilon}(x)-\frac{d}{dx}u_{2,\varepsilon}(x))\right| \\ &\le T^{1-\alpha}\sup_{x\in[0,T]}\left|\frac{d}{dx}u_{1,\varepsilon}(x)-\frac{d}{dx}u_{2,\varepsilon}(x)\right|\end{aligned}$$

Since $(u_{1,\varepsilon})_{\varepsilon>0}$ and $(u_{2,\varepsilon})_{\varepsilon>0}$ represent the same Colombeau generalized u we have that $\sup_{x\in[0,T]}\left|\frac{d}{dx}u_{1,\varepsilon}(x)-\frac{d}{dx}u_{2,\varepsilon}(x)\right|$ is negligible. Therefore, $\sup_{x\in[0,T]}\left|D^{\alpha}u_{1,\varepsilon}(x) - D^{\alpha}u_{2,\varepsilon}(x)\right|$ is negligible, too.

After proving the previous two lemmas we are able to introduce the α-fractional derivative of a Colombeau generalized on $[0,\infty)$.

Definition 3 Let $u\in\mathcal{G}([0,\infty))$ be a Colombeau generalized on $[0,\infty)$ The α-fractional derivative of u, in notation $D^{\alpha}u(x)=[D^{\alpha}u_{\varepsilon}(x)]$ is an element of $\mathcal{G}^{0}([0,\infty))$ satisfying (2).

Remark 1 For $0<\alpha<1$ the first-order derivative of $\frac{d}{dx}D^{\alpha}u_{\varepsilon}(x)=(1-\alpha)x^{-\alpha}\frac{d}{dx}u_{\varepsilon}(x)+x^{1-\alpha}\frac{d^2}{dx^2}u_{\varepsilon}(x)$ and it does not reach its upper limit. In general, the k-th order derivative $\frac{d^k}{dx^k}D^{\alpha}u_{\varepsilon}(x)$ it does not reach its upper limit on $[0,\infty)$

The new fractional derivative of a Colombeau generalized $u\in\mathcal{G}([0,\infty))$ If one wants this to be of $\mathcal{G}([0,\infty))$, then the regularization of the fractional derivative has to be done.

Definition 4 Let $(u_{\varepsilon})_{\varepsilon>0}$ be a representative of a Colombeau generalized $u\in\mathcal{G}([0,\infty))$. The regularized of new fractional derivative of $(u_{\varepsilon})_{\varepsilon>0}$, is defined by:

$$\tilde{D}^{\alpha}u_{\varepsilon}(x)=\begin{cases}(D^{\alpha}u_{\varepsilon}*\varphi_{\varepsilon})(x), 0<\alpha<1\\ u_{\varepsilon}^{'}(x)=\frac{d}{dx}u_{\varepsilon}(x)\end{cases} \tag{3}$$

where $D^{\alpha}u_{\varepsilon}(x)$ is given by (2) and $\varphi_{\varepsilon}(x)$ is given by (2.1).
The convolution in (3) is $D^{\alpha}u_{\varepsilon}*\varphi_{\varepsilon}(x)=\int_0^{\infty}D^{\alpha}u_{\varepsilon}(s)\varphi_{\varepsilon}(x-s)ds$

Lemma 7 *Let $(u_{\varepsilon})_{\varepsilon>0}$ be a representative of $u\in\mathcal{G}([0,\infty))$.*
Then, for every $\alpha>0$ and every $k\in\{0,1,2,\ldots\}$, $\sup_{x\in[0,T]}\left|\frac{d^k}{dx^k}\tilde{D}^{\alpha}u_{\varepsilon}(x)\right|$ has a moderate bound.

Proof Let $\varepsilon\in(0,1)$. For $\alpha\in\mathbb{N}$, $\tilde{D}^{\alpha}u_{\varepsilon}(x)$ is the usual derivative of order α of $u_{\varepsilon}(x)$ and the assertion immediately follows. In case when $0<\alpha<1$, we have

$$\begin{aligned}
\sup_{x\in[0,T]}\left|\tilde{D}^\alpha u_\varepsilon(x)\right| &= \sup_{x\in[0,T]}\left|(D^\alpha u_\varepsilon * \varphi_\varepsilon)(x)\right| \\
&\leq \sup_{x\in[0,T]}\left|\int_0^\infty D^\alpha u_\varepsilon(s)\varphi_\varepsilon(x-s)ds\right| \\
&\leq \sup_{s\in K}\left|D^\alpha u_\varepsilon(s)\right| \sup_{x\in[0,T]}\left|\int_K \varphi_\varepsilon(x-s)ds\right| \\
&\leq C \sup_{s\in K}\left|D^\alpha u_\varepsilon(s)\right|
\end{aligned}$$

for some constant $C > 0$

Since, according to Lemma 3.1, $\sup_{s\in K}\left|D^\alpha u_\varepsilon(s)\right|$ has a moderate bound, for every $\alpha > 0$, it follows that, $\sup_{x\in[0,T]}\left|\tilde{D}^\alpha u_\varepsilon(x)\right|$ has a moderate bound, too.
For arbitrary order derivative, we have

$$\begin{aligned}
\sup_{x\in[0,T]}\left|\frac{d^k}{dx^k}\tilde{D}^\alpha u_\varepsilon(x)\right| &\leq \sup_{s\in K}\left|D^\alpha u_\varepsilon(s)\right| \sup_{x\in[0,T]}\left|\int_K \frac{d^k}{dx^k}\varphi_\varepsilon(x-s)ds\right| \\
&\leq \frac{C}{\varepsilon^k}\sup_{s\in K}\left|D^\alpha u_\varepsilon(s)\right|
\end{aligned}$$

$k \in \mathbb{N}$ for some constant $C > 0$ according to Lemma 3.1, $\sup_{s\in K}\left|D^\alpha u_\varepsilon(s)\right|$ has a moderate bound.
Therefore $\sup_{x\in[0,T]}\left|\frac{d^k}{dx^k}\tilde{D}^\alpha u_\varepsilon(x)\right|$ has a moderate bound, too.

Lemma 8 *Let $(u_{1,\varepsilon})_{\varepsilon>0}$ and $(u_{2,\varepsilon})_{\varepsilon>0}$ be two different representative of $u \in \mathscr{G}([0,\infty))$.*
Then, for every $\alpha > 0$, and every $k \in \{0, 1, 2, \ldots\}$, $\sup_{x\in[0,T]}\left|\frac{d^k}{dx^k}(\tilde{D}^\alpha u_{1,\varepsilon}(x) - \tilde{D}^\alpha u_{2,\varepsilon}(x))\right|$ is negligible.

Proof

$$\begin{aligned}
&\sup_{x\in[0,T]}\left|\frac{d^k}{dx^k}(\tilde{D}^\alpha u_{1,\varepsilon}(x) - \tilde{D}^\alpha u_{2,\varepsilon}(x))\right| \\
&= \sup_{x\in[0,T]}\left|\frac{d^k}{dx^k}((D^\alpha u_{1,\varepsilon} * \varphi_\varepsilon)(x) - (D^\alpha u_{2,\varepsilon} * \varphi_\varepsilon)(x))\right| \\
&= \sup_{x\in[0,T]}\left|\frac{d^k}{dx^k}((D^\alpha u_{1,\varepsilon} - D^\alpha u_{2,\varepsilon}) * \varphi_\varepsilon)(x)\right| \\
&= \sup_{x\in[0,T]}\left|((D^\alpha u_{1,\varepsilon} - D^\alpha u_{2,\varepsilon}) * \frac{d^k}{dx^k}\varphi_\varepsilon)(x)\right| \\
&\leq \sup_{s\in K}\left|(D^\alpha u_{1,\varepsilon} - D^\alpha u_{2,\varepsilon})(s)\right| \sup_{x\in[0,T]}\left|\int_K \frac{d^k}{dx^k}\varphi_\varepsilon(x-s)ds\right| \\
&\leq C \sup_{s\in K}\left|(D^\alpha u_{1,\varepsilon} - D^\alpha u_{2,\varepsilon})(s)\right|
\end{aligned}$$

Lemma 3.2 $\sup_{s\in K}\left|(D^{\alpha}u_{1,\varepsilon}-D^{\alpha}u_{2,\varepsilon})(s)\right|$ is negligible then $\sup_{x\in[0,T]}$ $\left|\frac{d^k}{dx^k}(\tilde{D}^{\alpha}u_{1,\varepsilon}(x)-\tilde{D}^{\alpha}u_{2,\varepsilon}(x))\right|$ is negligible.

Now we introduce the regularized new fractional derivative of a Colombeau generalized on $[0,\infty)$ in the following way.

Definition 5 Let $u\in\mathcal{G}([0,\infty))$ be a Colombeau generalized on $[0,\infty)$ The αth fractional derivative of u, in notation $\tilde{D}^{\alpha}u(x)=[\tilde{D}^{\alpha}u_{\varepsilon}(x)]$, is the element of $\mathcal{G}([0,\infty))$ satisfying (3).

4 Fractional Integral of Colombeau Generalized

Let $(u_{\varepsilon})_{\varepsilon>0}$ be a representative of a Colombeau generalized $u\in\mathcal{G}([0,\infty))$. The fractional integral of $(u_{\varepsilon})_{\varepsilon>0}$, is defined by

$$I^{\alpha}u_{\varepsilon}(x)=\int_0^x(s^{\alpha-1}u_{\varepsilon}(s))ds \tag{4}$$

Lemma 9 *Let $(u_{\varepsilon})_{\varepsilon>0}$ be a representative of $u\in\mathcal{G}([0,\infty))$. Then, for every $\alpha>0$, $\sup_{x\in[0,T]}\left|I^{\alpha}u_{\varepsilon}(x)\right|$ has a moderate bound.*

Proof

$$\begin{aligned}\sup_{x\in[0,T]}\left|I^{\alpha}u_{\varepsilon}(x)\right| &= \sup_{x\in[0,T]}\left|\int_0^x(s^{\alpha-1}u_{\varepsilon}(s))ds\right|\\ &\leq \sup_{s\in[0,T]}\left|u_{\varepsilon}(s)\right|\int_0^x(s^{\alpha-1})ds\\ &\leq \frac{T^{\alpha}}{\alpha}\sup_{s\in[0,T]}\left|u_{\varepsilon}(s)\right|\\ &\leq C_{\alpha,T}\varepsilon^{-N}\end{aligned}$$

Lemma 10 *Let $(u_{1,\varepsilon})_{\varepsilon>0}$ and $(u_{2,\varepsilon})_{\varepsilon>0}$ be two different representative of $u\in\mathcal{G}([0,\infty))$.*
Then, for every $\alpha>0$, $\sup_{x\in[0,T]}\left|I^{\alpha}u_{1,\varepsilon}(x)-I^{\alpha}u_{2,\varepsilon}(x)\right|$ is negligible.

Proof

$$\begin{aligned}
\sup_{x\in[0,T]}\left|I^{\alpha}u_{1,\varepsilon}(x)-I^{\alpha}u_{2,\varepsilon}(x)\right| &= \sup_{x\in[0,T]}\left|\int_0^x (s^{\alpha-1}u_{1,\varepsilon}(s))ds-\int_0^x (s^{\alpha-1}u_{2,\varepsilon}(s))ds\right|\\
&= \sup_{x\in[0,T]}\left|\int_0^x (s^{\alpha-1}(u_{1,\varepsilon}(s)-u_{2,\varepsilon}(s))ds\right|\\
&\le \sup_{s\in[0,T]}\left|u_{1,\varepsilon}(s)-u_{2,\varepsilon}(s)\right|\left|\int_0^x (s^{\alpha-1})ds\right|\\
&\le \frac{T^{\alpha}}{\alpha}\sup_{s\in[0,T]}\left|u_{1,\varepsilon}(s)-u_{2,\varepsilon}(s)\right|
\end{aligned}$$

Since $(u_{1,\varepsilon})_{\varepsilon>0}$ and $(u_{2,\varepsilon})_{\varepsilon>0}$ represent the same Colombeau generalized u we have that

$$\sup_{x\in[0,T]}\left|u_{1,\varepsilon}(x)-u_{2,\varepsilon}(x)\right|$$

is negligible. Therefore,

$$\sup_{x\in[0,T]}\left|I^{\alpha}u_{1,\varepsilon}(x)-I^{\alpha}u_{2,\varepsilon}(x)\right|$$

is negligible, too.

After proving the previous two lemmas we are able to introduce the α fractional integral of a Colombeau generalized on $[0,\infty)$.

Definition 6 Let $u\in\mathcal{G}([0,\infty))$ be a Colombeau generalized on $[0,\infty)$ The αth fractional integral of u, in notation $I^{\alpha}u(x)=[I^{\alpha}u_{\varepsilon}(x)]$ is an element of $\mathcal{G}^0([0,\infty))$ satisfying (4).

Remark 2 For $0<\alpha<1$ the first-order derivative of $\frac{d}{dx}I^{\alpha}u_{\varepsilon}(x)=x^{\alpha-1}u_{\varepsilon}(x)$ and it does not reach its upper limit. In general, the kth order derivative $\frac{d^k}{dx^k}I^{\alpha}u_{\varepsilon}(x)$ it does not reach its upper limit on $[0,\infty)$.

The new fractional integral of a Colombeau generalized $u\in\mathcal{G}([0,\infty))$ If one wants this to be of $\mathcal{G}([0,\infty))$, then the regularization of the fractional integral has to be done.

Definition 7 Let $(u_{\varepsilon})_{\varepsilon>0}$ be a representative of a Colombeau generalized $u\in\mathcal{G}([0,\infty))$. The regularized new fractional integral of $(u_{\varepsilon})_{\varepsilon>0}$, is defined by:

$$\tilde{I}^{\alpha}u_{\varepsilon}(x)=(I^{\alpha}u_{\varepsilon}*\varphi_{\varepsilon})(x),0<\alpha<1 \tag{5}$$

where $I^{\alpha}u_{\varepsilon}(x)$ is given by (4) and $\varphi_{\varepsilon}(x)$is given by (2.1). The convolution in (5) is $I^{\alpha}u_{\varepsilon}*\varphi_{\varepsilon}(x)=\int_0^{\infty}I^{\alpha}u_{\varepsilon}(s)\varphi_{\varepsilon}(x-s)ds$.

Lemma 11 *Let $(u_\varepsilon)_{\varepsilon>0}$ be a representative of $u \in \mathcal{G}([0,\infty))$.*
Then, for every $\alpha > 0$ and every $k \in \{0, 1, 2, \ldots\}$,

$$\sup_{x\in[0,T]}\left|\frac{d^k}{dx^k}\tilde{I}^\alpha u_\varepsilon(x)\right|$$

has a moderate bound.

Proof Let $\varepsilon \in (0, 1)$ $0 < \alpha < 1$, we have

$$\begin{aligned}
\sup_{x\in[0,T]}\left|\tilde{I}^\alpha u_\varepsilon(x)\right| &= \sup_{x\in[0,T]}\left|(I^\alpha u_\varepsilon * \varphi_\varepsilon)(x)\right| \\
&\le \sup_{x\in[0,T]}\left|\int_0^\infty I^\alpha u_\varepsilon(s)\varphi_\varepsilon(x-s)ds\right| \\
&\le \sup_{s\in K}\left|I^\alpha u_\varepsilon(s)\right| \sup_{x\in[0,T]}\left|\int_K \varphi_\varepsilon(x-s)ds\right| \\
&\le C \sup_{s\in K}\left|I^\alpha u_\varepsilon(s)\right|
\end{aligned}$$

for some constant $C > 0$.
Since, according to Lemma 3.1, $\sup_{s\in K}\left|I^\alpha u_\varepsilon(s)\right|$ has a moderate bound, for every $\alpha > 0$, it follows that, $\sup_{x\in[0,T]}\left|\tilde{I}^\alpha u_\varepsilon(x)\right|$ has a moderate bound, too.
For arbitrary order derivative, we have

$$\begin{aligned}
\sup_{x\in[0,T]}\left|\frac{d^k}{dx^k}\tilde{I}^\alpha u_\varepsilon(x)\right| &\le \sup_{s\in K}\left|I^\alpha u_\varepsilon(s)\right| \sup_{x\in[0,T]}\left|\int_K \frac{d^k}{dx^k}\varphi_\varepsilon(x-s)ds\right| \\
&\le \frac{C}{\varepsilon^k}\sup_{s\in K}\left|I^\alpha u_\varepsilon(s)\right|
\end{aligned}$$

$k \in \mathbb{N}$ for some constant $C > 0$.
According to Lemma 3.1, $\sup_{s\in K}\left|I^\alpha u_\varepsilon(s)\right|$ has a moderate bound. Therefore $\sup_{x\in[0,T]}$ $\left|\frac{d^k}{dx^k}\tilde{I}^\alpha u_\varepsilon(x)\right|$ has a moderate bound, too.

Lemma 12 *Let $(u_{1,\varepsilon})_{\varepsilon>0}$ and $(u_{2,\varepsilon})_{\varepsilon>0}$ be two different representative of $u \in \mathcal{G}([0,\infty))$.*
Then, for every $\alpha > 0$, and every $k \in \{0, 1, 2, \ldots\}$,

$$\sup_{x\in[0,T]}\left|\frac{d^k}{dx^k}(\tilde{I}^\alpha u_{1,\varepsilon}(x) - \tilde{I}^\alpha u_{2,\varepsilon}(x))\right|$$

is negligible.

Proof

$$
\begin{aligned}
&\sup_{x\in[0,T]}\left|\frac{d^k}{dx^k}(\tilde{I}^\alpha u_{1,\varepsilon}(x) - \tilde{I}^\alpha u_{2,\varepsilon}(x))\right| \\
&= \sup_{x\in[0,T]}\left|\frac{d^k}{dx^k}((I^\alpha u_{1,\varepsilon} * \varphi_\varepsilon)(x) - (I^\alpha u_{2,\varepsilon} * \varphi_\varepsilon)(x))\right| \\
&= \sup_{x\in[0,T]}\left|\frac{d^k}{dx^k}((I^\alpha u_{1,\varepsilon} - I^\alpha u_{2,\varepsilon}) * \varphi_\varepsilon)(x)\right| \\
&= sup_{x\in[0,T]}\left|((I^\alpha u_{1,\varepsilon} - I^\alpha u_{2,\varepsilon}) * \frac{d^k}{dx^k}\varphi_\varepsilon)(x)\right| \\
&\leq \sup_{s\in K}\left|(I^\alpha u_{1,\varepsilon} - I^\alpha u_{2,\varepsilon})(s)\right| \sup_{x\in[0,T]}\left|\int_K \frac{d^k}{dx^k}\varphi_\varepsilon(x-s)ds\right| \\
&\leq C \sup_{s\in K}\left|(I^\alpha u_{1,\varepsilon} - I^\alpha u_{2,\varepsilon})(s)\right|
\end{aligned}
$$

By to Lemma 4.2 $\sup_{s\in K}\left|(I^\alpha u_{1,\varepsilon} - I^\alpha u_{2,\varepsilon})(s)\right|$ is negligible.
Then, $\sup_{x\in[0,T]}\left|\frac{d^k}{dx^k}(\tilde{I}^\alpha u_{1,\varepsilon}(x) - \tilde{I}^\alpha u_{2,\varepsilon}(x))\right|$ is negligible.

Now we introduce the regularized new fractional integral of a Colombeau generalized on $[0, \infty)$ in the following way.

Definition 8 Let $u \in \mathcal{G}([0, \infty))$ be a Colombeau generalized on $[0, \infty)$ The αth fractional integral of u, in notation

$$\tilde{I}^\alpha u(x) = [\tilde{I}^\alpha u_\varepsilon(x)]$$

is the element of $\mathcal{G}([0, \infty))$ satisfying (4).

Proposition 1 *Let $u \in \mathcal{G}([0, \infty))$, then*

$$D^\alpha I^\alpha u = u$$

Proof Let $(u_\varepsilon)_{\varepsilon>0}$ be a representative of a Colombeau generalized of $u \in \mathcal{G}([0, \infty))$

$$
\begin{aligned}
D^\alpha I^\alpha(u_\varepsilon)(x) &= (x^{1-\alpha})\frac{d}{dx}I^\alpha(u_\varepsilon(x)) \\
&= (x^{1-\alpha})\frac{d}{dx}\int_0^x (s^{\alpha-1}u_\varepsilon(s))ds \\
&= (x^{1-\alpha})(x^{\alpha-1}u_\varepsilon(x)) \\
&= u_\varepsilon(x)
\end{aligned}
$$

Proposition 2 *Let $u \in \mathcal{G}([0, \infty))$, then*

$$I^\alpha D^\alpha u = u_\varepsilon(x) - u_\varepsilon(0) \quad and \quad I^\alpha D^\alpha u \in \mathcal{G}([0, \infty))$$

Proof Let $(u_\varepsilon)_{\varepsilon>0}$ be a representative of a Colombeau generalized of $u \in \mathscr{G}([0,\infty))$

$$
\begin{aligned}
I^\alpha D^\alpha(u_\varepsilon)(x) &= \int_0^x (s^{\alpha-1}) D^\alpha(u_\varepsilon(s))ds \\
&= \int_0^x (s^{\alpha-1}(s^{1-\alpha})\frac{d}{ds}u_\varepsilon(s))ds \\
&= \int_0^x \frac{d}{ds}u_\varepsilon(s)ds \\
&= u_\varepsilon(x) - u_\varepsilon(0)
\end{aligned}
$$

then

$$I^\alpha D^\alpha u \in \mathscr{G}([0,\infty))$$

References

1. R. Khalil, M. Al Horani, A. Yousef, M. Sababheh, A new definition of fractional derivative. J. Comput. Appl. Math. **264**, 65–70 (2014)
2. M. Stojanovic, Extension Of Colombeau algebra to derivatives of arbitrary order $D^\alpha; \alpha \in \mathbb{R}_+ \cup \{0\}$: application to ODEs and PDEs with entire and fractional derivatives. Nonlinear Anal. **71**, 5458–5475 (2009)
3. D. Rajterc Ciric, M. Stojanovic, Convolution-type derivatives and transforms Of Colombeau generalized stochastic processes. Integr. Transforms Spec. Funct. **22**(4–5), 319–326 (2011)
4. T. Abdeljawad, On conformable fractional calculus (Submitted For Publication)
5. J.F. Colombeau, *Elementary Introduction in New Generalized Functions* (North Holland, Amsterdam, 1985)
6. M. Stojanovic, Fondation of the fractional calculus in generalized function algebras. Anal. Appl. **10**(4), 439–467 (2012)
7. M. Oberguggenberger, Generalized functions in nonlinear models—a survey. Nonlinear Anal. **47**, 5029–5040 (2001)
8. D. Rajterc Ciric, M. Stojanovic, Fractional derivatives of multidimensional Colombeau generalized stochastic processes. Fractional Calc. Appl. Anal. **16**(4), 949–961 (2013)
9. D. Rajter-Ciric, A note on fractional derivatives of Colombeau generalized stochastic processes. Novi Sad J. Math. **40**(1), 111–121 (2010)

Controlled Fuzzy Evolution Equations

Said Melliani, A. El Allaoui and L. S. Chadli

Abstract This paper is concerned with controlled fuzzy nonlinear evolution equations of the form

$$u'(t) = A\,u(t) + f\Big(t, u(t), u(\rho(t))\Big) + B(t)c(t), \quad t \in [t_0, t_1]; \quad u(t_0) = u_0.$$

Where $c(t) \in E^1$ is a control, A generate a fuzzy semigroup and $B : [t_0, t_1] \longrightarrow \mathscr{L}(E^1)$. We use the fuzzy strongly continuous semigroups theory to prove the existence, uniqueness and some properties of mild solutions.

1 Introduction

The theory of fuzzy sets has lately years been an object of increasing interest because of its vast applicability in several fields include mechanics, electrical engineering, processing signals and in more and more fields. Therefore, it draws a wide attention of the researchers in the recent years.

In this paper, we consider the controlled fuzzy nonlinear evolution equations of the form

$$\begin{cases} u'(t) = A\,u(t) + f\Big(t, u(t), u(\rho(t))\Big) + B(t)c(t), & t \in I \\ u(t_0) = u_0 \end{cases} \tag{1}$$

where $I = [t_0, t_1]$ be an interval of the real line and E^1 be the fuzzy metric space.

S. Melliani (✉) · A. El Allaoui · L. S. Chadli
Sultan Moulay Slimane University, BP 523 Beni Mellal, Morocco
e-mail: s.melliani@usms.ma

A. El Allaoui
e-mail: elallaoui199@gmail.com

L. S. Chadli
e-mail: sa.chadli@yahoo.fr

S. Melliani and O. Castillo (eds.), *Recent Advances in Intuitionistic Fuzzy Logic Systems*, Studies in Fuzziness and Soft Computing 372,
https://doi.org/10.1007/978-3-030-02155-9_10

Provided some conditions on the functions $f : I \times E^1 \times E^1 \to E^1$, $\rho : I \longrightarrow I$, $c(t) \in E^1$ a control, $B \in \mathscr{C}(I, \mathscr{L}(E^1))$ and A is the generator of a strongly continuous fuzzy semigroup. The aim of our paper is to study the existence of mild solution of (1) based on fuzzy strongly continuous semigroups theory (see [3, 10]).

Note that Kaleva [9] discussed the properties of differentiable fuzzy set-valued mappings by means of the concept of $H-$differentiability due to Puri and Ralescu [14], gave the existence and uniqueness theorem for a solution of the fuzzy differential equation

$$u'(t) = f\Big(t, u(t)\Big); \quad u(0) = u_0$$

when $f : I \times E^n \to E^n$ satisfies the Lipschitz condition.

In [10], Said Melliani, El Hassan Eljaoui and Lalla Saadia Chadli studied, with more details, the existence and uniqueness of mild solution for the fuzzy differential equation

$$u'(t) = A\,u(t) + f(t, u(h(t))); \quad u(0) = u_0 + (-1)g(u)$$

In [4] Bhaskar Dubey and Raju K. George studied the linear time-invariant systems with fuzzy initial condition

$$u'(t) = Au(t) + Bc(t), \quad u(t_0) = u_0.$$

where $c(t) \in (E^1)^m$ a control and A, B, are $n \times n$, $n \times m$ real matrices, respectively, $t_0 \geq 0$. Other parallel results [5, 6].

2 Preliminaries

Let $\mathscr{P}_K(\mathbb{R}^n)$ denote the family of all nonempty compact convex subsets of $\mathbb{R}^n$ and define the addition and scalar multiplication in $\mathscr{P}_K(\mathbb{R}^n)$ as usual. Let A and B be two nonempty bounded subsets of $\mathbb{R}^n$. The distance between A and B is defined by the Hausdorf metric,

$$d(A, B) = max\left\{\sup_{a\in A}\inf_{b\in B}\|a - b\|, \sup_{b\in B}\inf_{a\in A}\|a - b\|\right\}$$

where $\|\|$ denotes the usual Euclidean norm in $\mathbb{R}^n$.

Then it is clear that $(\mathscr{P}_K(\mathbb{R}^n), d)$ becomes a complete and separable metric space (see [14]).

Denote

$$E^n = \left\{u : \mathbb{R}^n \longrightarrow [0, 1] \mid \text{u satisfies (i)–(iv) below}\right\},$$

where

(i) u is normal i.e there exists an $x_0 \in \mathbb{R}^n$ such that $u(x_0) = 1$,
(ii) u is fuzzy convex,
(iii) u is upper semicontinuous,
(iv) $[u]^0 = cl\{x \in \mathbb{R}^n / u(x) > 0\}$ is compact.

For $0 < \alpha \leq 1$, denote $[u]^\alpha = \{t \in \mathbb{R}^n \,/\, u(t) \geq \alpha\}$. Then from (i)–(iv), it follows that the α-level set $[u]^\alpha \in \mathscr{P}_K(\mathbb{R}^n)$ for all $0 \leq \alpha \leq 1$.

According to Zadeh's extension principle, we have addition and scalar multiplication in fuzzy number space E^n as follows:

$$[u+v]^\alpha = [u]^\alpha + [v]^\alpha, \quad [ku]^\alpha = k[u]^\alpha$$

where $u, v \in E^n$, $k \in \mathbb{R}^n$ and $0 \leq \alpha \leq 1$.

Define $D : E^n \times E^n \to \mathbb{R}^+$ by the equation

$$D(u, v) = \sup_{0 \leq \alpha \leq 1} d\Big([u]^\alpha , [v]^\alpha\Big)$$

where d is the Hausdorff metric for non-empty compact sets in $\mathbb{R}^n$.

Then it is easy to see that D is a metric in E^n. Using the results in [14], we know that

1. (E^n, D) is a complete metric space;
2. $D(u+w, v+w) = D(u, v)$ for all $u, v, w \in E^n$;
3. $D(k\,u, k\,v) = |k|\; D(u, v)$ for all $u, v \in E^n$ and $k \in \mathbb{R}^n$

If we denote $\|u\|_{\mathscr{F}} = D\left(u, \tilde{0}\right)$, $u \in E^n$, then $\|u\|_{\mathscr{F}}$ has the properties of an usual norm on E^n (see [7]),

1. $\|u\|_{\mathscr{F}} = 0$ iff $u = \tilde{0}$;
2. $\|\lambda u\|_{\mathscr{F}} = |\lambda| \|u\|_{\mathscr{F}}$ for all $u \in E^n$, $\lambda \in \mathbb{R}$;
3. $\|u+v\|_{\mathscr{F}} \leq \|u\|_{\mathscr{F}} + \|v\|_{\mathscr{F}}$ for all $u, v \in E^n$;
4. $D(\alpha u, \beta u) \leq |\alpha - \beta| D(u, \tilde{0})$, for all $\alpha, \beta \geq 0$ or $\alpha, \beta \leq 0$, $u \in E^n$.

On E^n, we can define the subtraction $\ominus$, called the H-difference (see [8]) as follows $u \ominus v$ has sense if there exists $w \in E^n$ such that $u = v + w$.

Denote $\mathscr{C}(I, E^n) = \{f : I \longrightarrow E^n; f$ is continuous on $I\}$, endowed with the metric

$$H(u, v) = \sup_{t \in I} D(u(t), v(t))$$

Then $(\mathscr{C}, H)$ is a complete metric space.

Lets $a, b \in \mathbb{R}, f \in \mathscr{C}(I, E^n)$, if we denote $\|f\| = H\left(f, \tilde{0}\right)$, then $\|f\|$ has the properties of an usual norm on E^n (see [7]),

1. $\|f\| = 0$ if $f = \tilde{0}$;
2. $\|\lambda f\| = |\lambda| \|f\|$ for all $f \in \mathscr{C}(I, E^n)$, $\lambda \in \mathbb{R}$;

3. $\|f + g\| \leq \|f\| + \|g\|$ for all $f, g \in \mathscr{C}(I, E^n)$;
4. $H(\alpha f, \beta f) \leq |\alpha - \beta| H(f, \tilde{0})$, for all $\alpha, \beta \geq 0$ or $\alpha, \beta \leq 0, f \in \mathscr{C}(I, E^n)$.

Definition 1 A mapping $F : I \to E^n$ is Hukuhara differentiable at $t_0 \in I$ if there exists a $F'(t) \in E^n$ such that the following limits

$$\lim_{h \to 0^+} \frac{F(t+h) \ominus F(t)}{h} \quad \text{and} \quad \lim_{h \to 0^+} \frac{F(t) \ominus F(t-h)}{h}$$

exist and equal to $F'(t)$.

We recall some measurability, integrability properties for fuzzy set-valued mappings (see [9]).

Definition 2 A mapping $F : I \to E^n$ is strongly mesurable if for all $\alpha \in [0, 1]$ the set-valued function $F_\alpha : I \to \mathscr{P}_K(\mathbb{R}^n)$ defined by $F_\alpha(t) = [F(t)]^\alpha$ is Lebesgue mesurable.

A mapping $F : I \to E^n$ is called integrably bounded if there exists an integrable function k such that $||x|| \leq k(t)$ for all $x \in F_0(t)$.

Definition 3 Let $F : I \to E^n$. Then the integral of F over I denoted by $\int_I F(t)dt$ or $\int_I F(t)dt$, is defined by the equation

$$\left[\int_I F(t)dt\right]^\alpha = \int_I F_\alpha(t)dt = \left\{\int_I f(t)dt / f : I \to \mathbb{R}^n \textbf{ is a mesurable selection for } F_\alpha\right\}$$

for all $\alpha \in]0, 1]$.

Also, a strongly mesurable and integrably bounded mapping $F : I \to E^n$ is said to be integrable over I if

$$\int_I F(t)dt \in E^n$$

Proposition 1 (Aumann [1]) *If $F : I \to E^n$ is strongly measurable and integrably bounded, then F is integrable.*

Proposition 2 [9] *Let $F, G : I \to E^n$ be integrable and $\lambda \in \mathbb{R}$. Then*

(i) $\int_I (F(t) + G(t))dt = \int_I F(t)dt + \int_I G(t)dt$,
(ii) $\int_I \lambda F(t)dt = \lambda \int_I F(t)dt$,
(iii) $D(F, G)$ *is integrable,*
(iv) $D(\int_I F(t)dt, \int_I G(t)dt) \leq \int_I D(F, G)(t)dt$.

2.1 Operator Theory

We give here a definition of linear operator, which is similar to that given by Gal and Gal in [7].

Definition 4 $A : E^n \longrightarrow E^n$ is a linear operator if

$$\begin{cases} A(x+y) = Ax + Ay \\ A(\lambda x) = \lambda A(x) \end{cases}$$

for all $x, y \in E^n$, $\lambda \in \mathbb{R}$.

Remark 1 If $A : E^n \longrightarrow E^n$ is linear and continuous at $\tilde{0} \in E^n$, then the latter does not imply the continuity of A at each $x \in E^n$, because in general, we cannot write $x_0 = (x_0 \ominus x) + x$.

However, we can prove the following theorems, which is similar to that given in [7].

Theorem 1 *If $A : E^n \longrightarrow E^n$ is linear, then it is continuous at $\tilde{0} \in E^n$ if and only if there exists $M > 0$ such that*

$$\|A(x)\|_{\mathscr{F}} \leq M \|x\|_{\mathscr{F}}, \ \forall x \in E^n.$$

Now, for $A : E^n \longrightarrow E^n$ linear and continuous at $\tilde{0}$, let us denote

$$\mathscr{M}_A := \left\{M > 0; \|A(x)\|_{\mathscr{F}} \leq M \|x\|_{\mathscr{F}}, \forall x \in E^n\right\}$$

Furthermore, in the both cases, denote $|\|A\||_{\mathscr{F}} = \inf_M \mathscr{M}_A$.

We have the following:

Theorem 2 *If $A : E^n \longrightarrow E^n$ is linear and continuous at $\tilde{0}$, then*

$$\|A(x)\|_{\mathscr{F}} \leq |\|A\||_{\mathscr{F}} \|x\|_{\mathscr{F}}.$$

for all $x \in E^n$ and

$$|\|A\||_{\mathscr{F}} = \sup\left\{\|A(x)\|_{\mathscr{F}}; x \in E^n, \|x\|_{\mathscr{F}} \leq 1\right\}.$$

Corollary 1 *If $A : E^n \longrightarrow E^n$ is additive, positive homogeneous and continuous at $\tilde{0}$, then*

$$\|A(x)\|_{\mathscr{F}} \leq |\|A\||_{\mathscr{F}} \|x\|_{\mathscr{F}}, \quad \forall x \in E^n.$$

Next, let us denote

$$\mathscr{L}_0^+(E^n) = \left\{A : E^n \longrightarrow E^n; A \text{ is additive, positive homogeneous and continuous at } \tilde{0}\right\},$$
$$\mathscr{L}^+(E^n) = \left\{A \in \mathscr{L}_0^+(E^n); A \text{ is continuous at each } x \in E^n\right\},$$
$$\mathscr{L}_0(E^n) = \left\{A : E^n \longrightarrow E^n; A \text{ is linear and continuous at } \tilde{0}\right\},$$

and

$$\mathscr{L}(E^n) = \left\{A \in \mathscr{L}_0(E^n); A \quad \text{is continuous at each } x \in E^n\right\},$$

We consider the metric $\Phi : \mathscr{L}_0^+(E^n) \times \mathscr{L}_0^+(E^n) \longrightarrow \mathbb{R}^+$ by

$$\Phi(A, B) = \sup\left\{D(A(x), B(x)); \|x\|_{\mathscr{F}} \leq 1\right\}, A, B \in \mathscr{L}_0^+(E^n).$$

Then it is easy to see that $\Phi(A, \tilde{O}) = |||A|||_{\mathscr{F}}$, $A \in \mathscr{L}_0^+(E^n)$, where $\tilde{O} : E^n \longrightarrow E^n$ is given by $\tilde{O}(x) = \tilde{0}$, $\forall x \in E^n$.

We can prove the following theorem, which is similar to that given by Gal and Gal in [7].

Theorem 3 *1.* $\left(\mathscr{L}_0^+(E^n), \Phi\right)$, $(\mathscr{L}_0(E^n), \Phi)$, $\left(\mathscr{L}^+(E^n), \Phi\right)$ *and* $(\mathscr{L}(E^n), \Phi)$ *are complete metric spaces.*
2. $\Phi(A + B, C + D) \leq \Phi(A, C) + \Phi(B, D)$,
3. $\Phi(kA, kB) = |k|\Phi(A, B)$,
4. $\Phi(A, B) \leq |||A|||_{\mathscr{F}} + |||B|||_{\mathscr{F}}$,
5. $\Phi(A + B, C) \leq \Phi(A, C) + \Phi(B, C)$,
6. $\Phi(A + B, \tilde{O}) \leq |||A|||_{\mathscr{F}} + |||B|||_{\mathscr{F}}$.

for all $A, B, C \in \mathscr{L}_0^+(E^n)$.

2.2 *Fuzzy Strongly Continuous Semigroups*

We give here a definition of a fuzzy semigroups, which is similar to that given in [2, 10, 13].

Definition 5 A family $\{T(t), t \geq 0\}$ of operators from E^n into itself is a fuzzy strongly continuous semigroup if

(i) $T(0) = i$, the identity mapping on E^n,
(ii) $T(t + s) = T(t)T(s)$ for all $t, s \geq 0$,
(iii) the function $g : [0, \infty[\rightarrow E^n$, defined by $g(t) = T(t)x$ is continuous at $t = 0$ for all $x \in E^n$ i.e

$$\lim_{t \to 0^+} T(t)x = x$$

(iv) There exist two constants $R > 0$ and ω such that

$$D\Big(T(t)x, T(t)y\Big) \leq R\, e^{\omega t}\, D(x, y), \quad for\; t \geq 0, x, y \in E^n$$

In particular if $R = 1$ and $\omega = 0$, we say that $\{T(t), t \geq 0\}$ is a contraction fuzzy semigroup.

Remark 2 The condition (iii) implies that the function $t \longrightarrow T(t)x$ is continuous on $[0, \infty[$ for all $x \in E^n$.

Definition 6 Let $\{T(t), t \geq 0\}$ be a fuzzy strongly continuous semigroup on E^n and $x \in E^n$. If for $h > 0$ sufficiently small, the Hukuhara difference $T(h)x \ominus x$ exits, we define

$$Ax = \lim_{h \to 0^+} \frac{T(h)x \ominus x}{h}$$

whenever this limit exists in the metric space (E^n, D). Then the operator A defined on

$$D(A) = \left\{ x \in E^n : \lim_{h \to 0^+} \frac{T(h)x \ominus x}{h} \text{ exists} \right\} \subset E^n$$

is called the infinitesimal generator of the fuzzy semigroup $\{T(t), t \geq 0\}$.

Lemma 1 *Let A be the generator of a fuzzy semigroup $\{T(t), t \geq 0\}$ on E^n, then for all $x \in E^n$ such that $T(t)x \in D(A)$ for all $t \geq 0$, the mapping $t \to T(t)x$ is differentiable and*

$$\frac{d}{dt}\Big(T(t)\,x\Big) = A\,T(t)\,x, \quad \forall t \geq 0$$

Example 1 We define on E^n the family of operator $\{T(t), t \geq 0\}$ by

$$T(t)x = e^{kt}x, \quad k \in \mathbb{R}.$$

For $k \geq 0$, $\{T(t), t \geq 0\}$ is a fuzzy strongly continuous semigroup on E^n, and the linear operator A defined by $Ax = kx$ is the infinitesimal generator of this fuzzy semigroup.

Proposition 3 *Let $\{T(t), t \geq 0\}$ be a fuzzy strongly continuous semigroup on E^n. Then for all $t \geq 0$ we have $T(t) \in \mathscr{L}(E^n)$.*

Proof Lets $x, y \in E^n$, the condition (iv) in the Definition 2.5 implies that, there exist two constants $M > 0$ and ω such that

$$D\Big(T(t)x, T(t)y\Big) \leq R\, e^{\omega t}\, D(x, y)$$

Or, $D(x, y) \longrightarrow 0$ as $x \to y$. Then $D\Big(T(t)x, T(t)y\Big) \longrightarrow 0$ as $x \to y$.
which implies $T(t) \in \mathscr{L}(E^n)$, for all $t \geq 0$.

3 Main Results

To begin our discussion, we need to introduce the concept of mild solution for the problem (1), provided $f : [t_0, t_1] \times E^1 \to E^1, \rho : [t_0, t_1] \longrightarrow [t_0, t_1], B : [t_0, t_1] \longrightarrow \mathscr{L}(E^1)$, A is the generator of a strongly continuous fuzzy semigroup. the input $c(t) \in E^1$ for each $t \in [t_0, t_1]$ and $c(.)$ is fuzzy-integrable (see [9, 10]) in $[t_0, t_1]$. Therefore, it is important to understand the structure of the solutions of (1).

Definition 7 We say that u is a mild solution of the equation (1) if

(i) $u \in \mathscr{C}([t_0, t_1], E^1)$, $u(t) \in D(A)$ for all $t \in [t_0, t_1]$;

(ii) and $u(t) = T(t - t_0)u_0 + \int_{t_0}^{t} T(t - s)\Big(f(s, u(s), u(\rho(s)) + B(s)c(s)\Big)ds,$

for all $t \in [t_0, t_1]$.

We give here a definition of the Controllability, which is similar to that given by Dubey et al. in [4].

Definition 8 The problem (1) with fuzzy initial condition $u(0) = u_0 \in E^1$ is said to be controllable to a fuzzy-state $u_1 \in E^1$ at $t_2 > t_0$ if there exists a fuzzy-integrable control $c(t) \in E^1$ for $t \in [t_0, t_1]$ such that the solution of problem (1) with this control satisfies $u(t_2) = u_1$.

Let $N = \sup_{t\in[t_0,t_1]} Re^{\omega t}$.

We first study the existence and uniqueness of mild solutions using the fixed point argument.

Suppose the assumptions:

(H$_0$) A is the infinitesimal generator of a strongly continuous fuzzy semigroup $\{T(t), t \geq 0\}$ on E^1.

(H$_1$). $f : [t_0, t_1] \times E^1 \to E^1$ is continuous and there exist two constants $L_1,\ L_2 > 0$ such that

$$D\Big(f(t, x, y), f(t, x', y')\Big) \leq L_1 D(x, y) + L_2 D(x', y'), \qquad \forall t \in [t_0, t_1], \quad x, y, x', y' \in E^1.$$

(H$_2$). $\rho : [t_0, t_1] \longrightarrow [t_0, t_1]$ is continuous and $\rho(t) \leq t$ for all $t \in [t_0, t_1]$.

(H$_3$). for $t \in [t_0, t_1]$, $c(t) \in E^1$ is a control and $c(.)$ is fuzzy-integrable.

(H$_4$). $B : [t_0, t_1] \longrightarrow \mathscr{L}(E^1)$is continuous.

(H$_5$). There exist a constant $M > 0$ such that
$D(B(t)x, B(t)y) \leq MD(x, y)$ for all $x, y \in E^1, t \in [t_0, t_1]$.

Theorem 4 *Assume that the conditions* $(\boldsymbol{H}_0) - (\boldsymbol{H}_4)$ *are satisfied. Then for all* $u_0 \in E^1$ *such that* $T(t)u_0 \in E^1$ *for all* $t \geq 0$*, the problem (1) has a unique mild solution on* $[t_0, t_1]$.

Proof Transform the problem (1) into a fixed point problem. Consider the operator $\Gamma : \mathscr{C}([t_0, t_1]; E^1) \longrightarrow \mathscr{C}([t_0, t_1]; E^1)$ defined as

$$\Gamma u(t) = T(t - t_0)u_0 + \int_{t_0}^{t} T(t - s)\Big(f(s, u(s), u(\rho(s))) + B(s)u(s)\Big).$$

Γ is well defined and maps $\mathscr{C}([t_0, t_1]; E^1)$ into itself. Indeed
For $u \in E^1$, $t \in [t_0, t_1]$ and h very small, we have

$$
\begin{aligned}
&D(\Gamma u(t+h), \Gamma u(t)) \\
&= D\Bigg(T(t+h-t_0)u_0 + \int_{t_0}^{t+h} T(t+h-s)\Big(f(s, u(s), u(\rho(s))) + B(s)c(s)\Big)ds\,, \\
&T(t-t_0)u_0 + \int_{t_0}^{t} T(t-s)\Big(f(s, u(s), u(\rho(s))) + B(s)c(s)\Big)ds\Bigg) \\
&\leq D\left(T(t+h-t_0)u_0, T(t-t_0)u_0\right) \\
&+ D\Bigg(\int_{t_0}^{t+h} T(t+h-s)\Big(f(s, u(s), u(\rho(s))) + B(s)c(s)\Big)ds\,, \\
&\int_{t_0}^{t} T(t-s)\Big(f(s, u(s), u(\rho(s))) + B(s)c(s)\Big)ds\Bigg) \\
&\leq ND\left(T(h)u_0, u_0\right) + \int_{t_0}^{t_0+h} D\left(T(t-h+s)\Big(f(s, u(s), u(\rho(s))) + B(s)c(s)\Big)ds, \tilde{0}\right) ds \\
&+ N\int_{t_0}^{t} D\Big(f(s+h, u(s+h), u(\rho(s+h))), f(s, u(s), u(\rho(s)))\Big)ds \\
&+ N\int_{t_0}^{t} D\Big(B(s+h)c(s+h), B(s)c(s)\Big)ds.
\end{aligned}
$$

It is obviously that

$$
\begin{aligned}
&D\left(T(h)u_0, u_0\right) \to 0, \\
&\int_{t_0}^{t_0+h} D\left(T(t-h+s)\Big(f(s, u(s), u(\rho(s))) + B(s)c(s)\Big)ds, \tilde{0}\right) ds \to 0.
\end{aligned}
$$

And by the dominated convergence theorem we have

$$
\int_{t_0}^{t} D\Big(f(s+h, u(s+h), u(\rho(s+h))), f(s, u(s), u(\rho(s)))\Big)ds \to 0, \quad .
$$

and $\int_{t_0}^{t} D\Big(B(s+h)c(s+h), B(s)c(s)\Big)ds \to 0$
From above, we infer that $\Gamma u \in \mathscr{C}([t_0, t_1]; E^1)$, for all $u \in \mathscr{C}([t_0, t_1]; E^1)$.
For $t \in [t_0, t_1]$, $u, v \in E^1$, we have

$$\begin{aligned}
D(\Gamma u(t), \Gamma v(t)) &= D\Big(T(t-t_0)u_0 + \int_{t_0}^{t} T(t-s)\Big(f(s, u(s), u(\rho(s))) + B(s)c(s)\Big)ds,\\
&\quad T(t-t_0)u_0 + \int_{t_0}^{t} T(t-s)\Big(f(s, v(s), v(\rho(s))) + B(s)c(s)\Big)ds\Big)\\
&\le D\Big(\int_{t_0}^{t} T(t-s)\Big(f(s, u(s), u(\rho(s))) + B(s)c(s)\Big)ds,\\
&\quad \int_{t_0}^{t} T(t-s)\Big(f(s, v(s), v(\rho(s))) + B(s)c(s)\Big)ds\Big)\\
&\le N\int_{t_0}^{t} D\Big(f(s, u(s), u(\rho(s)))\\
&\quad + B(s)c(s), f(s, v(s), v(\rho(s))) + B(s)c(s)\Big)ds\\
&\le N\int_{t_0}^{t} D\Big(f(s, u(s), u(\rho(s))), f(s, v(s), v(\rho(s)))\Big)ds\\
&\le N\int_{t_0}^{t} \Big(L_1 D(u(s), v(s)) + L_2 D(u(\rho(s)), v(\rho(s)))\Big)ds\\
&\le (L_1 + L_2)N(t-t_0)H(u, v).
\end{aligned}$$

Therefore, we have

$$\begin{aligned}
D(\Gamma^2 u(t), \Gamma^2 v(t)) &\le N\int_{t_0}^{t} \Big(L_1 D(\Gamma u(s), \Gamma v(s)) + L_2 D(\Gamma u(\rho(s)), \Gamma v(\rho(s)))\Big)ds\\
&\le N\int_{t_0}^{t} \Big(L_1(L_1 + L_2)N(s-t_0)H(u, v)\\
&\quad + L_2(L_1 + L_2)N(\rho(s) - t_0)H(u, v)\Big)ds\\
&\le (L_1 + L_2)N(t-t_0)H(u, v).
\end{aligned}$$

Since $\rho(s) \le s$, then

$$\begin{aligned}
D(\Gamma^2 u(t), \Gamma^2 v(t)) &\le N\int_{t_0}^{t} \Big(L_1 D(\Gamma u(s), \Gamma v(s)) + L_2 D(\Gamma u(\rho(s)), \Gamma v(\rho(s)))\Big)ds\\
&\le N\int_{t_0}^{t} \Big(L_1(L_1 + L_2)N(s-t_0)H(u, v)\\
&\quad + L_2(L_1 + L_2)N(s - t_0)H(u, v)\Big)ds\\
&= (L_1 + L_2)^2 N^2 H(u, v)\int_{t_0}^{t} (s-t_0)ds\\
&= \frac{(L_1 + L_2)^2 N^2 (t-t_0)^2}{2} H(u, v).
\end{aligned}$$

By inference, we have for all $m > 0$

$$D\left(\Gamma^m u(t), \Gamma^m v(t)\right) \leq \frac{\left((L_1+L_2)N(t-t_0)\right)^m}{m!} H(u,v), \forall t \in [t_0, t_1].$$

Which means that

$$H\left(\Gamma^m u, \Gamma^m v\right) \leq \frac{\left((L_1+L_2)N(t-t_0)\right)^m}{m!} H(u,v), \forall t \in [t_0, t_1].$$

Since

$$\lim_{m\to+\infty} \frac{\left((L_1+L_2)N(t-t_0)\right)^m}{m!} = 0.$$

Then there exists $p > 0$ such that

$$\frac{\left((L_1+L_2)N(t-t_0)\right)^p}{p!} < 1.$$

which shows that Γ^p is a contraction. So there exists a unique fixed point $u \in \mathscr{C}([t_0, t_1]; E^1)$ such that $\Gamma^p u = u$.
Which implies that $\Gamma^p(\Gamma u) = \Gamma(\Gamma^p u) = \Gamma u$. Since u is unique, then $\Gamma u = u$. It follows that u is the unique mild solution of the problem (1).

The next theorem provides a correspondence between the continuity of solution and the continuity of control.

Theorem 5 *Let $\{c_n, n \geq 0\} \subset E^1$ be a sequence of controls as in ($\boldsymbol{H}_3$) such that $\lim_{n\to+\infty} c_n = c$. Suppose that the conditions ($\boldsymbol{H}_0$) – ($\boldsymbol{H}_5$) hold. Let u^{c_n} be the mild solution for (1) corresponding to c_n and u^c be the mild solution corresponding to c. If $N(L_1+L_2)(t_1-t_0) < 1$, then $\lim_{n\to+\infty} u^{c_n} = u^c$.*

Proof For $t \in [t_0, t_1]$ we have

$$\begin{aligned}
D\left(u^{c_n}(t), u^c(t)\right) &= D\Big(T(t-t_0)u_0 + \int_{t_0}^t T(t-s)\Big(f(s, u^{c_n}(s), u^{c_n}(\rho(s))) + B(s)c_n(s)\Big)ds, \\
&\quad T(t-t_0)u_0 + \int_{t_0}^t T(t-s)\Big(f(s, u^c(s), u^c(\rho(s))) + B(s)c(s)\Big)ds\Big) \\
&\leq D\Big(\int_{t_0}^t T(t-s)\Big(f(s, u^{c_n}(s), u^{c_n}(\rho(s))) + B(s)c_n(s)\Big)ds, \\
&\quad \int_{t_0}^t T(t-s)\Big(f(s, u^c(s), u^c(\rho(s))) + B(s)c(s)\Big)ds\Big) \\
&\leq N \int_{t_0}^t D\big(f(s, u^{c_n}(s), u^{c_n}(\rho(s))) \\
&\quad + B(s)c_n(s), f(s, u^c(s), u^c(\rho(s))) + B(s)c(s)\big)\, ds
\end{aligned}$$

$$\leq N \int_{t_0}^{t} \Big(L_1 D\left(u^{c_n}(s), u^c(s)\right) + L_2 D\left(u^{c_n}(\rho(s)), u^c(\rho(s))\right) \Big) ds$$
$$+ N \int_{t_0}^{t} D\Big(B(s)c_n(s), B(s)c(s)\Big) ds$$
$$\leq N(L_1 + L_2)(t_1 - t_0) H(u^{c_n}, u^c) + MN(t_1 - t_0) H(c_n, c).$$

Which implies that

$$H(u^{c_n}, u^c) \leq \frac{MN(t_1 - t_0)}{1 - N(L_1 + L_2)(t_1 - t_0)} H(c_n, c).$$

Since $c_n \to c$, hence $u^{c_n} \to u^c$ as $n \to +\infty$.

we establish the following result about continuous dependence of a mild Solution.

Theorem 6 *Suppose that the conditions* $(\boldsymbol{H}_0) - (\boldsymbol{H}_4)$ *hold. Lets* u *and* v *be mild solutions of (1) on* $[t_0, t_1]$ *corresponding to* u_0 *and* v_0 *respectively.*
If $L_2 N(t_1 - t_0) e^{L_1 N(t_1 - t_0)} < 1$*, then*

$$H(u, v) \leq \frac{N}{e^{L_1 N(t_0 - t_1)} + L_2 N(t_0 - t_1)} D(u_0, v_0).$$

Proof for $t \in [t_0, t_1]$, we have

$$D\left(u(t), v(t)\right) = D\Bigg(T(t - t_0)u_0 + \int_{t_0}^{t} T(t - s)\Big(f(s, u(s), u(\rho(s))) + B(s)c(s)\Big) ds,$$
$$T(t - t_0)v_0 + \int_{t_0}^{t} T(t - s)\Big(f(s, v(s), v(\rho(s))) + B(s)c(s)\Big)\Bigg) ds$$
$$\leq N D(u_0, v_0) + N \int_{t_0}^{t} D\Big(f(s, u(s), u(\rho(s))), f(s, v(s), v(\rho(s)))\Big) ds$$
$$\leq N D(u_0, v_0) + N \int_{t_0}^{t} \Big(L_1 D(u(s), v(s)) + L_2 D(u(\rho(s)), v(\rho(s))) \Big) ds$$
$$\leq N D(u_0, v_0) + L_2 N(t_1 - t_0) H(u, v) + L_1 N \int_{t_0}^{t} D(u(s), v(s)) ds$$

By using Gronwall inequality, we find that

$$D\left(u(t), v(t)\right) \leq \Big(N D(u_0, v_0) + L_2 N(t_1 - t_0) H(u, v) \Big) e^{L_1 N(t - t_0)}$$
$$\leq \Big(N D(u_0, v_0) + L_2 N(t_1 - t_0) H(u, v) \Big) e^{L_1 N(t_1 - t_0)}, \qquad \forall t \in [t_0, t_1].$$

which implies that

$$H(u,v) \leq \Big(ND(u_0,v_0) + L_2N(t_1-t_0)H(u,v)\Big)e^{L_1N(t_1-t_0)}.$$

Finally, we get

$$H(u,v) \leq \frac{N}{e^{L_1N(t_0-t_1)} + L_2N(t_0-t_1)}D(u_0,v_0).$$

4 Conclusion

In order to describe a random evolution of the temperature of the rode using a control, we consider a controlled fuzzy evolution equation. By using operator semigroup of fuzzy sets theory, we obtain existence results. In addition, future work includes expanding the idea signalized in this work and introducing observability and generalize other works [11, 12]. This is a fertile field with vast research projects, which can lead to numerous theories and applications. We plan to devote significant attention to this direction. And we intend to investigate the applications which are based on experimental data (real world problems) of the proposed theory.

References

1. R.J. Aumann, Integrals of set-valued functions. J. Math. Anal. Appl. 1–12 (1965)
2. A. El Allaoui, S. Melliani, L.S. Chadli, Fuzzy dynamical systems and Invariant attractor sets for fuzzy strongly continuous semigroups. J. Fuzzy Set Valued Anal. **2016**(2), 148–155 (2016)
3. A. El Allaoui, S. Melliani, L.S. Chadli, Fuzzy α-semigroups of operators. Gen. Lett. Math. **2**(2), 42–49 (2017)
4. B. Dubey, R.K. George, Controllability of linear time-invariant dynamical systems with fuzzy initial condition, in *Proceedings of the World Congress on Engineering and Computer Science*, pp. 23–25 (2013)
5. B. Dubey, R.K. George, *Estimation of Controllable Initial Fuzzy States of Linear Time-Invariant Dynamical Systems, Communications in Computer and Information Science* (Springer, Berlin, 2012), pp. 316–324
6. Y. Feng, L. Hua, On the quasi-controllability of continuous-time dynamic fuzzy control systems. Chaos, Solitons Fractals 177–188 (2006)
7. C.G. Gal, S.G. Gal, Semigroups of operators on spaces of fuzzy-number-valued functions with applications to fuzzy differential equations, arXiv:1306.3928v1 (2013)
8. M. Hukuhara, Integration des applications measurables dont la valeur est un compact convexe. Funk. Ekvacioj 207–223 (1967)
9. O. Kaleva, Fuzzy differentiel equations. Fuzzy Sets Syst. **24**, 301–317 (1987)
10. S. Melliani, A. El Allaoui, L.S. Chadli, Fuzzy differential equation with nonlocal conditions and fuzzy semigroups. Adv. Differ. Equ. (2016)
11. S. Melliani, L.S. Chadli, A. El Allaoui, Periodic boundary value problems for controlled nonlinear impulsive evolution equations on Banach spaces. Int. J. Nonlinear Anal. Appl. **8**(1), 301–314 (2017)

12. S. Melliani, A. El Allaoui, L.S. Chadli, A general class of periodic boundary value problems for controlled nonlinear impulsive evolution equations on Banach spaces. Adv. Diff. Equ. 290 (2016)
13. S. Melliani, A. El Allaoui, L.S. Chadli, Relation between fuzzy semigroups and fuzzy dynamical systems. Nonlinear Dyn. Syst. Theory **17**(1), 60–69 (2017)
14. M.L. Puri, D.A. Ralescu, Fuzzy random variables. J. Math. Anal. Appl. **114**, 409–422 (1986)

Multiplication Operation and Powers of Trapezoidal Fuzzy Numbers

E. Eljaoui, Said Melliani and L. S. Chadli

Abstract In this work, we explore and exhibe some numerical methods to calculate two types of the product (respectively powers) of trapezoidal fuzzy numbers. As a particular case, we study the multiplication operation on triangular fuzzy numbers, we report and correct errors in an article dealing with the same topic.

1 Introduction

The fuzzy numbers theory is used in different domains, in linguistics, statistics, engineering, physics, biology and experimental sciences. In 1965, Zadeh introduced the fuzzy sets theory. Then, many works was devoted to the study of arithmetic operations on fuzzy numbers. Especially, Dubois and Prade [1] defined a fuzzy number as a subset of the real line [2].

In [3], Bansal studied the arithmetic behavior of trapezoidal fuzzy numbers and formulated the basic mathematical operations on this type of fuzzy numbers.

Gao et al. [2] used the extension principle to develop nonlinear programming method, analytical method, computer drawing method and computer simulation method for solving multiplication operation of two fuzzy numbers.

In [4], Mahanta et al. introduced an alternative method to evaluate the arithmetic operations on triangular fuzzy numbers.

In this paper, we report and correct some errors in the article [2], and we develop numerical methods to calculate the inner and canonical products of trapezoidal and triangular fuzzy numbers.

E. Eljaoui · S. Melliani (✉) · L. S. Chadli
Sultan Moulay Slimane University, BP 523 Beni Mellal, Morocco
e-mail: s.melliani@usms.ma

E. Eljaoui
e-mail: eljaouihass@gmail.com

L. S. Chadli
e-mail: sa.chadli@yahoo.fr

S. Melliani and O. Castillo (eds.), *Recent Advances in Intuitionistic Fuzzy Logic Systems*, Studies in Fuzziness and Soft Computing 372,
https://doi.org/10.1007/978-3-030-02155-9_11

In Sect. 2, we give some preliminaries on fuzzy numbers and trapezoidal fuzzy numbers and we correct an error in [2]. In Sect. 3, we apply analytical method, computer drawing method and computer simulation method for solving multiplication operation of two trapezoidal fuzzy numbers. In Sect. 4, we study the square and the nth power of trapezoidal fuzzy numbers using inner and canonical multiplication.

2 Preliminaries

A fuzzy number is a fuzzy set on the real axis, i.e. a mapping $u : \mathbb{R} \longrightarrow [0, 1]$ which satisfies the following four conditions:

(i) u is normal, i.e. $\exists x_0 \in \mathbb{R}$ for which $u(x_0) = 1$,
(ii) u is fuzzy convex, i.e. $u(\lambda x + (1-\lambda)y) \geq \min(u(x), u(y))$ for any $x, y \in \mathbb{R}$ and $\lambda \in [0, 1]$,
(iii) u is upper semi-continuous,
(iv) The closure $[u]^0 = \overline{\{x \in \mathbb{R} | u(x) > 0\}}$ of the support $\text{supp}u = \{x \in \mathbb{R} | u(x) > 0\}$ of u is compact.

We denote the set of all fuzzy numbers on $\mathbb{R}$ by E and called it as the space of fuzzy numbers. For $0 < \alpha \leq 1$, denote $[u]_\alpha = \{x \in \mathbb{R} \mid u(x) \geq \alpha\}$.
Then, from (i) to (iv), it follows that the α-level set $[u]_\alpha$ is a compact interval $[u]_\alpha = [u^-(\alpha), u^+(\alpha)]$, for all $0 \leq \alpha \leq 1$.
It is well known that the following properties are true

$$[u + v]_\alpha = [u]_\alpha + [v]_\alpha\,, \quad [ku]_\alpha = k\,[u]_\alpha\,,$$

and

$$[uv]_\alpha = [u]_\alpha\,[v]_\alpha\,,$$

for each $u, v \in E$, $k \in \mathbb{R}$ and $\alpha \in [0, 1]$.

The following theorem known as "Representation Theorem" is very important.

Theorem 1 *Let $[u]_\alpha = [u^-(\alpha), u^+(\alpha)]$, for $u \in E$ and for all $0 \leq \alpha \leq 1$. Then the following statements hold:*

(i) $u^-(\alpha)$ is a bounded non-decreasing left continuous function in $(0, 1]$.
(ii) $u^+(\alpha)$ is a bounded non-increasing left continuous function in $(0, 1]$.
(iii) $u^-(\alpha)$ and $u^+(\alpha)$ are right continuous at $\alpha = 0$.
(iv) $u^-(1) \leq u^+(1)$.

A crisp number k is simply represented by $u^-(\alpha) = u^+(\alpha) = k$, for all $0 \leq \alpha \leq 1$. A triangular fuzzy number is represented by three reals as follows $\widetilde{A} = (a_1, a_2, a_3)$ and its membership function $\mu_{\widetilde{A}}$ is given by

$$\mu_{\widetilde{A}}(x) = \begin{cases} 0 & si \quad x < a_1 \\ \frac{x-a_1}{a_2-a_1} & si \; a_1 \leq x \leq a_2 \\ \frac{a_3-x}{a_3-a_2} & si \; a_2 \leq x \leq a_3 \\ 0 & si \quad x > a_3 \end{cases}$$

Please notice that the membership function of a triangular fuzzy number is piecewise affine.

The α-cut of a triangular fuzzy number $\widetilde{A} = (a_1, a_2, a_3)$ is given by $A_\alpha = [a_L^{(\alpha)}, a_R^{(\alpha)}] = [a_1 + (a_2 - a_1)\alpha, a_3 - (a_3 - a_2)\alpha]$, with $\alpha \in [0, 1]$.

Definition 1 A trapezoidal fuzzy number is represented by four real numbers as follows $\widetilde{A} = (a, b, c, d)$, with $a \leq b \leq c \leq d$ and its membership function $\mu_{\widetilde{A}}$ is given by

$$\mu_{\widetilde{A}}(x) = \begin{cases} \frac{x-a}{b-a} & if \;\; a \leq x \leq b \\ 1 & if \;\; b \leq x \leq c \\ \frac{d-x}{d-c} & si \;\; c \leq x \leq d \\ 0 & , otherwise \end{cases}$$

This fuzzy number is said to be positive (respectively negative) if $a \geq 0$ (respectively $d \leq 0$).

A trapezoidal fuzzy number $\widetilde{A} = (a, b, c, d)$ such that $b = c$ is a triangular fuzzy number.

Definition 2 (*Operations on trapezoidal fuzzy numbers*)
Let $\widetilde{A} = (a, b, c, d)$ and $\widetilde{B} = (e, f, g, h)$ be two trapezoidal fuzzy numbers, we define

(i) $\widetilde{A} \oplus \widetilde{B} = (a + e, b + f, c + g, d + h)$.
(ii) $\widetilde{A} \ominus \widetilde{B} = (a - h, b - g, c - f, d - e)$.
(iii) $-\widetilde{A} = (-d, -c, -b, -a)$.

We define two type of multiplication for trapezoidal fuzzy numbers:

- the first type is based on Zadeh's extension principle, which we will denote by $\times$,
- and the second one is defined in a way to results in a trapezoidal fuzzy number, which we will denote by $\otimes$.

Definition 3 (*Canonical multiplication*)
Let $\widetilde{A} = (a, b, c, d)$ and $\widetilde{B} = (e, f, g, h)$ be two trapezoidal fuzzy numbers. Using Zadeh's extension principle, we define their canonical product $\widetilde{Q} = \widetilde{A} \times \widetilde{B}$ by its α-cuts

$$\widetilde{Q}_\alpha = \widetilde{A}_\alpha \times \widetilde{B}_\alpha.$$

Definition 4 (*Inner multiplication*)
Let $\widetilde{A} = (a, b, c, d)$ and $\widetilde{B} = (e, f, g, h)$ be two trapezoidal fuzzy numbers, we define their inner product $\widetilde{A} \otimes \widetilde{B}$ by

$$\widetilde{A} \otimes \widetilde{B} = (\min\{ae, ah, de, dh\}, \min\{bf, cf, cg, bg\}, \max\{bf, cf, cg, bg\}, \max\{ae, ah, de, dh\}).$$

3 Multiplication Operation on Trapezoidal Fuzzy Numbers

The procedure of addition or substraction is simple, but the procedure of multiplication or division is difficult and complex.
Now we extend some numerical methods presented in [2] for solving multiplication operation of two triangular fuzzy numbers, which we apply for the multiplication of two trapezoidal fuzzy numbers.

3.1 *Analytical Method*

3.1.1 Canonical Multiplication

Let $\widetilde{A} = (a, b, c, d)$, $\widetilde{B} = (e, f, g, h)$ be two trapezoidal fuzzy numbers and $\widetilde{Q} = \widetilde{A} \times \widetilde{B}$. We have $\widetilde{A}_\alpha = [a^L_\alpha, a^R_\alpha] = [a + \alpha(b - a), d + \alpha(c - d)]$ and $\widetilde{B}_\alpha = [b^L_\alpha, b^R_\alpha] = [e + \alpha(f - e), h + \alpha(g - h)]$. Then, $\widetilde{Q}_\alpha = [a^L_\alpha, a^R_\alpha] \times [b^L_\alpha, b^R_\alpha] = [q^L_\alpha, q^R_\alpha]$, where $q^L_\alpha = \min\{a^L_\alpha b^L_\alpha, a^L_\alpha b^R_\alpha, a^R_\alpha b^L_\alpha, a^R_\alpha b^R_\alpha\}$ and $q^R_\alpha = \max\{a^L_\alpha b^L_\alpha, a^L_\alpha b^R_\alpha, a^R_\alpha b^L_\alpha, a^R_\alpha b^R_\alpha\}$. We suppose that the trapezoidal fuzzy numbers $\widetilde{A}$ and $\widetilde{B}$ are positive, then $a^L_\alpha, a^R_\alpha, b^L_\alpha, b^R_\alpha$ are positive. Therefore

$$
\begin{aligned}
q^L_\alpha &= a^L_\alpha b^L_\alpha \\
&= (a + \alpha(b - a))\,(e + \alpha(f - e)) \\
q^L_\alpha &= (b - a)(f - e)\alpha^2 + (af + be - 2ae)\alpha + ae
\end{aligned}
$$

substituting $q^L_\alpha = z$, we get

$$
\alpha = \frac{-(af + be - 2ae) + \sqrt{(af - be)^2 + 4(b - a)(f - e)z}}{2(b - a)(f - e)}, \quad (Omit\ \alpha < 0).
$$

Similarly, we have

$$
\begin{aligned}
q^R_\alpha &= a^R_\alpha b^R_\alpha \\
&= (d + \alpha(c - d))\,(h + \alpha(g - h)) \\
q^R_\alpha &= (c - d)(g - h)\alpha^2 + (dg + ch - 2dh)\alpha + dh
\end{aligned}
$$

substituting $q^R_\alpha = z$, we get

$$
\alpha = \frac{-(dg + ch - 2dh) - \sqrt{(dg - ch)^2 + 4(c - d)(g - h)z}}{2(c - d)(g - h)}, \quad (Omit\ \alpha > 1).
$$

Therefore, the membership function of $\widetilde{Q} = \widetilde{A} \times \widetilde{B}$ is

$$\mu_{\widetilde{Q}}(z) = \begin{cases} \frac{-(af+be-2ae)+\sqrt{(af-be)^2+4(b-a)(f-e)z}}{2(b-a)(f-e)} & , ae \leq z \leq bf \\ 1 & , bf \leq z \leq cg \\ \frac{-(dg+ch-2dh)-\sqrt{(dg-ch)^2+4(c-d)(g-h)z}}{2(c-d)(g-h)} & , cg \leq z \leq dh \\ 0 & , otherwise \end{cases}$$

Note that the core $[bf, cg]$ of $\widetilde{Q}$ is obtained by using interval methods.
So, we can formulate the following result.

Proposition 1 *Let $\widetilde{A} = (a, b, c, d)$, $\widetilde{B} = (e, f, g, h)$ be two positive trapezoidal fuzzy numbers, then the membership function of their canonical product $\widetilde{Q} = \widetilde{A} \times \widetilde{B}$ is given by*

$$\mu_{\widetilde{Q}}(z) = \begin{cases} \frac{-(af+be-2ae)+\sqrt{(af-be)^2+4(b-a)(f-e)z}}{2(b-a)(f-e)} & , ae \leq z \leq bf \\ 1 & , bf \leq z \leq cg \\ \frac{-(dg+ch-2dh)-\sqrt{(dg-ch)^2+4(c-d)(g-h)z}}{2(c-d)(g-h)} & , cg \leq z \leq dh \\ 0 & , otherwise \end{cases}$$

Example 1 Let $\widetilde{A} = (1, 4, 6, 8)$, $\widetilde{B} = (2, 3, 7, 9)$ and $\widetilde{Q} = \widetilde{A} \times \widetilde{B}$, then by the analytic method we get

$$\mu_{\widetilde{Q}}(z) = \begin{cases} \frac{-7+\sqrt{25+12z}}{6} & , 2 \leq z \leq 12 \\ 1 & , 12 \leq z \leq 42 \\ \frac{17-\sqrt{1+4z}}{4} & , 42 \leq z \leq 72 \\ 0 & , otherwise \end{cases}$$

3.1.2 Inner Multiplication

Let $\widetilde{A} = (a, b, c, d)$, $\widetilde{B} = (e, f, g, h)$ be two trapezoidal fuzzy numbers and $\widetilde{Q} = \widetilde{A} \otimes \widetilde{B}$. We set $u = \min\{ae, ah, de, dh\}$, $v = \min\{bf, cf, cg, bg\}$, $w = \max\{bf, cf, cg, bg\}$ and $t = \max\{ae, ah, de, dh\}$. Then, the membership function of $\widetilde{Q}$ is the following piecewise affine mapping

$$\mu_{\widetilde{Q}}(z) = \begin{cases} \frac{z-u}{v-u} & , u \leq z \leq v \\ 1 & , v \leq z \leq w \\ \frac{t-z}{t-w} & , w \leq z \leq t \\ 0 & , otherwise \end{cases}$$

In particular, if $a \geq 0$ and $e \geq 0$, then $u = ae, v = bf, w = cg, t = dh$. Therefore

$$\mu_{\widetilde{Q}}(z) = \begin{cases} \frac{z-ae}{bf-ae} & , ae \leq z \leq bf \\ 1 & , bf \leq z \leq cg \\ \frac{dh-z}{dh-cg} & , cg \leq z \leq dh \\ 0 & , otherwise \end{cases}$$

Example 2 **Inner product**
Let $\widetilde{A} = (1, 4, 6, 8), \widetilde{B} = (2, 3, 7, 9)$ and $\widetilde{Q} = \widetilde{A} \otimes \widetilde{B}$, then $\widetilde{Q} = (2, 12, 42, 72)$ and

$$\mu_{\widetilde{Q}}(z) = \begin{cases} \frac{z-2}{10} & , 2 \leq z \leq 12 \\ 1 & , 12 \leq z \leq 42 \\ \frac{72-z}{30} & , 42 \leq z \leq 72 \\ 0 & , otherwise \end{cases}$$

Remark 1 **Product of two triangular fuzzy numbers**
Let $\widetilde{A} = (a, b, d), \widetilde{B} = (e, f, h)$ be two triangular fuzzy numbers, which can be regarded as particular trapezoidal fuzzy numbers $\widetilde{A} = (a, b, c, d)$ and $\widetilde{B} = (e, f, g, h)$, with $b = c$ and $f = g$. Let $\widetilde{Q} = \widetilde{A} \otimes \widetilde{B}$. We set $u = \min\{ae, ah, de, dh\}$, $v = bf = w$ and $t = \max\{ae, ah, de, dh\}$.
Then, the membership function of $\widetilde{Q}$ is the following piecewise affine mapping

$$\mu_{\widetilde{Q}}(z) = \begin{cases} \frac{z-u}{v-u} & , u \leq z \leq v \\ \frac{t-z}{t-v} & , v \leq z \leq t \\ 0 & , otherwise \end{cases}$$

In particular, if $a \geq 0$ and $e \geq 0$, then $u = ae, v = bf = w, t = dh$. Therefore

$$\mu_{\widetilde{Q}}(z) = \begin{cases} \frac{z-ae}{bf-ae} & , ae \leq z \leq bf \\ \frac{dh-z}{dh-bf} & , bf \leq z \leq dh \\ 0 & , otherwise \end{cases}$$

Example 3 **Product of triangular fuzzy numbers**
Let $\widetilde{A} = (2, 3, 5), \widetilde{B} = (3, 5, 6)$ be two triangular fuzzy numbers (see [2]), regarded as trapezoidal fuzzy numbers i.e $\widetilde{A} = (2, 3, 3, 5), \widetilde{B} = (3, 5, 5, 6)$ and $\widetilde{Q} = \widetilde{A} \times \widetilde{B}$, then by the analytic method above we get the same result as in [2]:

$$\mu_{\widetilde{Q}}(z) = \begin{cases} \frac{-7+\sqrt{1+8z}}{4} & , 6 \leq z \leq 15 \\ \frac{17-\sqrt{49+8z}}{4} & , 15 \leq z \leq 30 \\ 0 & , otherwise \end{cases}$$

Please notice that contrarily of the membership function of a triangular fuzzy number, this mapping is not piecewise affine. So, the canonical product of two triangular fuzzy

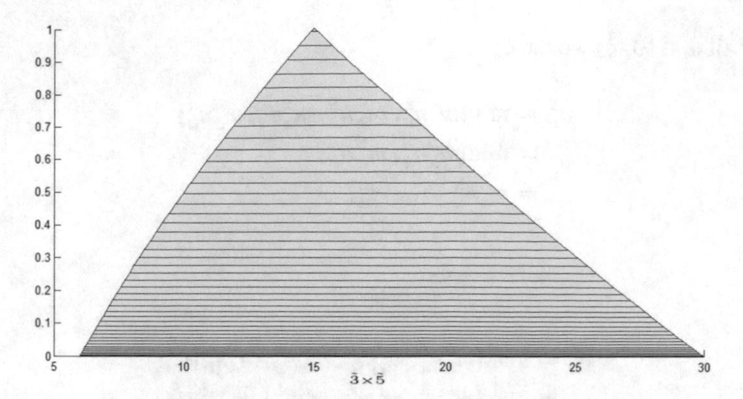

Fig. 1 Membership function of canonical product $\tilde{3} \times \tilde{5}$ by analytical method

numbers is not a triangular fuzzy number. And the canonical (or usual) multiplication does not preserve the shape of the operands (see [5]).
In the other hand, the membership function of the inner product $\widetilde{Q} = \widetilde{3} \otimes \widetilde{5}$ is given by the following (Fig. 1).

$$\mu_{\widetilde{Q}}(z) = \begin{cases} \frac{z-6}{9} & 6 \leq z \leq 15 \\ \frac{30-z}{15} & 15 \leq z \leq 30 \\ 0 & otherwise \end{cases},$$

which is piecewise affine. So the inner product of two triangular fuzzy numbers is a triangular fuzzy number. And the inner multiplication preserves the shape of the operands (see [5]).

Now let us recall the invalid results in [2]:
Let $\widetilde{M} = (a_1, b_1, c_1), \widetilde{N} = (a_2, b_2, c_2)$ be two triangular fuzzy numbers and $\widetilde{Q} = \widetilde{M} \times \widetilde{N}$. We have $\widetilde{M}_\alpha = [m_\alpha^L, m_\alpha^R] = [a_1 + \alpha(b_1 - a_1), c_1 + \alpha(b_1 - c_1)]$ and $\widetilde{N}_\alpha = [n_\alpha^L, n_\alpha^R] = [a_2 + \alpha(b_2 - a_2), c_2 + \alpha(b_2 - c_2)]$. Then, $\widetilde{Q}_\alpha = [m_\alpha^L, m_\alpha^R] \times [n_\alpha^L, n_\alpha^R] = [q_\alpha^L, q_\alpha^R]$, where $q_\alpha^L = \min\{m_\alpha^L n_\alpha^L, m_\alpha^L n_\alpha^R, m_\alpha^R n_\alpha^L, m_\alpha^R n_\alpha^R\}$ and $q_\alpha^R = \max\{m_\alpha^L n_\alpha^L, m_\alpha^L n_\alpha^R, m_\alpha^R n_\alpha^L, m_\alpha^R n_\alpha^R\}$.
The error appear clearly when the authors in [2], suppose that $c_1 \geq b_1 \geq a_1$ and $c_2 \geq b_2 \geq a_2$, and conclude that $q_\alpha^L = a_\alpha^L b_\alpha^L$ and $q_\alpha^R = a_\alpha^R b_\alpha^R$.
Their assertion is not correct, as can be shown by the following counter example:

Example 4 Let $\widetilde{M} = (-6, -4, -3), \widetilde{N} = (-2, -1, 2)$, then

$$m_\alpha^L = 2\alpha - 6 < 0, m_\alpha^R = -\alpha - 3 < 0, n_\alpha^L = \alpha - 2 < 0, n_\alpha^R = -3\alpha + 2,$$

with $n_\alpha^R > 0$ for all $\alpha \in [0, \frac{3}{2}]$. Then, for all $\alpha \in [0, \frac{3}{2}]$ we have

$$m_\alpha^L n_\alpha^L > 0, m_\alpha^L n_\alpha^R < 0, m_\alpha^R n_\alpha^L > 0, m_\alpha^R n_\alpha^R < 0.$$

So, for all $\alpha \in [0, \frac{3}{2}]$ we have

$$\begin{aligned} q_\alpha^L &= \min\{m_\alpha^L n_\alpha^L, m_\alpha^L n_\alpha^R, m_\alpha^R n_\alpha^L, m_\alpha^R n_\alpha^R\} \\ &= \min\{m_\alpha^L n_\alpha^R, m_\alpha^R n_\alpha^R\} \\ &= m_\alpha^L n_\alpha^R \neq m_\alpha^L n_\alpha^L, \end{aligned}$$

and

$$\begin{aligned} q_\alpha^R &= \max\{m_\alpha^L n_\alpha^L, m_\alpha^L n_\alpha^R, m_\alpha^R n_\alpha^L, m_\alpha^R n_\alpha^R\} \\ &= \max\{m_\alpha^L n_\alpha^L, m_\alpha^L n_\alpha^R, m_\alpha^R n_\alpha^L\} \neq m_\alpha^R n_\alpha^R, \end{aligned}$$

Precisely, for $\alpha = 0$, we obtain

$$\begin{aligned} q_0^L = \min\{12, -12, 6, -6\} = -12; \; q_0^R = \max\{12, -12, 6, -6\} = 12; \\ m_0^L n_0^L = 12; \; m_0^R n_0^R = -6. \end{aligned}$$

Then it is obvious that $q_0^L \neq m_0^L n_0^L$ and $q_0^R \neq m_0^R n_0^R$.

The authors must suppose that the triangular fuzzy numbers $\widetilde{M}$ and $\widetilde{N}$ are positive that is $c_1 \geq b_1 \geq a_1 \geq 0$ and $c_2 \geq b_2 \geq a_2 \geq 0$. In this case $m_\alpha^L \geq 0, m_\alpha^R \geq 0, n_\alpha^L \geq 0, n_\alpha^R \geq 0$, then

$$q_\alpha^R = m_\alpha^R n_\alpha^R \quad and \quad q_\alpha^L = m_\alpha^L n_\alpha^L.$$

3.2 Computer Drawing Method

The aim of this method is to calculate the product of two trapezoidal (respectively triangular) fuzzy numbers $\widetilde{A}, \widetilde{B}$, it is based on the calculus of α-cuts.
Here we develop and generalize the method used in [2] for solving multiplication of two triangular fuzzy numbers.

3.2.1 Canonical Multiplication

Let $\widetilde{A} = (a, b, c, d), \widetilde{B} = (e, f, g, h)$ be two trapezoidal fuzzy numbers and $\widetilde{Q} = \widetilde{A} \otimes \widetilde{B}$. The method based on using the formula $\widetilde{Q}_\alpha = \widetilde{A}_\alpha \times \widetilde{B}_\alpha$, is as follows:

Step 1: Set $\alpha = 0.01$;
Step 2: If $\alpha > 1$, stop. The grey sector is the membership function of $\widetilde{Q}$. Otherwise, go to step 3;
Step 3: Calculate α-cuts $[a_\alpha^L, a_\alpha^R]$ and $[b_\alpha^L, b_\alpha^R]$ of $\widetilde{A}$ and $\widetilde{B}$ respectively;

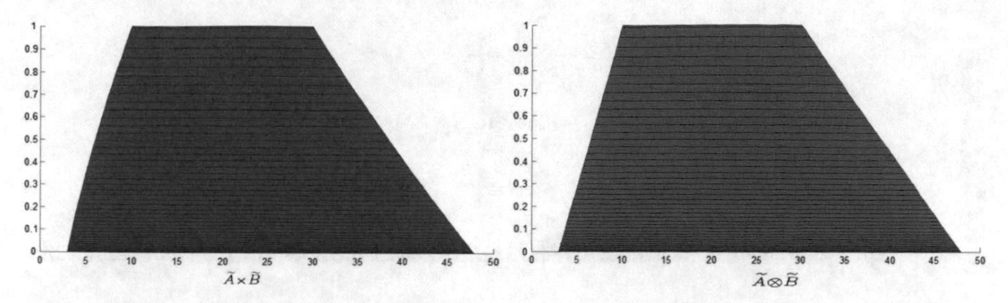

Fig. 2 Canonical and inner product $\widetilde{A} \times \widetilde{B}, \widetilde{A} \otimes \widetilde{B}$ by computer drawing method

Step 4: Calculate $q_\alpha^L = \min\{a_\alpha^L b_\alpha^L, a_\alpha^L b_\alpha^R, a_\alpha^R b_\alpha^L, a_\alpha^R b_\alpha^R\}$ and $q_\alpha^R = \max\{a_\alpha^L b_\alpha^L, a_\alpha^L b_\alpha^R, a_\alpha^R b_\alpha^L, a_\alpha^R b_\alpha^R\}$;
Step 5: Set Q rows axis and set membership function of $\widetilde{Q}$ column axis, then create a line from (q_α^L, α) to (q_α^R, α);
Step 6: $\alpha \leftarrow \alpha + 0.01$, then go to step 2.

Example 5 Let $\widetilde{A} = (1, 2, 5, 6), \widetilde{B} = (3, 5, 6, 8)$, then the membership function of $\widetilde{Q} = \widetilde{A} \times \widetilde{B}$ is given in Fig. 2.

3.2.2 Inner Multiplication

Let $A = (a, b, c, d), B = (e, f, g, h)$ be two trapezoidal fuzzy numbers and $\widetilde{Q} = \widetilde{A} \otimes \widetilde{B}$. The method based on the calculus above is as follows:

Step 1: Set $\alpha = 0.01$;
Step 2: If $\alpha > 1$, stop. The grey sector is the membership function of $\widetilde{Q}$. Otherwise, go to step 3;
Step 3: Calculate $u = \min\{ae, ah, de, dh\}$, $v = \min\{bf, cf, cg, bg\}$, $w = \max\{bf, cf, cg, bg\}$ and $t = \max\{ae, ah, de, dh\}$;
Step 4: Calculate $q_\alpha^L = u + \alpha(v - u)$ and $q_\alpha^R = t + \alpha(w - t)$;
Step 5: Set $\widetilde{Q}$ rows axis and set membership function of $\widetilde{Q}$ column axis, then create a line from (q_α^L, α) to (q_α^R, α);
Step 6: $\alpha \leftarrow \alpha + 0.01$, then go to step 2.

Example 6 Let $\widetilde{A} = (1, 2, 5, 6), \widetilde{B} = (3, 5, 6, 8)$ and $\widetilde{Q} = \widetilde{A} \otimes \widetilde{B}$, then $\widetilde{Q} = (3, 10, 25, 48)$ and

$$\mu_{\widetilde{Q}}(z) = \begin{cases} \frac{z-3}{7} & , 3 \leq z \leq 10 \\ 1 & , 10 \leq z \leq 25 \\ \frac{48-z}{23} & , 25 \leq z \leq 48 \\ 0 & , \textit{otherwise} \end{cases}$$

Remark 2 To get the correct method in the case of triangular fuzzy numbers, it is sufficient to take $b = c$ and $f = g$.

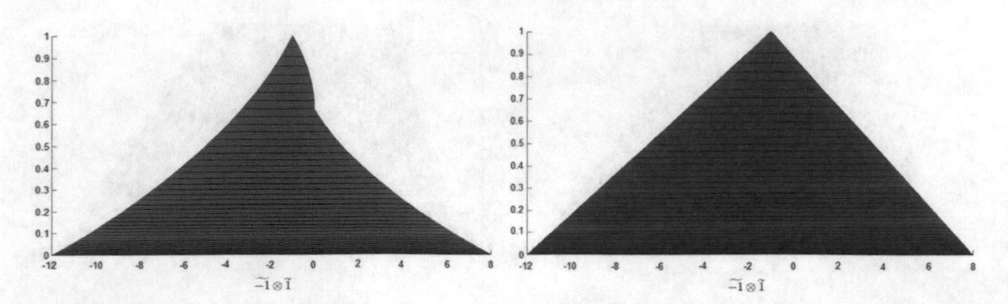

Fig. 3 Canonical and inner products $\widetilde{-1} \times \widetilde{1}$, $\widetilde{-1} \otimes \widetilde{1}$ by computer drawing method

Letting $\widetilde{M} = (a_1, b_1, c_1), \widetilde{N} = (a_2, b_2, c_2)$ be two triangular fuzzy numbers (we adopt the same notations in [2]). The computer drawing algorithm to calculate $\widetilde{Q} = \widetilde{M} \otimes \widetilde{N}$ is as follows:

Step 1: Set $\alpha = 0.01$;
Step 2: If $\alpha > 1$, stop. The grey sector is the membership function of $\widetilde{Q}$. Otherwise, go to step 3;
Step 3: Calculate $u = \min\{a_1b_1, a_1b_3, a_3b_1, a_3b_3\}$, $v = a_2b_2$ and $w = \max\{a_1b_1, a_1 b_3, a_3b_1, a_3b_3\}$;
Step 4: Calculate $q_\alpha^L = u + \alpha(v - u)$ and $q_\alpha^R = w + \alpha(v - w)$;
Step 5: Set $\widetilde{Q}$ rows axis and set membership function of $\widetilde{Q}$ column axis, then create a line from (q_α^L, α) to (q_α^R, α);
Step 6: $\alpha \leftarrow \alpha + 0.01$, then go to step 2.

Example 7 Let $\widetilde{-1} = (-3, -1, 2) = (-3, -1, -1, 2), \widetilde{1} = (-2, 1, 4) = (-2, 1, 1, 4)$ and $\widetilde{Q} = \widetilde{-1} \otimes \widetilde{1}$, then $\widetilde{Q} = (-12, -1, 0, 8)$ and

$$\mu_{\widetilde{Q}}(z) = \begin{cases} \frac{17-\sqrt{1-24z}}{12} & , -12 \leq z \leq -1 \\ \frac{2+\sqrt{-z}}{3} & , -1 \leq z \leq 0 \\ \frac{3-\sqrt{1+z}}{3} & , 0 \leq z \leq 8 \\ 0 & , \textit{otherwise} \end{cases}$$

This figure is most precise and smooth than the following figure obtained by the same method in [2].

Here we draw booth of the membership functions for canonical and inner product in the same figure, obtained by computer drawing method (Fig. 3).

3.3 Computer Simulation Method

The computer simulation method needn't calculate α-cuts interval, it is as follows (Fig. 4):

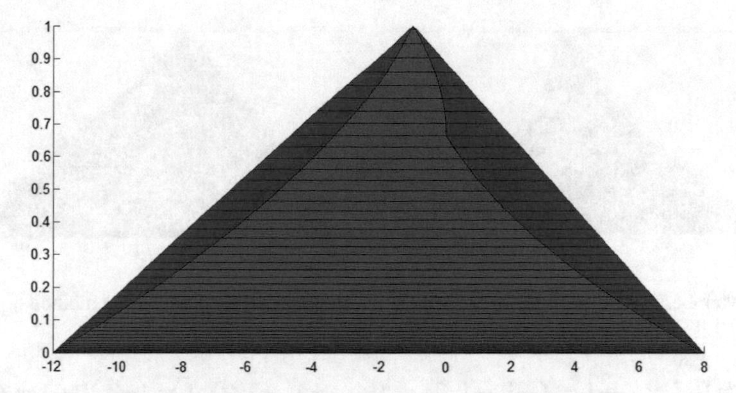

Fig. 4 Membership functions of $\widetilde{-1} \times \widetilde{1}$ and $\widetilde{-1} \otimes \widetilde{1}$ by computer drawing method

Step 1: Set $i = 1$ and simulation times N;
Step 2: If $i > N$, stop. The grey sector is the membership function of $\widetilde{Q}$. Otherwise, go to step 3;
Step 3: Generate two random numbers x_i on the interval $[a, d]$ and y_i on the interval $[e, h]$. Calculate $z_i = x_i y_i$ and $v_i = \min\{\mu_{\widetilde{A}}(x_i), \mu_{\widetilde{B}}(y_i)\}$
Step 4: Create a line from $(z_i, 0)$ to (z_i, v_i), then go to step 2.

Example 8 Let $\widetilde{A} = (1, 2, 5, 6)$, $\widetilde{B} = (3, 5, 6, 8)$ and $\widetilde{Q} = \widetilde{A} \times \widetilde{B}$, then $\widetilde{Q} = (3, 10, 25, 48)$ and

$$\mu_{\widetilde{Q}}(z) = \begin{cases} \frac{-5+\sqrt{1+8z}}{4} & , 3 \le z \le 10 \\ 1 & , 10 \le z \le 30 \\ \frac{10-\sqrt{4+2z}}{2} & , 30 \le z \le 48 \\ 0 & , otherwise \end{cases}$$

(Fig. 5).

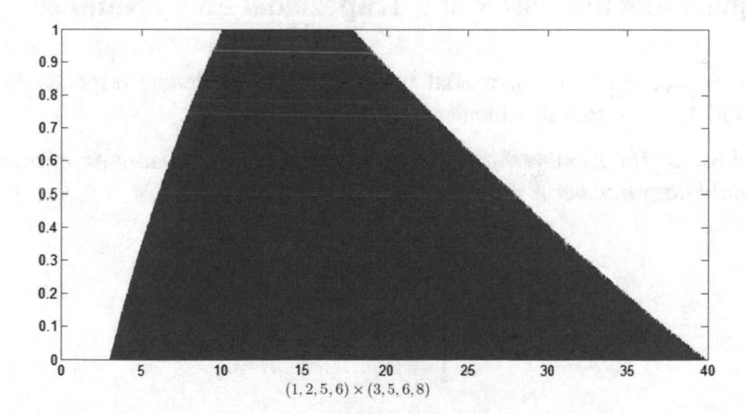

Fig. 5 Membership function of $\widetilde{Q} = (1, 2, 5, 6) \times (3, 5, 6, 8)$ by the computer simulation method

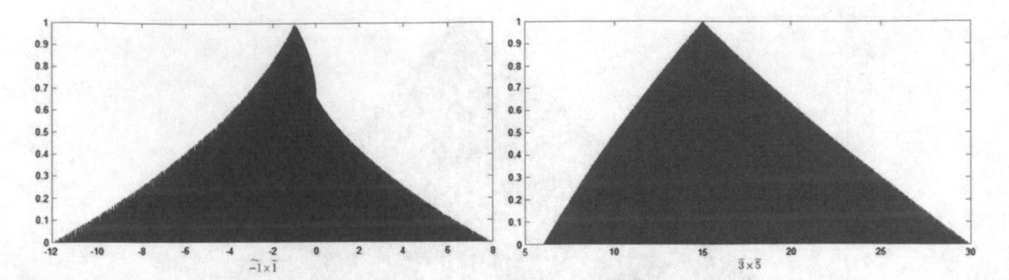

Fig. 6 Membership function of $\widetilde{-1} \times \widetilde{1}$ and $\widetilde{3} \times \widetilde{5}$ by the computer simulation method

Example 9 Let $\widetilde{-1} = (-3, -1, 2) = (-3, -1, -1, 2)$, $\widetilde{1} = (-2, 1, 4) = (-2, 1, 1, 4)$ and $\widetilde{Q} = \widetilde{-1} \otimes \widetilde{1}$, then $\widetilde{Q} = (-12, -1, 0, 8)$ and

$$\mu_{\widetilde{Q}}(z) = \begin{cases} \frac{17-\sqrt{1-24z}}{12} & , -12 \leq z \leq -1 \\ \frac{2+\sqrt{-z}}{3} & , -1 \leq z \leq 0 \\ \frac{3-\sqrt{1+z}}{3} & , 0 \leq z \leq 8 \\ 0 & , \textit{otherwise} \end{cases}$$

Example 10 Let $\widetilde{3} = (2, 3, 5) = (2, 3, 3, 5)$, $\widetilde{5} = (3, 5, 6) = (3, 5, 5, 6)$ and $\widetilde{Q} = \widetilde{-1} \otimes \widetilde{1}$, then $\widetilde{Q} = (3, 15, 15, 30)$ and

$$\mu_{\widetilde{Q}}(z) = \begin{cases} \frac{z-3}{12} & , 3 \leq z \leq 15 \\ \frac{30-z}{15} & , 15 \leq z \leq 30 \\ 0 & , \textit{otherwise} \end{cases}$$

(Fig. 6).

4 Square and nth Power of a Trapezoidal Fuzzy Number

Let $\widetilde{A} = (a, b, c, d)$ be a trapezoidal fuzzy number, we denote respectively $\widetilde{A}^2 = \widetilde{A} \otimes \widetilde{A}$ and $\widetilde{A}^{(2)} = \widetilde{A} \otimes \widetilde{A}$ its canonical and inner square.

Proposition 2 *The membership function of the canonical square $\widetilde{A}^2$ of a positive trapezoidal fuzzy number $\widetilde{A} = (a, b, c, d)$ is given by*

$$\mu_{\widetilde{A}^2}(z) = \begin{cases} \frac{-a+\sqrt{z}}{b-a} & , a^2 \leq z \leq b^2 \\ 1 & , b^2 \leq z \leq c^2 \\ \frac{d-\sqrt{z}}{d-c} & , c^2 \leq z \leq d^2 \\ 0 & , \textit{otherwise} \end{cases}$$

Proof It results immediately from the Proposition 1.

Corollary 1 *The membership function of the canonical square $\widetilde{A}^2$ of a positive triangular fuzzy number $\widetilde{A} = (a, b, c)$ is given by*

$$\mu_{\widetilde{A}^2}(z) = \begin{cases} \frac{-a+\sqrt{z}}{b-a} & , a^2 \leq z \leq b^2 \\ \frac{c-\sqrt{z}}{c-b} & , b^2 \leq z \leq c^2 \\ 0 & , otherwise \end{cases}$$

Proof It results obviously from the Proposition 2.

Proposition 3 *The membership function of the inner square $\widetilde{A}^{(2)}$ of a positive trapezoidal fuzzy number $\widetilde{A} = (a, b, c, d)$ is given by*

$$\mu_{\widetilde{A}^{(2)}}(z) = \begin{cases} \frac{z-a^2}{b^2-a^2} & , a^2 \leq z \leq b^2 \\ 1 & , b^2 \leq z \leq c^2 \\ \frac{d^2-z}{d^2-c^2} & , c^2 \leq z \leq d^2 \\ 0 & , otherwise \end{cases}$$

It results immediately from the results in paragraph 3.1.2.

Corollary 2 *The membership function of the inner square $\widetilde{A}^{(2)}$ of a positive triangular fuzzy number $\widetilde{A} = (a, b, c)$ is given by*

$$\mu_{\widetilde{A}^{(2)}}(z) = \begin{cases} \frac{z-a^2}{b^2-a^2} & , a^2 \leq z \leq b^2 \\ \frac{c^2-z}{c^2-b^2} & , b^2 \leq z \leq c^2 \\ 0 & , otherwise \end{cases}$$

Proof It results obviously from the Proposition 3.

Example 11 Let $\widetilde{A} = (2, 3, 5, 6)$, then its inner square $\widetilde{A}^{(2)} = (4, 9, 25, 36)$ and the membership function of its canonical square is given by

$$\mu_{\widetilde{A}^2}(z) = \begin{cases} \sqrt{z} - 2 & , 4 \leq z \leq 9 \\ 1 & , 9 \leq z \leq 25 \\ 6 - \sqrt{z} & , 25 \leq z \leq 36 \\ 0 & , otherwise \end{cases}$$

Example 12 Let $\widetilde{A} = (2, 3, 5)$, then its inner square $\widetilde{A}^{(2)} = (4, 9, 25)$ and the membership function of its canonical square is given by

$$\mu_{\widetilde{A}^2}(z) = \begin{cases} \sqrt{z} - 2 & , 4 \leq z \leq 9 \\ \frac{5-\sqrt{z}}{2} & , 9 \leq z \leq 25 \\ 0 & , otherwise \end{cases}$$

(Fig. 7).

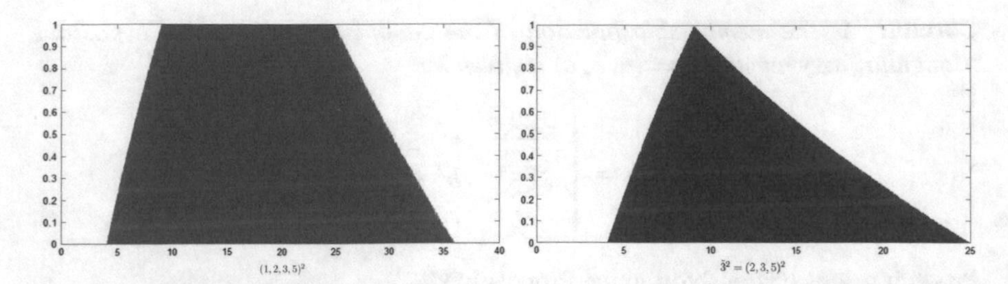

Fig. 7 Canonical squares of (1, 2, 3, 5) and $\widetilde{3} = (2, 3, 5)$ by the computer simulation method

Let $\widetilde{A} = (a, b, c, d)$ be a trapezoidal fuzzy number and $n \geq$ be a positive integer. By recurrence, we define respectively $\widetilde{A}^n = \widetilde{A}^{n-1} \otimes \widetilde{A}$ and $\widetilde{A}^{(n)} = \widetilde{A}^{(n-1)} \otimes \widetilde{A}$ its canonical and inner nth power.

Example 13 Let $\widetilde{A} = (2, 3, 5, 6)$, then its inner nth power is the trapezoidal fuzzy number $\widetilde{A}^{(n)} = (2^n, 3^n, 5^n, 6^n)$. Using the analytical method, we get the membership function of its canonical nth power:

$$\mu_{\widetilde{A}^2}(z) = \begin{cases} \sqrt[n]{z} - 2 \; , 2^n \leq z \leq 3^n \\ \quad 1 \qquad , 3^n \leq z \leq 5^n \\ 6 - \sqrt[n]{z} \; , 5^n \leq z \leq 6^n \\ \quad 0 \qquad , \textit{otherwise} \end{cases}$$

5 Conclusion

We have presented three numerical methods for solving inner and canonical multiplication on trapezoidal fuzzy numbers. As it is remarked in [2], the analytical method is more precise and the computer drawing method is simple, but there are both based on calculus of α-cuts interval, which seems difficult and hard in certain cases. The computer simulation method is the most simple, but it gives a membership function rough and not very smooth. For future research, one can apply these methods to solve nonlinear fuzzy differential equations, which involves the powers of trapezoidal or triangular fuzzy numbers.

References

1. D. Dubois, H. Prade, Operations on fuzzy numbers. Int. J. Syst. Sci. **9**, 613–626 (1978)
2. S. Gao, Z. Zhang, C. Cao, Multiplicatin operation on fuzzy numbers. J. Softw. **2009**, 331–338 (2009)
3. A. Bansal, Trapezoidal fuzzy numbers (a, b, c, d): arithmetic behavior. Int. J. Phys. Math. Sci. **2011**, 39–44 (2011)

4. S. Mahanta, R. Chuta, H.K. Baruah, Fuzzy arithmetic without using the method of α-Cuts. Int. J. Latest Trents Comput. **1**, 73–80 (2010)
5. J. Fodor, B. Bede, Arithmetics With Fuzzy Numbers: A Comparative Overview
6. S. Heilpern, Representation and application of fuzzy numbers. Fuzzy Sets Syst. **91**, 259–268 (1997)
7. G.J. Klir, Fuzzy arithmetic with requisite constraints. Fuzzy Sets Syst. **91**, 165–175 (1997)
8. G.J. Klir, Fuzzy sets: an overview of fundmentals, applications and personal views. Beijing Norm. Univ. Press **2000**, 44–49 (2000)
9. H.M. Lee, L. Lin , Using weighted triangular fuzzy numbers to evaluate the rate of aggregative risk in software developpement, in Proceeding of 7th International Conference of Machine Learning and Cybernetics, **7** (2008), pp. 3762–3767
10. N. Gani, S.N.M. Assarudeen, A new operation on triangular fuzzy number for solving linear programming problem. Appl. Math. Sci. **6**, 525–532 (2012)
11. L.A. Zadeh, Fuzzy set. Inf. Control **8**, 339–353 (1965)

Intuitioinistic Fuzzy Hilbert Space

Said Melliani, M. Elomari and L. S. Chadli

Abstract In the present paper we benefit the generalized Hukuhar's difference in order to built an intuitionistic fuzzy vector space and a Hilbert space on the set of all intuitionistic fuzzy numbers. Also we give a proof of the existence and uniqueness of an approximation triangular of an intuitionistic fuzzy number.

1 Introduction

In this paper we introduce the notion of intuitionistic fuzzy Hilbert space. Intuitionistic fuzzy sets were defined in [1] as a model for inexact concepts and subjective judgments. Such a model can be used in situations when deterministic and/or probabilistic models do not provide a realistic description of the process under study. Techniques of intuitionistic fuzzy sets and systems theory were applied in various domains. A concept of fuzzy linear space was defined by [2]. In [3] Chi-Tsuen Yeh use the properties of Hilbert space in order to define the approximation of a fuzzy number and propose a generalization by the name of weighted trapezoidal approximation. From this idea we proposed the concept of the intuitionistic fuzzy Hilbert space and we answer the question about the existence of approximation intuitionistic fuzzy triangular number.

This paper is organized as follows In Sect. 2 we recall some concept concerning the fuzzy numbers. The concept of inner product take place in Sect. 3. A proof of existence and uniqueness of a triangular approximation is given in Sect. 4. We finished by a example in Sect. 5.

S. Melliani (✉) · M. Elomari · L. S. Chadli
Sultan Moulay Slimane University, BP 523 Beni Mellal, Morocco
e-mail: s.melliani@usms.ma

M. Elomari
e-mail: m.elomari@usms.ma

L. S. Chadli
e-mail: sa.chadli@yahoo.fr

S. Melliani and O. Castillo (eds.), *Recent Advances in Intuitionistic Fuzzy Logic Systems*, Studies in Fuzziness and Soft Computing 372,
https://doi.org/10.1007/978-3-030-02155-9_12

2 Preliminaries

In this section, we present some definitions and introduce the necessary notation, which will be used throughout the paper.

We denote $\mathbb{F}_1$ the class of function defined as follows:

$$\mathbb{F}_1 = \Big\{ < u, v >: \mathbb{R} \to [0,1]^2, \quad 0 \le u + v \le 1 \quad \text{satisfies } (1-4) \text{ below} \Big\}$$

1. $< u, v >$ is normal, i.e. there is a $x_0, x_1 \in \mathbb{R}$ such that $u(x_0) = 1, v(x_1) = 1$;
2. $< u, v >$ is an intuitionistic fuzzy convex set;
3. u is upper semi-continuous, and v is lower semi-continuous;
4. Closure of $\{x \in \mathbb{R}^n, \quad v(x) < 1\}$ is compact

For all $\alpha, \beta \in L^* = \big\{(x, y) \in [0, 1]^2, x + y \le 1\big\}$ the α, β-cut of an element of $\mathbb{F}_1$ is defined by

$$< u, v >^{\alpha,\beta} = \Big\{ x \in \mathbb{R}, \ u(x) \ge \alpha, \quad v(x) \le \beta \Big\}$$

By the previous properties we can write

$$\Big[< u, v > \Big]^{\alpha,\beta} = \Big[\underline{< u, v >}(\alpha, \beta), \overline{< u, v >}(\alpha, \beta) \Big]$$

By the extension principal of Zadeh we have

$$\Big[< u, v > + < u', v' > \Big]^{\alpha,\beta} = \Big[< u, v > \Big]^{\alpha,\beta} + \Big[< u', v' > \Big]^{\alpha,\beta}$$
$$\Big[\lambda < u, v > \Big]^{\alpha,\beta} = \lambda \Big[< u, v > \Big]^{\alpha,\beta}$$

For all $< u, v >, < u', v' > \in \mathbb{F}_1$ and $\lambda \in \mathbb{R}$, the distance between two element of $\mathbb{F}_1$ is given by (see [4]).

Theorem 1 *[4] On $\mathbb{F}_1$ we can define the metric*

$$d_\infty\Big(\langle u, v \rangle, \langle z, w \rangle \Big) = \frac{1}{4} \sup_{0<\alpha\le 1} \Big| \Big[\langle u, v \rangle \Big]_r^+ (\alpha) - \Big[\langle z, w \rangle \Big]_r^+ (\alpha) \Big|$$
$$+ \frac{1}{4} \sup_{0<\alpha\le 1} \Big| \Big[\langle u, v \rangle \Big]_l^+ (\alpha) - \Big[\langle z, w \rangle \Big]_l^+ (\alpha) \Big|$$
$$+ \frac{1}{4} \sup_{0<\alpha\le 1} \Big| \Big[\langle u, v \rangle \Big]_r^- (\alpha) - \Big[\langle z, w \rangle \Big]_r^- (\alpha) \Big|$$
$$+ \frac{1}{4} \sup_{0<\alpha\le 1} \Big| \Big[\langle u, v \rangle \Big]_l^- (\alpha) - \Big[\langle z, w \rangle \Big]_l^- (\alpha) \Big|$$

and

$$d_p\left(<u,v>,<u',v'>\right)=\left(\frac{1}{4}\int_0^1\left|[<u,v>]_l^+(\alpha)-[<u',v'>]_l^+(\alpha)\right|^p d\alpha\right.$$
$$+\frac{1}{4}\int_0^1\left|[<u,v>]_r^+(\alpha)-[<u',v'>]_r^+(\alpha)\right|^p d\alpha$$
$$+\frac{1}{4}\int_0^1\left|[<u,v>]_l^-(\alpha)-[<u',v'>]_l^-(\alpha)\right|^p d\alpha$$
$$\left.+\frac{1}{4}\int_0^1\left|[<u,v>]_r^-(\alpha)-[<u',v'>]_r^-(\alpha)\right|^p d\alpha\right)^{\frac{1}{p}}$$

For $p\in[1,\infty)$, we have $\left(\mathbb{F}_1,d_p\right)$ is a complete metric space.

For $p=2$ we have

$$d\Big(<u,v>,<u',v'>\Big)=\Big[\int_0^1\left(\underline{<u,v>}(\alpha,1-\alpha)-\underline{<u',v'>}(\alpha,1-\alpha)\right)^2 d\alpha$$
$$\int_0^1\left(\overline{<u,v>}(\alpha,1-\alpha)-\overline{<u',v'>}(\alpha,1-\alpha)\right)^2 d\alpha\Big]^{\frac{1}{2}}$$

The metric space $\left(\mathbb{F}_1,d\right)$ is complete, separable (see [4]) and the following properties for metric d are valid

1. $d\Big(<u,v>+<u',v'>,<u,v>+<w,z>\Big)=d\Big(<u,v>,<w,z>\Big)$;
2. $d\Big(\lambda<u,v>,\lambda<w,z>\Big)=|\lambda|d\Big(<u,v>,<w,z>\Big)$;
3. $d\Big(<u,v>+<u',v'>,<w,z>+<w',z'>\Big)\le d\Big(<u,v>,<w,z>\Big)+d\Big(<u',v'>,<w',z'>\Big)$;

Definition 1 [5] The generalized Hukuhara difference of two fuzzy numbers $<u,v>,<u',v'>\in\mathbb{F}_1$ is defined as follows

$$<u,v>\ominus_{gH}<u',v'>=<w,z>$$
$$\Leftrightarrow\begin{cases}<u,v>=<u',v'>+<z,w>\\ \text{or}\quad <u',v'>=<u,v>+(-1)<w,z>\end{cases}$$

In terms of α, β-levels we have

$$\left(< u, v > \ominus_{gH} < u', v' >\right)^{\alpha,\beta}$$
$$= \Big[\min\left\{\underline{< u, v >}(\alpha, \beta) - \underline{< u', v' >}(\alpha, \beta)\overline{< u, v >}(\alpha, \beta) - \overline{< u', v' >}(\alpha, \beta)\right\},$$
$$\max\left\{\underline{< u, v >}(\alpha, \beta) - \underline{< u', v' >}(\alpha, \beta), \overline{< u, v >}(\alpha, \beta) - \overline{< u', v' >}(\alpha, \beta)\right\}\Big]$$

and the conditions for the existence of $< w, z >=< u, v > \ominus_{gH} < u', v' >\in \mathbb{F}_1$ are

$$\text{case (i)} \quad \begin{cases} \underline{< w, z >}(\alpha, 1-\alpha) \\ = \underline{< u, v >}(\alpha, 1-\alpha) - \underline{< u', v' >}(\alpha, 1-\alpha) \text{ and } \overline{< w, z >}(\alpha, 1-\alpha) \\ = \overline{< u, v >}(\alpha, 1-\alpha) - \overline{< u', v' >}(\alpha, 1-\alpha) \\ \text{with } \underline{< w, z >}(\alpha, 1-\alpha) \text{ increasing, } \overline{< w, z >}(\alpha, 1-\alpha) \text{ decreasing,} \\ \underline{< w, z >}(\alpha, 1-\alpha) \leq \overline{< w, z >}(\alpha, 1-\alpha) \end{cases}$$

$$\text{case (ii)} \quad \begin{cases} \underline{< w, z >}(\alpha, 1-\alpha) \\ = \overline{< u, v >}(\alpha, 1-\alpha) - \overline{< u', v' >}(\alpha, 1-\alpha) \text{ and } \overline{< w, z >}(\alpha, 1-\alpha) \\ = \underline{< u, v >}(\alpha, 1-\alpha) - \underline{< u', v' >}(\alpha, 1-\alpha) \\ \text{with } \underline{< w, z >}(\alpha, 1-\alpha) \text{ increasing, } \overline{< w, z >}(\alpha, 1-\alpha) \text{ decreasing,} \\ \underline{< w, z >}(\alpha, 1-\alpha) \leq \overline{< w, z >}(\alpha, 1-\alpha) \end{cases}$$

for all $\alpha \in [0, 1]$.

3 Intuitionistic Fuzzy Inner Product

For all $< u, v >, < u', v' >\in \mathbb{F}_1$,

$$\left(< u, v > \oplus < u', v' >\right)(y) = \left\langle \sup_{x\in\mathbb{R}} u(x) \wedge u'(y-x), \inf_{x\in\mathbb{R}} v(x) \wedge v'(y-x)\right\rangle$$

The inverse of $\oplus$ is defined by

$$\ominus : \begin{cases} \mathbb{F}_1 \times \mathbb{F}_1 \to \mathbb{F}_1 \\ (< u, v >, < u', v' >) \to \begin{cases} < z, w >, \; if \quad < u, v >\neq< u', v' > \\ \widetilde{0}, \; if \quad < u, v >=< u', v' > \end{cases} \end{cases}$$

where

$$< z, w >^{\alpha,\beta}$$
$$= \Big[\min\{\underline{< u, v >}(\alpha,\beta) - \underline{< u', v' >}(\alpha,\beta)\overline{< u, v >}(\alpha,\beta) - \overline{< u', v' >}(\alpha,\beta)\},$$
$$\max\{\underline{< u, v >}(\alpha,\beta) - \underline{< u', v' >}(\alpha,\beta), \overline{< u, v >}(\alpha,\beta) - \overline{< u', v' >}(\alpha,\beta)\}\Big]$$

It is clear that $\ominus$ define a low on $\mathbb{F}_1$.

Lemma 1 *($\mathbb{F}_1, \oplus$) is a commutat if group*

Proof Let $< u, v > \in \mathbb{F}_1$, with

$$< u, v >^{\alpha,\beta} =$$
$$\Big[\min\{\underline{< u, v >}(\alpha,\beta) - \underline{< u', v' >}(\alpha,\beta)\overline{< u, v >}(\alpha,\beta) - \overline{< u', v' >}(\alpha,\beta)\},$$
$$\max\{\underline{< u, v >}(\alpha,\beta) - \underline{< u', v' >}(\alpha,\beta), \overline{< u, v >}(\alpha,\beta) - \overline{< u', v' >}(\alpha,\beta)\}\Big]$$

First

$$[0, 0] \ominus < u, v >^{\alpha,\beta} =$$
$$\Big[-\max\{\underline{< u, v >}(\alpha,\beta) - \underline{< u', v' >}(\alpha,\beta), \overline{< u, v >}(\alpha,\beta) - \overline{< u', v' >}(\alpha,\beta)\},$$
$$-\min\{\underline{< u, v >}(\alpha,\beta) - \underline{< u', v' >}(\alpha,\beta)\overline{< u, v >}(\alpha,\beta) - \overline{< u', v' >}(\alpha,\beta)\}\Big]$$

By the conditions (1) it is clear that the inverse of $< u, v >$ belong to $\mathbb{F}_1$. Also $\widetilde{0} \in \mathbb{F}_1$. If $< u, v >, < u', v' > \in \mathbb{F}_1$ by definition we get $< u, v > \oplus < u', v' > \in \mathbb{F}_1$.

Lemma 2 *For all $\lambda, \beta \in \mathbb{R}$ and $< u, v >, < u', v' > \in \mathbb{F}_1$ we have*

1. $1. < u, v > = < u, v >$;
2. $(\lambda + \beta) < u, v > = \lambda < u, v > + \beta < u, v >$
3. $\lambda(< u, v > + < u', v' >) = \lambda < u, v > + \lambda < u', v' >$

Proof 1. Let $< u, v > \in \mathbb{F}_1$, by Zadeh's extension $(1. < u, v >)(x) = < u, v > (x), \quad \forall x \in \mathbb{R}$;

2. For example suppose that $\alpha \geq 0$ and $\beta \leq 0$

$$\begin{aligned}(\lambda + \beta) < u, v > &= (\lambda - (-\beta)) < u, v > \\ &= \lambda < u, v > \ominus(-\beta) < u, v >\end{aligned}$$

in the other hand

$$\begin{aligned}\lambda < u, v > + \beta < u, v > &= \lambda < u, v > -(-\beta) < u, v > \\ &= \lambda < u, v > \ominus(-\beta) < u, v >\end{aligned}$$

Hence $(\lambda + \beta) < u, v >= \lambda < u, v > +\beta < u, v >$;

3. By Zadeh's extension

$$\left[\lambda(< u, v > + < u', v' >)\right]^{\alpha,\beta} = \lambda < u, v >^{\alpha,\beta} + < u', v' >^{\alpha,\beta}$$

For all $\alpha \in [0, 1]$, which implies the equality.

By the previous lemma we get

Theorem 2 *($\mathbb{F}_1, +, .$) is a vector space.*

Let the map $\Big\langle . \mid . \Big\rangle : \mathbb{F}_1 \times \mathbb{F}_1 \longrightarrow \mathbb{R}$ defined by

$$\Big\langle < u, v > \mid < u', v' > \Big\rangle = \int_0^1 \underline{< u, v >}(\alpha, 1-\alpha)\underline{< u', v' >}(\alpha, 1-\alpha)d\alpha$$
$$\int_0^1 \overline{< u, v >}(\alpha, 1-\alpha)\overline{< u', v' >}(\alpha, 1-\alpha)d\alpha$$

It is easy to show that

Lemma 3 *The previous map is well defined.*

Theorem 3 *The map $\langle .|. \rangle$ defined a inner product on $\mathbb{F}_1$*

Proof 1. It is clear that $\langle .|. \rangle$ is symetric.

2. $\langle .|. \rangle$ is bilinear.
   Let $\lambda \in \mathbb{R}$, $u, v, \omega \in E^1$, we have

$$\begin{aligned} &\langle \lambda < u, v > + < u', v' > |\omega\rangle \\ &= \int_0^1 \big(\lambda\underline{< u, v >}(\alpha, 1-\alpha) + \underline{< u', v' >}(\alpha)\big)\,\underline{\omega}(\alpha)d\alpha \\ &+ \int_0^1 \big(\lambda\overline{< u, v >}(\alpha) + \overline{< u', v' >}(\alpha)\big)\,\omega(\alpha)d\alpha \end{aligned}$$

   by linearity of integral we have

$$\langle \lambda < u, v > + < u', v' > |\omega\rangle = \lambda\,\langle < u, v > |\omega\rangle + \langle < u', v' > |\omega\rangle$$

3. By definition it is clear that $\langle < u, v > \mid < u, v > \rangle \geq 0$;
4. Let $< u, v > \in \mathbb{F}_1$ such that $\langle < u, v > \mid < u, v > \rangle = 0$, then $\int_0^1 \underline{< u, v >}^2(\alpha, 1-\alpha)d\alpha = 0$ and $\int_0^1 \overline{< u, v >}^2(\alpha, 1-\alpha)d\alpha = 0$ which implies that $\underline{< u, v >}(\alpha, 1-\alpha) = \overline{u}(\alpha) = 0$, a.e. set $J \subset [0, 1]$ on which $\underline{u}$ and $\overline{u}$ non-zero, but $[0, 1] \setminus J$ is dense in $[0, 1]$. Now let $\alpha \in J$ and let $\alpha_n \in [0, 1] \setminus J$ converge to α, we get $[\underline{< u, v >}(\alpha, 1-\alpha), \overline{< u, v >}(\alpha, 1-\alpha)] =$

$\bigcap_n [\underline{<u,v>}(\alpha_n), \overline{<u,v>}(\alpha_n, 1-\alpha_n)] = [0,0]$, thus $\underline{<u,v>}(\alpha, 1-\alpha) = \overline{<u,v>}(\alpha, 1-\alpha) = 0$, which prove that $<u,v>=\widetilde{0}$.

Remark 1 For all $<u,v>, <u',v'>\in \mathbb{F}_1$

$$\begin{aligned}\| <u,v> \ominus <u',v'> \| &= \sqrt{\langle <u,v> \ominus <u',v'> \mid <u,v> \ominus_{gH} <u',v'> \rangle} \\ &= d\left(<u,v>, <u',v'>\right)\end{aligned}$$

Example 1 If $u = (l, u, x, y; 1-l, 1-u, 1-x, 1-y)$ and $v = (l', u', x', y'; 1-l', 1-u', 1-x', 1-y')$ two trapezoidal intuitionistic fuzzy numbers (see [3]), with $a = b = 1, c = d = \frac{1}{12}$ and $\omega_L = \omega_U = \frac{1}{2}$. We have

$$\begin{aligned}\langle <u,v> \mid <u',v'> \rangle &= \Big[\textstyle\int_0^1 \left(l + x(\alpha - \frac{1}{2})\right)\left(l' + x'(\alpha - \frac{1}{2})\right) d\alpha \\ &\quad + \textstyle\int_0^1 \left(u - y(\alpha - \frac{1}{2})\right)\left(u' - y'(\alpha - \frac{1}{2})\right) d\alpha\Big]^{\frac{1}{2}} \\ &= ll' + uu' + \tfrac{1}{12}xx' + \tfrac{1}{12}yy'\end{aligned}$$

Remark 2 We can find the same result as [3] in the weighted case.

4 Projection on the Set of All Triangular Intuitionistic Fuzzy Numbers

Let the set

$$\mathbb{F}_T = \left\{<u,v>\in \mathbb{F}_1, \quad <u,v>^{\alpha,1-\alpha} = [t_1 + \alpha t_2, t_3 + \alpha t_4], \; t_1 = t_1 + t_2 \le t_3 + t_4 \le t_3\right\}$$

Lemma 4 *$\mathbb{F}_T$ is convex closed subset of $\mathbb{F}_1$.*

Proof Let $\gamma \in [0,1]$ and $<u,v>, <u',v'>\in \mathbb{F}_T$ with

$$<u,v>^{(\alpha,1-\alpha)} = [t_1 + \alpha t_2, t_3 + \alpha t_4] \text{ and } <u',v'>^{\alpha,1-\alpha} = [s_1 + \alpha s_2, s_3 + \alpha s_4]$$

we have

$$\begin{aligned}&\left(\gamma <u,v> + (1-\gamma) <u',v'>\right)^{\alpha,1-\alpha} = \\ &\left[(\gamma t_1 + (1-\gamma)s_1) + (\gamma t_2 + (1-\gamma)s_2)\alpha, (\gamma t_3 + (1-\gamma)s_3) + (\gamma t_4 + (1-\gamma)s_4)\alpha\right]\end{aligned}$$

but

$$(\gamma t_1 + (1-\gamma)s_1) \leq \gamma t_1 + (1-\gamma)s_1) + (\gamma t_2 + (1-\gamma)s_2 \leq \\ \gamma t_3 + (1-\gamma)s_3) + (\gamma t_4 + (1-\gamma)s_4 \leq \gamma t_3 + (1-\gamma)s_3$$

which implies that $\mathbb{F}_T$ is a convex subset of $\mathbb{F}_1$.

Now let $< u_n, v_n > \in \mathbb{F}_T \longrightarrow < u, v >$ with respect d, then

$$\underline{< u_n, v_n >}(\alpha, 1-\alpha) \rightarrow \underline{< u, v >}(\alpha, 1-\alpha)$$

and

$$\overline{< u_n, v_n >}(\alpha, 1-\alpha) \rightarrow \overline{< u, v >}(\alpha, 1-\alpha)$$

in L^2, By Fisher's theorem [6] there is a subsequence such that

$$\underline{< u_{n_k}, v_{n_k} >}(\alpha, 1-\alpha) \rightarrow \underline{u}(\alpha, 1-\alpha)$$

and

$$\overline{u_{n_k}}(\alpha, 1-\alpha) \rightarrow \overline{u}(\alpha, 1-\alpha)$$

using the result in [7], we have

$$\sup_{\alpha \in (0,1]} d_H(< u_n, v_n >^{(\alpha,1-\alpha)}, < u, v >^{(\alpha,1-\alpha)}) \longrightarrow 0$$

but $(\mathbb{F}_1, d_\infty)$ is a complete metric space, thus $< u, v > \in \mathbb{F}_T$ then $\mathbb{F}_T$ is a closed subset of $\mathbb{F}_1$.

By the Lemma (4) the existence and uniqueness of projection in the Hilbert space we have.

Theorem 4 *For each $< u, v > \in \mathbb{F}_1$ there is a unique $< u', v' > \in \mathbb{F}_T$ such that*

1. *$d\left(< u, v >, < u', v' >\right) = d\left(< u, v >, \mathbb{F}_T\right)$;*
2. *the map $< u, v > \longrightarrow < u', v' >$ is a linear and injective mapping.*

Example 2 We consider the following $< u, v > \in \mathbb{F}_1$ where

$$u(x) = \begin{cases} 1 - 2x^2, & |x| \leq \frac{1}{2} \\ 0, & |x| > \frac{1}{2} \end{cases}$$

$$v(x) = \begin{cases} x^2, & |x| \leq 1 \\ 0, & |x| > 1 \end{cases}$$

His projection on the intuitionistic fuzzy triangular space is given by

$$u_t(x) = \begin{cases} 1 + 2x, & \frac{-1}{2} \leq x \leq 0 \\ 1 - 2x, & 0 \leq x \leq \frac{1}{2} \\ 0, & |x| > \frac{1}{2} \end{cases}$$

$$v_t(x) = \begin{cases} x, & 0 \leq x \leq 1 \\ -x, & -1 \leq x0 \\ 0, & |x| > 1 \end{cases}$$

We have

$$[< u, v >]^\alpha = [-\sqrt{\frac{1-\alpha}{2}}, \sqrt{\frac{1-\alpha}{2}}]$$

$$[< u, v >]_\alpha = [-\sqrt{1-\alpha}, \sqrt{1-\alpha}]$$

and

$$[< u, v >]^\alpha = [\frac{1-\alpha}{2}, \frac{1-\alpha}{2}]$$

$$[< u, v >]_\alpha = [-\alpha, \alpha]$$

So

$$d(< u, v >, < u_t, v_t >) = \inf_{< z, w > \in E_T} d(< u, v >, < z, w >)$$

5 Conclusion

In this paper we will try to give a sens of Hilbert space on the set of all fuzzy numbers, and we studied the existence and uniqueness of an approximation triangular fuzzy number of a fuzzy number.

References

1. K.T. Atanassov, Intuitionistic fuzzy sets. Fuzzy Sets Syst. **20**, 87–96 (1986)
2. M.L. Puri, D.A. Ralescu, Fuzzy random variables. J. Math. Anal. Appl. **114**, 409–422 (1986)
3. Chi-Tsuen Yeh, Weighted trapezoidal and triangular approximations of fuzzy numbers. Fuzzy Sets Syst. **160**, 3059–3079 (2009)
4. S. Melliani, M. Elomari, L.S. Chadli, R. Ettoussi, Intuitionistic fuzzy metric space. Notes Intuitionistic Fuzzy Sets **21**(1), 43–53 (2015)
5. M. Melliani, S. Elomari, L.S. Chadli, R. Ettoussi, Extension of Hukuhara difference in intuitionist fuzzy set theory. Notes on Intuitionistic Fuzzy Sets **21**(4), 34–47 (2015)

6. H. Brezis, *Functional Analysis, Sobolev Spaces and Partial Differential Equations* (Springer, Berlin, 2010)
7. O. Kaleva, S. Seikkala, On the convergence of fuzzy sets. Fuzzy Sets Syst. **17**, 53–65 (1985)
8. L. Stefanini, B. Bede, Generalized hukuhara differentiability of interval-valued functions and interval differential equations. Nonlinear Anal. **71**(34), 1311–1328 (2009)
9. R. Ettoussi, S. Melliani, L.S. Chadli, Differential equation with intuitionistic fuzzy parameters. Notes Intuitionistic Fuzzy Sets **23**(4), 46–61 (2017)
10. L.A. Zadeh, Fuzzy sets. Inf. Control **8**, 338–353 (1965)

On Intuitionistic Fuzzy Laplace Transforms for Second Order Intuitionistic Fuzzy Differential Equations

R. Ettoussi, Said Melliani and L. S. Chadli

Abstract In this paper, we introduce a result for the intuitionistic fuzzy Laplace transforms. The related theorems and properties are proved in detail and we propose a procedure for solving second-order intuitionistic fuzzy differential equations by using the intuitionistic fuzzy Laplace transform method. Finally, we present an example to illustrate this work.

1 Introduction

In 1965, Zadeh [14] first introduced the fuzzy set theory. Later many researchers have applied this theory to the well known results in the classical set theory.

The idea of intuitionistic fuzzy set was first published by Atanassov [1–3] as a generalization of the notion of fuzzy set. The notions of differential and integral calculus for intuitionistic fuzzy-set-valued are given using Hukuhara difference in intuitionistic Fuzzy theory [7]. The authors of papers [10, 11] are discussed differential and partial differential equations under intuitionistic fuzzy environment respectively. The existence and uniqueness of the solution of intuitionistic fuzzy differential equations by using successive approximations method have been discussed in [7], while in [6] the theorem of the existence and uniqueness of the solution for differential equations with intuitionistic fuzzy data are proved by using the theorem of fixed point in the complete metric space, also the explicit formula of the solution are given by using the α-cuts method.

Laplace transform is a very useful apparatus to solve differential equation. Laplace transforms give the solution of a differential equations satisfying the initial condition

R. Ettoussi · S. Melliani (✉) · L. S. Chadli
Sultan Moulay Slimane University, BP 523 Beni Mellal, Morocco
e-mail: saidmelliani@gmail.com

R. Ettoussi
e-mail: razika.imi@gmail.com

L. S. Chadli
e-mail: sa.chadli@yahoo.fr

S. Melliani and O. Castillo (eds.), *Recent Advances in Intuitionistic Fuzzy Logic Systems*, Studies in Fuzziness and Soft Computing 372,
https://doi.org/10.1007/978-3-030-02155-9_13

directly without use the general solution of the differential equation. Fuzzy Laplace Transform (FLT) was first introduced by Allahviranloo and Ahmadi [2]. Here first order fuzzy differential equation with fuzzy initial condition is solved by FLT. Tolouti and Ahmadi [12] applied the FLT in 2nd order FDE. FLT also used to solve many areas of differential equation [1].

This paper is organized as follows: in Sect. 2 we give preliminaries which we will use throughout this work. in Sect. 3 a result for intuitionistic fuzzy Laplace transform is presented. In Sect. 4 we construct a procedure for solving intuitionistic fuzzy initial value problems (IFIVPs) of the second-order. In the last section, we present an example for illustrate this work.

2 Preliminaries

Let us $T = [a, b] \subset \mathbb{R}$ be a compact interval.

Definition 1 we denote by

$$\mathrm{IF}_1 = \mathrm{IF}(\mathbb{R}) = \left\{ \langle u, v \rangle \ : \ \mathbb{R} \rightarrow [0, 1]^2 , |\forall\, x \in \mathbb{R}| 0 \leq u(x) + v(x) \leq 1 \right\}$$

An element $\langle u, v \rangle$ of IF_1 is said an intuitionistic fuzzy number if it satisfies the following conditions

(i) $\langle u, v \rangle$ is normal i.e there exists $x_0, x_1 \in \mathbb{R}$ such that $u(x_0) = 1$ and $v(x_1) = 1$.
(ii) u is fuzzy convex and v is fuzzy concave.
(iii) u is upper semi-continuous and v is lower semi-continuous
(iv) $supp\, \langle u, v \rangle = \mathrm{cl}\{x \in \mathbb{R} \ : | \ v(x) < 1\}$ is bounded.

so we denote the collection of all intuitionistic fuzzy number by IF_1.

Definition 2 [8] An intuitionistic fuzzy number $\langle u, v \rangle$ in parametric form is a pair $\langle u, v \rangle = \Big((\underline{\langle u, v \rangle}^+, \overline{\langle u, v \rangle^+}), (\underline{\langle u, v \rangle}^-, \overline{\langle u, v \rangle}^-) \Big)$ of functions $\underline{\langle u, v \rangle}^-(\alpha)$, $\overline{\langle u, v \rangle}^-(\alpha)$, $\underline{\langle u, v \rangle}^+(\alpha)$ and $\overline{\langle u, v \rangle}^+(\alpha)$, which satisfies the following requirements:

1. $\underline{\langle u, v \rangle}^+(\alpha)$ is a bounded monotonic increasing continuous function,
2. $\overline{\langle u, v \rangle}^+(\alpha)$ is a bounded monotonic decreasing continuous function,
3. $\underline{\langle u, v \rangle}^-(\alpha)$ is a bounded monotonic increasing continuous function,
4. $\overline{\langle u, v \rangle}^-(\alpha)$ is a bounded monotonic decreasing continuous function,
5. $\underline{\langle u, v \rangle}^-(\alpha) \leq \overline{\langle u, v \rangle}^-(\alpha)$ and $\underline{\langle u, v \rangle}^+(\alpha) \leq \overline{\langle u, v \rangle}^+(\alpha)$, for all $0 \leq \alpha \leq 1$.

Example A Triangular Intuitionistic Fuzzy Number (TIFN) $\langle u, v \rangle$ is an intuitionistic fuzzy set in $\mathbb{R}$ with the following membership function u and non-membership function v:

$$u(x) = \begin{cases} \frac{x - a_1}{a_2 - a_1} & \text{if} a_1 \le x \le a_2 \\ \frac{a_3 - x}{a_3 - a_2} & \text{if} a_2 \le x \le a_3, \\ 0 & \text{otherwise} \end{cases}$$

$$v(x) = \begin{cases} \frac{a_2 - x}{a_2 - a_1'} & \text{if } a_1' \le x \le a_2 \\ \frac{x - a_2}{a_3' - a_2} & \text{if } a_2 \le x \le a_3', \\ 1 & \textit{otherwise.} \end{cases}$$

where $a_1' \le a_1 \le a_2 \le a_3 \le a_3'$

This TIFN is denoted by $\langle u, v\rangle = \langle a_1, a_2, a_3; a_1', a_2, a_3'\rangle$.
Its parametric form is

$$\underline{\langle u, v\rangle}^+(\alpha) = a_1 + \alpha(a_2 - a_1), \ \overline{\langle u, v\rangle}^+(\alpha) = a_3 - \alpha(a_3 - a_2)$$

$$\underline{\langle u, v\rangle}^-(\alpha) = a_1' + \alpha(a_2 - a_1'), \ \overline{\langle u, v\rangle}^-(\alpha) = a_3' - \alpha(a_3' - a_2)$$

For $\alpha \in [0, 1]$ and $\langle u, v\rangle \in \mathrm{IF}_1$, the upper and lower α-cuts of $\langle u, v\rangle$ are defined by

$$[\langle u, v\rangle]^\alpha = \{x \in \mathbb{R} : v(x) \le 1 - \alpha\}$$

and

$$[\langle u, v\rangle]_\alpha = \{x \in \mathbb{R} : u(x) \ge \alpha\}$$

Remark 1 If $\langle u, v\rangle \in \mathrm{IF}_1$, so we can see $[\langle u, v\rangle]_\alpha$ as $[u]^\alpha$ and $[\langle u, v\rangle]^\alpha$ as $[1 - v]^\alpha$ in the fuzzy case.

We define $0_{\langle 1,0\rangle} \in \mathrm{IF}_1$ as

$$0_{\langle 1,0\rangle}(t) = \begin{cases} \langle 1, 0\rangle & t = 0 \\ \langle 0, 1\rangle & t \ne 0 \end{cases}$$

For $\langle u, v\rangle$, $\langle z, w\rangle \in \mathrm{IF}_1$ and $\lambda \in \mathbb{R}$, the addition and scaler-multiplication are defined as follows:

$$\Big[\langle u,v\rangle \oplus \langle z,w\rangle\Big]^{\alpha} = \Big[\langle u,v\rangle\Big]^{\alpha} + \Big[\langle z,w\rangle\Big]^{\alpha}, \quad \Big[\lambda\,\langle z,w\rangle\Big]^{\alpha} = \lambda\Big[\langle z,w\rangle\Big]^{\alpha}$$
$$\Big[\langle u,v\rangle \oplus \langle z,w\rangle\Big]_{\alpha} = \Big[\langle u,v\rangle\Big]_{\alpha} + \Big[\langle z,w\rangle\Big]_{\alpha}, \quad \Big[\lambda\,\langle z,w\rangle\Big]_{\alpha} = \lambda\Big[\langle z,w\rangle\Big]_{\alpha}.$$

Definition 3 Let $\langle u, v\rangle$ an element of IF_1 and $\alpha \in [0, 1]$, we define the following sets:

$$\Big[\langle u,v\rangle\Big]_{l}^{+}(\alpha) = \inf\{x \in \mathbb{R} \mid u(x) \geq \alpha\},$$
$$\Big[\langle u,v\rangle\Big]_{r}^{+}(\alpha) = \sup\{x \in \mathbb{R} \mid u(x) \geq \alpha\}$$
$$\Big[\langle u,v\rangle\Big]_{l}^{-}(\alpha) = \inf\{x \in \mathbb{R} \mid v(x) \leq 1-\alpha\},$$
$$\Big[\langle u,v\rangle\Big]_{r}^{-}(\alpha) = \sup\{x \in \mathbb{R} \mid v(x) \leq 1-\alpha\}$$

Remark 2

$$\Big[\langle u,v\rangle\Big]_{\alpha} = \Big[\Big[\langle u,v\rangle\Big]_{l}^{+}(\alpha), \Big[\langle u,v\rangle\Big]_{r}^{+}(\alpha)\Big]$$
$$\Big[\langle u,v\rangle\Big]^{\alpha} = \Big[\Big[\langle u,v\rangle\Big]_{l}^{-}(\alpha), \Big[\langle u,v\rangle\Big]_{r}^{-}(\alpha)\Big]$$

On the space IF_1 we will consider the following metric,

$$\begin{aligned} d_{\infty}\Big(\langle u,v\rangle, \langle z,w\rangle\Big) = &\frac{1}{4}\sup_{0<\alpha\leq 1}\Big|\Big[\langle u,v\rangle\Big]_{r}^{+}(\alpha) - \Big[\langle z,w\rangle\Big]_{r}^{+}(\alpha)\Big| \\ &+ \frac{1}{4}\sup_{0<\alpha\leq 1}\Big|\Big[\langle u,v\rangle\Big]_{l}^{+}(\alpha) - \Big[\langle z,w\rangle\Big]_{l}^{+}(\alpha)\Big| \\ &+ \frac{1}{4}\sup_{0<\alpha\leq 1}\Big|\Big[\langle u,v\rangle\Big]_{r}^{-}(\alpha) - \Big[\langle z,w\rangle\Big]_{r}^{-}(\alpha)\Big| \\ &+ \frac{1}{4}\sup_{0<\alpha\leq 1}\Big|\Big[\langle u,v\rangle\Big]_{l}^{-}(\alpha) - \Big[\langle z,w\rangle\Big]_{l}^{-}(\alpha)\Big| \end{aligned}$$

Theorem 1 [9] *The metric space* (IF_1, d_{∞}) *is complete.*

Proposition 1 *For all* $\alpha, \beta \in [0, 1]$ *and* $\langle u, v\rangle \in IF_1$

(i) $\Big[\langle u,v\rangle\Big]_{\alpha} \subset \Big[\langle u,v\rangle\Big]^{\alpha}$

(ii) $\Big[\langle u,v\rangle\Big]_{\alpha}$ *and* $\Big[\langle u,v\rangle\Big]^{\alpha}$ *are nonempty compact convex sets in* $\mathbb{R}$

(iii) if $\alpha \leq \beta$ *then* $\Big[\langle u,v\rangle\Big]_{\beta} \subset \Big[\langle u,v\rangle\Big]_{\alpha}$ *and* $\Big[\langle u,v\rangle\Big]^{\beta} \subset \Big[\langle u,v\rangle\Big]^{\alpha}$

(iv) If $\alpha_n \nearrow \alpha$ *then* $\Big[\langle u, v\rangle\Big]_{\alpha} = \bigcap_n \Big[\langle u, v\rangle\Big]_{\alpha_n}$ *and* $\Big[\langle u, v\rangle\Big]^{\alpha} = \bigcap_n \Big[\langle u, v\rangle\Big]^{\alpha_n}$

Let M any set and $\alpha \in [0, 1]$ we denote by

$$M_\alpha = \{x \in \mathbb{R} : u(x) \geq \alpha\} \quad \text{and} \quad M^\alpha = \{x \in \mathbb{R} : v(x) \leq 1 - \alpha\}$$

Lemma 1 [9] *Let* $\Big\{M_\alpha,\ \alpha \in [0, 1]\Big\}$ *and* $\Big\{M^\alpha,\ \alpha \in [0, 1]\Big\}$ *two families of subsets of* $\mathbb{R}$ *satisfies (i)–(iv) in Proposition 1, if u and v define by*

$$u(x) = \begin{cases} 0 & \text{if } x \notin M_0 \\ \sup\{\alpha \in [0, 1] : x \in M_\alpha\} & \text{if } x \in M_0 \end{cases}$$

$$v(x) = \begin{cases} 1 & \text{if } x \notin M^0 \\ 1 - \sup\{\alpha \in [0, 1] : x \in M^\alpha\} & \text{if } x \in M^0 \end{cases}$$

Then $\langle u, v\rangle \in IF_1$

Definition 4 [6] Let $F : T \to \mathrm{IF}_1$ be an intuitionistic fuzzy valued mapping and $t_0 \in T$. Then F is called intuitionistic fuzzy continuous in t_0 iff:

$$(\forall \varepsilon > 0)(\exists \delta > 0)\Big(\forall t \in T \text{such as} \mid t - t_0 \mid < \delta\Big) \Rightarrow d_\infty(F(t), F(t_0)) < \varepsilon.$$

Definition 5 [6] A mapping $F : T \to \mathrm{IF}_1$ is said to be differentiable at $t_0 \in (a, b)$ if there exist $F'(t_0) \in \mathrm{IF}_1$ such that limits:

$$\lim_{\Delta t \to 0^+} \frac{F(t_0 + \Delta t) \ominus F(t_0)}{\Delta t} \quad \textit{and} \quad \lim_{\Delta t \to 0^+} \frac{F(t_0) \ominus F(t_0 - \Delta t)}{\Delta t}$$

exist and they are equal to $F'(t_0)$.

If $F : T \to \mathrm{IF}_1$ is differentiable at $t_0 \in T$, then we say that $F'(t_0)$ is the intuitionistic fuzzy derivative of $F(t)$ at the point t_0.

Definition 6 A mapping $F : T \to \mathrm{IF}_1$ is said to be differentiable of second order at $t_0 \in (a, b)$ if there exist $F''(t_0) \in \mathrm{IF}_1$ such that limits:

$$\lim_{\Delta t \to 0^+} \frac{F'(t_0 + \Delta t) \ominus F'(t_0)}{\Delta t} \quad \textit{and} \quad \lim_{\Delta t \to 0^+} \frac{F'(t_0) \ominus F'(t_0 - \Delta t)}{\Delta t}$$

exists and they are equal to $F''(t_0)$.

Here the limit is taken in the metric space $(\mathrm{IF}_1, d_\infty)$. At the end points of T we consider only the one-sided derivatives.

Theorem 2 *Let $F(t)$ and $F'(t)$ are differentiable. Denote*

$$[F'(t)]_\alpha = [F_l^{'-}(t,\alpha), F_r^{'-}(t,\alpha)],\ [F'(t)]^\alpha = [F_l^{'+}(t,\alpha), F_r^{'+}(t,\alpha)].$$

Then $F_l^{'-}(t,\alpha), F_r^{'-}(t,\alpha), F_l^{'+}(t,\alpha)$ and $F_r^{'+}(t,\alpha)$ are differentiable and we have

$$[F''(t)]_\alpha = [F_l^{''-}(t,\alpha), F_r^{''-}(t,\alpha)],\ [F''(t)]^\alpha = [F_l^{''+}(t,\alpha), F_r^{''+}(t,\alpha)]$$

Proof Since $F(t)$ is differentiable then we get

$$[F'(t)]_\alpha = [F_l^{'-}(t,\alpha), F_r^{'-}(t,\alpha)],\ [F'(t)]^\alpha = [F_l^{'+}(t,\alpha), F_r^{'+}(t,\alpha)].$$

also $F'(t)$ is differentiable then by the Definition 6 we have
$\Big[F'(t+h)\ominus F'(t)\Big]_\alpha = \Big[F_l^{'-}(t+h,\alpha) - F_l^{'-}(t,\alpha), F_r^{'-}(t+h,\alpha) - F_r^{'-}(t,\alpha)\Big]$,
and $\Big[F'(t)\ominus F'(t-h)\Big]_\alpha = \Big[F_l^{'-}(t,\alpha) - F_l^{'-}(t-h,\alpha), F_r^{'-}(t,\alpha) - F_r^{'-}(t-h,\alpha)\Big]$.
and multiplying by $h > 0$ we get:

$$\frac{1}{h}\Big[F'(t+h)\ominus F'(t)\Big]_\alpha = \Big[\frac{F_l^{'-}(t+h,\alpha) - F_l^{'-}(t,\alpha)}{h}, \frac{F_r^{'-}(t+h,\alpha) - F_r^{'-}(t,\alpha)}{h}\Big]$$

and

$$\frac{1}{h}\Big[F'(t)\ominus F'(t-h)\Big]_\alpha = \Big[\frac{F_l^{'-}(t,\alpha) - F_l^{'-}(t-h,\alpha)}{h}, \frac{F_r^{'-}(t,\alpha) - F_r^{'-}(t-h,\alpha)}{h}\Big]$$

Finally, passing to the limit we find that

$$[F''(t)]_\alpha = [F_l^{''-}(t,\alpha), F_r^{''-}(t,\alpha)]$$

we use the same idea to prove that

$$[F''(t)]^\alpha = [F_l^{''+}(t,\alpha), F_r^{''+}(t,\alpha)].$$

Proposition 2 ([13], p. 82, Lebesgue's Theorem). *Let f be a bounded function on $[a,b]$. Then f is Riemann-integrable on $[a,b]$ if and only if f is continuous a.e. on $[a,b]$.*

Proposition 3 ([3], p. 276, Theorem 10.33). *Assume f is Riemann-integrable on $[a,b]$ for every $b \geq a$, and assume there is a positive constant M such that $\int_a^\infty |f(t)|dt \leq M$ for every $b \geq a$, Then both f and $|f|$ are improper Riemann integrable on $[a,\infty)$. Also, f is Lebesgue-integrable on $[a,\infty)$ and the Lebesgue integral of f is equal to the improper Riemann integral of f.*

3 Intuitionistic Fuzzy Laplace Transform

Definition 7 Let $f(t)$ be an intuitionistic fuzzy-valued function on $[a, b]$. Suppose that $f_l^+(t, \alpha), f_r^+(t, \alpha)$,
$f_l^-(t, \alpha)$ and $f_r^-(t, \alpha)$ are Riemann-integrable on $[a, b]$ for all $\alpha \in [0, 1]$.
Let

$$A_\alpha = \left[\int_a^b f_l^+(t, \alpha)dt, \int_a^b f_r^+(t, \alpha)dt \right]$$

and

$$A^\alpha = \left[\int_a^b f_l^-(t, \alpha)dt, \int_a^b f_r^-(t, \alpha)dt \right].$$

Then we say that $f(t)$ is intuitionistic fuzzy Riemann-integrable on $[a, b]$ denoted as $\mathbb{IFRI}$ on $[a, b]$ and the membership function and the nonmembership function of $\int_a^b f(t)dt$ are defined by,

$$u_{\int_a^b f(t)dt}(y) = \sup_{0\leq\alpha\leq1} \alpha.\chi_{A_\alpha}(y)$$

$$v_{\int_a^b f(t)dt}(y) = \inf_{0\leq\alpha\leq1} \alpha.\chi_{A^\alpha}(y)$$

for all $y \in A^{\mho}$.

Theorem 3 *Let $f(t)$ be an intuitionistic fuzzy-valued function on $[a, \infty)$ represented by* $\left(f_l^+(t, \alpha), f_r^+(t, \alpha), f_l^-(t, \alpha), f_r^-(t, \alpha)\right)$, *for any fixed $\alpha \in [0, 1]$, assume that $f_l^+(t, \alpha), f_r^+(t, \alpha), f_l^-(t, \alpha)$ and $f_r^-(t, \alpha)$ are Riemann-integrable on $[a, b]$ for every $b \geq a$, and assume there are four positive constants $M_l^+(\alpha)$, $M_r^+(\alpha)$, $M_l^-(\alpha)$ and $M_r^-(\alpha)$ such that*

$$\int_a^\infty |f_l^+(t, \alpha)|dt \leq M_l^+(\alpha), \quad \int_a^\infty |f_r^+(t, \alpha)|dt \leq M_r^+(\alpha)$$

$$\int_a^\infty |f_l^-(t, \alpha)|dt \leq M_l^-(\alpha), \quad \int_a^\infty |f_r^-(t, \alpha)|dt \leq M_r^-(\alpha)$$

Then $f(t)$ is an improper intuitionistic fuzzy Riemann integrable on $[a, \infty)$ and the improper intuitionistic fuzzy Riemann-integral is an intuitionistic fuzzy number. Furthermore, we have:

$$\int_a^{\infty} f(t,\alpha)dt =$$
$$\left(\int_a^{\infty} f_l^+(t,\alpha)dt, \int_a^{\infty} f_r^+(t,\alpha)dt, \int_a^{\infty} f_l^-(t,\alpha)dt, \int_a^{\infty} f_r^-(t,\alpha)dt\right)$$

Proof For $\alpha,\ \beta \in]0,1]$ we consider the following sets

$$A_\alpha = \left[\int_a^b f_l^+(t,\alpha)dt, \int_a^b f_r^+(t,\alpha)dt\right]$$

and

$$A^\alpha = \left[\int_a^b f_l^-(t,\alpha)dt, \int_a^b f_r^-(t,\alpha)dt\right]$$

Since $f(t)$ is intuitionistic fuzzy-valued function on $[a,\infty)$ and by Proposition 1, then we have $\left[f(t)\right]_\alpha \subseteq \left[f(t)\right]^\alpha$ So, $A_\alpha \subseteq A^\alpha$ the property (i) of Proposition 1 is holds.

In addition $\left[f(t)\right]_\beta \subseteq \left[f(t)\right]_\alpha$ and $\left[f(t)\right]^\beta \subseteq \left[f(t)\right]^\alpha$ this implies that $A_\beta \subseteq A_\alpha$ and $A^\beta \subseteq A^\alpha$ for $\alpha \leq \beta$ which means that the Property (ii) of Proposition 1 is verified.

Moreover, the function $f(t)$ is intuitionistic fuzzy-valued function on $[a,\infty)$ then it's verified the conditions of the Definition 1 which ensure that the sets A_α and A^α are nonempty compact convex sets in $\mathbb{R}$.

It remains to show property (iv) of the Proposition 1 i.e $A_\alpha = \bigcap_{k\geq 1} A_{\alpha_k}$ and $A^\alpha = \bigcap_{k\geq 1} A^{\alpha_k}$ for all increasing sequences $\alpha_k \uparrow \alpha$ converging to $\alpha \in]0,1]$. For $\alpha_k \uparrow \alpha$, we have
$f_l^+(t,0) \leq f_l^+(t,\alpha_k) \leq f_l^+(t,1)$ and $f_r^+(t,0) \leq f_r^+(t,\alpha_k) \leq f_r^+(t,1)$, thus

$$|f_l^+(t,\alpha_k)| \leq \max\{|f_l^+(t,0)|, |f_l^+(t,1)|\} :\equiv g_1(t)$$

and

$$|f_r^+(t,\alpha_k)| \leq \max\{|f_r^+(t,0)|, |f_r^+(t,1)|\} :\equiv g_2(t).$$

Since $f_l^+(t,0), f_l^+(t,1), f_r^+(t,0)$ and $f_r^+(t,1)$ are Riemann-integrable on $[a,b]$, then $|f_l^+(t,0)|$, $|f_l^+(t,1)|$, $g_1(t)$, $|f_r^+(t,0)|$, $|f_r^+(t,1)|$ and $g_2(t)$ are also Riemann-integrable on $[a,b]$ by Proposition 2.

Now, $0 \leq g_1(t) \leq |f_l^+(t,0)| + |f_l^+(t,1)|$, we have

$$\int_a^b |g_1(t)|dt = \int_a^b g_1(t)dt \leq \int_a^b |f_l^+(t,0)|dt$$
$$+ \int_a^b |f_l^+(t,1)|dt \leq M_l^+(0) + M_l^+(1) = M$$

for every $b \geq a$.

Also, $0 \leq g_2(t) \leq |f_r^+(t,0)| + |f_r^+(t,1)|$, we have

$$\int_a^b |g_2(t)|dt = \int_a^b g_2(t)dt \leq \int_a^b |f_r^+(t,0)|dt + \int_a^b |f_r^+(t,1)|dt \leq M_r^+(0) + M_r^+(1) = N$$

for every $b \geq a$. Then by Proposition 3, $g_1(t)$ and $g_2(t)$ are improper Riemann-integrable and also Lebesgue-integrable on $[a, \infty)$. By the Lebesgue's Dominated Convergence Theorem, we have

$$\lim_{k\to\infty} \int_a^\infty f_l^+(t,\alpha_k)dt = \int_a^\infty f_l^+(t,\alpha)dt$$

and

$$\lim_{k\to\infty} \int_a^\infty f_r^+(t,\alpha_k)dt = \int_a^\infty f_r^+(t,\alpha)dt$$

Thus, $A_\alpha = \bigcap_{k\geq 1} A_{\alpha_k}$ for $\alpha_k \uparrow \alpha$. Similarly we prove that $A^\alpha = \bigcap_{k\geq 1} A^{\alpha_k}$.

Finally By Lemma 1 the proof is complete.

Definition 8 Let $f(t)$ be continuous intuitionistic fuzzy-value function on $[0, \infty)$. Suppose that $f(t)e^{-st}$ is improper intuitionistic fuzzy Riemann-integrable on $[0, \infty)$, then $\int_0^\infty f(t)e^{-st}dt$ is called intuitionistic fuzzy Laplace transform and is denoted as

$$\mathbf{L}[f(t)] = \int_0^\infty f(t)e^{-st}dt, \; s > 0$$

We have

$$\int_0^\infty f(t,\alpha)e^{-st}dt =$$

$$\left(\int_0^\infty f_l^+(t,\alpha)e^{-st}dt, \int_0^\infty f_r^+(t,\alpha)e^{-st}dt, \int_0^\infty f_l^-(t,\alpha)e^{-st}dt, \int_0^\infty f_r^-(t,\alpha)e^{-st}dt\right)$$

Also by using the definition of classical Laplace transform

$$\mathscr{L}(f_l^+(t,\alpha)) = \int_0^\infty f_l^+(t,\alpha)e^{-st}dt, \;\; \mathscr{L}(f_r^+(t,\alpha)) = \int_0^\infty f_r^+(t,\alpha)e^{-st}dt$$

$$\mathscr{L}(f_l^-(t,\alpha)) = \int_0^\infty f_l^-(t,\alpha)e^{-st}dt, \;\; \mathscr{L}(f_r^-(t,\alpha)) = \int_0^\infty f_r^-(t,\alpha)e^{-st}dt$$

Then, we get

$$\mathbf{L}[f(t,\alpha)] = \Big(\mathscr{L}(f_l^+(t,\alpha)), \mathscr{L}(f_r^+(t,\alpha)), \mathscr{L}(f_l^-(t,\alpha)), \mathscr{L}(f_r^-(t,\alpha))\Big)$$

Theorem 4 *Suppose that $f(t)$ and $f'(t)$ are the continuous intuitionistic fuzzy valued functions on $[0,\infty)$ such that $f(t)e^{-st}$ and $f'(t)e^{-st}$ exists, are continuous and are improper intuitionistic fuzzy Riemann Integrable on $[0,\infty)$, then we have*

$$\mathbf{L}[f''(t)] = s^2\mathbf{L}[f(t)] \ominus sf(0) \ominus f'(0).$$

Proof First, we denote by $\Big(f_l^+(t,\alpha), f_r^+(t,\alpha), f_l^-(t,\alpha), f_r^-(t,\alpha)\Big)$, $\Big(f_l^{'+}(t,\alpha), f_r^{'+}(t,\alpha), f_l^{'-}(t,\alpha), f_r^{'-}(t,\alpha)\Big)$ and $\Big(f_l^{''+}(t,\alpha), f_r^{''+}(t,\alpha), f_l^{''-}(t,\alpha), f_r^{''-}(t,\alpha)\Big)$ the parametric forms respectively of the $f(t), f'(t)$ and $f''(t)$.

By the definition of intuitionistic fuzzy laplace transform we have

$$\begin{aligned}\mathbf{L}[f''(t)] &= \int_0^\infty f''(t)e^{-st}dt,\ s>0.\\ &= s\int_0^\infty f'(t)e^{-st}dt \ominus f'(0)\\ &= s\mathbf{L}[f'(t)] \ominus f'(0).\end{aligned}$$

Now for any arbitrary fixed $\alpha \in [0,1]$, by using the parametric form and classical transform, we get

$$\begin{cases}\mathscr{L}(f_l^{''+}(t,\alpha)) = s.[s\mathscr{L}(f_l^+(t,\alpha)) - f_l^+(0,\alpha)] - f_l^{'+}(0,\alpha),\\ \mathscr{L}(f_r^{''+}(t,\alpha)) = s.[s\mathscr{L}(f_r^+(t,\alpha)) - f_r^+(0,\alpha)] - f_r^{'+}(0,\alpha),\\ \mathscr{L}(f_l^{''-}(t,\alpha)) = s.[s\mathscr{L}(f_l^-(t,\alpha)) - f_l^-(0,\alpha)] - f_l^{'-}(0,\alpha),\\ \mathscr{L}(f_r^{''-}(t,\alpha)) = s.[s\mathscr{L}(f_r^-(t,\alpha)) - f_r^-(0,\alpha)] - f_r^{'-}(0,\alpha).\end{cases}$$

Thus,

$$\begin{cases} \mathscr{L}(f_l^{''+}(t,\alpha)) = s^2\mathscr{L}(f_l^{+}(t,\alpha)) - sf_l^{+}(0,\alpha) - f_l^{'+}(0,\alpha), \\ \mathscr{L}(f_r^{''+}(t,\alpha)) = s^2\mathscr{L}(f_r^{+}(t,\alpha)) - sf_r^{+}(0,\alpha) - f_r^{'+}(0,\alpha), \\ \mathscr{L}(f_l^{''-}(t,\alpha)) = s^2\mathscr{L}(f_l^{-}(t,\alpha)) - sf_l^{-}(0,\alpha) - f_l^{'-}(0,\alpha), \\ \mathscr{L}(f_r^{''-}(t,\alpha)) = s^2\mathscr{L}(f_r^{-}(t,\alpha)) - sf_r^{-}(0,\alpha) - f_r^{'-}(0,\alpha). \end{cases}$$

Since $\left(\mathscr{L}(f_l^{+}(t,\alpha)), \mathscr{L}(f_r^{+}(t,\alpha)), \mathscr{L}(f_l^{-}(t,\alpha)), \mathscr{L}(f_r^{-}(t,\alpha))\right)$ is the parametric form of the intuitionistic fuzzy laplace transform of $f(t)$ that is

$$\mathbf{L}[f(t,\alpha)] = \left(\mathscr{L}(f_l^{+}(t,\alpha)), \mathscr{L}(f_r^{+}(t,\alpha)), \mathscr{L}(f_l^{-}(t,\alpha)), \mathscr{L}(f_r^{-}(t,\alpha))\right)$$

Then we have

$$\mathbf{L}[f''(t)] = s^2\mathbf{L}[f(t)] \ominus sf(0) \ominus f'(0).$$

□

Theorem 5 *Let $f(t)$, $g(t)$ be continuous intuitionistic fuzzy-valued functions on $[a,b]$ and c_1, c_2 two real constants, then*

$$\mathbf{L}[c_1 f(t) \oplus c_2 g(t)] = c_1\mathbf{L}[f(t)] \oplus c_2\mathbf{L}[g(t)].$$

Proof By the definition of the intuitionistic fuzzy laplace transform of $f(t)$ we have

$$\begin{aligned} \mathbf{L}[c_1 f(t) \oplus c_2 g(t)] &= \int_0^{\infty} (c_1 f(t) \oplus c_2 g(t))e^{-st}dt, s > 0 \\ &= \int_0^{\infty} (c_1 f(t)e^{-st} \oplus c_2 g(t)e^{-st})dt \end{aligned}$$

By the Theorem 4.3 in [7], we have the linearity of the integral that allows us to write

$$\begin{aligned} \int_0^{\infty} (c_1 f(t) \oplus c_2 g(t))e^{-st}dt &= \int_0^{\infty} c_1 f(t)e^{-st}dt \oplus \int_0^{\infty} c_2 g(t)e^{-st}dt, s > 0 \\ &= c_1\int_0^{\infty} f(t)e^{-st}dt \oplus c_2\int_0^{\infty} g(t)e^{-st}dt \end{aligned}$$

Then, we get

$$\int_0^{\infty} (c_1 f(t) \oplus c_2 g(t))e^{-st}dt = c_1\mathbf{L}[f(t)] \oplus c_2\mathbf{L}[g(t)] \tag{1}$$

Thus,

$$\mathbf{L}[c_1 f(t) \oplus c_2 g(t)] = c_1 \mathbf{L}[f(t)] \oplus c_2 \mathbf{L}[g(t)].$$

□

4 Constructing Solutions via Second Order Intuitionistic Fuzzy Initial Value Problem

In this section,we consider the following second order intuitionistic fuzzy initial value problem in general form

$$\begin{cases} y''(t) = f(t, y(t), y'(t)) \\ y(0) = \left(y_l^+(0,\alpha), y_r^+(0,\alpha)0, y_l^-(0,\alpha), y_r^-(0,\alpha)\right) \\ y'(0) = \left(y_l^{'+}(0,\alpha), y_r^{'+}(0,\alpha), y_l^{'-}(0,\alpha), y_r^{'-}(0,\alpha)\right) \end{cases} \tag{2}$$

By using intuitionistic fuzzy Laplace transform method we have:

$$\mathbf{L}[y''(t)] = \mathbf{L}[f(t, y(t), y'(t))] \tag{3}$$

By using Theorem 4, Eq. (3) can be written as follows:

$$s^2\mathbf{L}[y(t)] \ominus sy(0) \ominus y'(0) = \mathbf{L}[f(t, y(t), y'(t))]$$

Thus,

$$\begin{cases} s^2\mathscr{L}[y_l^+(t,\alpha)] - sy_l^+(0,\alpha) - y_l^{'+}(0,\alpha) = \mathscr{L}[f_l^+(t, y(t), y'(t), \alpha)], \\ s^2\mathscr{L}[y_r^+(t,\alpha)] - sy_r^+(0,\alpha) - y_r^{'+}(0,\alpha) = \mathscr{L}[f_r^+(t, y(t), y'(t), \alpha)], \\ s^2\mathscr{L}[y_l^-(t,\alpha)] - sy_l^-(0,\alpha) - y_l^{'-}(0,\alpha) = \mathscr{L}[f_l^-(t, y(t), y'(t), \alpha)], \\ s^2\mathscr{L}[y_r^-(t,\alpha)] - sy_r^-(0,\alpha) - y_r^{'-}(0,\alpha) = \mathscr{L}[f_r^-(t, y(t), y'(t), \alpha)]. \end{cases} \tag{4}$$

To solve the linear system (4), for simplicity we assume that:

$$\begin{cases} \mathscr{L}[y_l^+(t,\alpha)] = H_1(s,\alpha), \\ \mathscr{L}[y_r^+(t,\alpha)] = H_2(s,\alpha), \\ \mathscr{L}[y_l^-(t,\alpha)] = K_1(s,\alpha), \\ \mathscr{L}[y_r^-(t,\alpha)] = K_2(s,\alpha). \end{cases} \tag{5}$$

where $H_1(s,\alpha)$, $H_2(s,\alpha)$, $K_1(s,\alpha)$ and $K_2(s,\alpha)$ are solutions of system (4).

By using inverse classical Laplace transform, $y_l^+(t,\alpha)$, $y_r^+(t,\alpha)$, $y_l^-(t,\alpha)$ and $y_r^-(t,\alpha)$ are computed as follows

$$\begin{cases} y_l^+(t,\alpha) = \mathscr{L}^{-1}[H_1(s,\alpha)], \\ y_r^+(t,\alpha) = \mathscr{L}^{-1}[H_2(s,\alpha)], \\ y_l^-(t,\alpha) = \mathscr{L}^{-1}[K_1(s,\alpha)], \\ y_r^-(t,\alpha) = \mathscr{L}^{-1}[K_2(s,\alpha)]. \end{cases} \tag{6}$$

Example We consider the following intuitionistic fuzzy initial value problem (IFIVP)

$$\begin{cases} y''(t) = 2y'(t) + 3y(t) \\ y(0) = \left(3+\alpha, 5-\alpha, 2+2\alpha, 6-2\alpha\right) \\ y'(0) = \left(2+\alpha, 4-\alpha, 1+2\alpha, 5-2\alpha\right) \end{cases} \tag{7}$$

the parametric form of the IFIVP is given by the following form:

$$y''(t,\alpha) = 2y'(t,\alpha) + 3y(t,\alpha)$$

i.e

$$\begin{cases} y_l''^+(t,\alpha) = 2y_l'^+(t,\alpha) + 3y_l^+(t,\alpha), \\ y_r''^+(t,\alpha) = 2y_r'^+(t,\alpha) + 3y_r^+(t,\alpha), \\ y_l''^-(t,\alpha) = 2y_l'^-(t,\alpha) + 3y_l^-(t,\alpha), \\ y_r''^-(t,\alpha) = 2y_r'^-(t,\alpha) + 3y_r^-(t,\alpha), \end{cases} \tag{8}$$

Applying intuitionistic fuzzy Laplace Transform

$$\mathbf{L}[y''(t)] = 2\mathbf{L}[y'(t)] + 3\mathbf{L}[y(t)]$$

i.e,

$$\begin{cases} \mathscr{L}[y_l^{''+}(t,\alpha)] = 2\mathscr{L}[y_l^{'+}(t,\alpha)] + 3\mathscr{L}[y_l^{+}(t,\alpha)], \\ \mathscr{L}[y_r^{''+}(t,\alpha)] = 2\mathscr{L}[y_r^{'+}(t,\alpha)] + 3\mathscr{L}[y_r^{+}(t,\alpha)], \\ \mathscr{L}[y_l^{''-}(t,\alpha)] = 2\mathscr{L}[y_l^{'-}(t,\alpha)] + 3\mathscr{L}[y_l^{-}(t,\alpha)], \\ \mathscr{L}[y_r^{''-}(t,\alpha)] = 2\mathscr{L}[y_r^{'-}(t,\alpha)] + 3\mathscr{L}[y_r^{-}(t,\alpha)]. \end{cases} \tag{9}$$

Therefore,

$$\begin{cases} s^2\mathscr{L}[y_l^{+}(t,\alpha)] - sy_l^{+}(0,\alpha) - y_l^{'+}(0,\alpha)) = 2s\mathscr{L}[y_l^{+}(t,\alpha)] - 2y_l^{+}(0,\alpha) \\ +3\mathscr{L}[y_l^{+}(t,\alpha)], \\ s^2\mathscr{L}[y_r^{+}(t,\alpha)] - sy_r^{+}(0,\alpha) - y_r^{'+}(0,\alpha)) = 2s\mathscr{L}[y_r^{+}(t,\alpha)] - 2y_r^{+}(0,\alpha) \\ +3\mathscr{L}[y_r^{+}(t,\alpha)], \\ s^2\mathscr{L}[y_l^{-}(t,\alpha)] - sy_l^{-}(0,\alpha) - y_l^{'-}(0,\alpha)) = 2s\mathscr{L}[y_l^{-}(t,\alpha)] - 2y_l^{-}(0,\alpha) \\ +3\mathscr{L}[y_l^{-}(t,\alpha)], \\ s^2\mathscr{L}[y_r^{-}(t,\alpha)] - sy_r^{-}(0,\alpha) - y_r^{'-}(0,\alpha)) = 2s\mathscr{L}[y_r^{r}(t,\alpha)] - 2y_r^{-}(0,\alpha) \\ +3\mathscr{L}[y_r^{-}(t,\alpha)]. \end{cases} \tag{10}$$

Thus,

$$\begin{cases} \mathscr{L}[y_l^{+}(t,\alpha)] = \frac{1}{s^2-2s-3} y_l^{'+}(0,\alpha) + \frac{s-2}{s^2-2s-3} y_l^{+}(0,\alpha), \\ \mathscr{L}[y_r^{+}(t,\alpha)] = \frac{1}{s^2-2s-3} y_r^{'+}(0,\alpha) + \frac{s-2}{s^2-2s-3} y_r^{+}(0,\alpha), \\ \mathscr{L}[y_l^{-}(t,\alpha)] = \frac{1}{s^2-2s-3} y_l^{'-}(0,\alpha) + \frac{s-2}{s^2-2s-3} y_l^{-}(0,\alpha), \\ \mathscr{L}[y_r^{-}(t,\alpha)] = \frac{1}{s^2-2s-3} y_r^{'-}(0,\alpha) + \frac{s-2}{s^2-2s-3} y_r^{-}(0,\alpha). \end{cases} \tag{11}$$

After that we replace the value of parametric form of $y'(0)$ and $y(0)$ in the system (11), we have

$$\begin{cases} \mathscr{L}[y_l^{+}(t,\alpha)] = \frac{(2+\alpha)+(s-2)(3+\alpha)}{s^2-2s-3}, \\ \mathscr{L}[y_r^{+}(t,\alpha)] = \frac{(4-\alpha)+(s-2)(5-\alpha)}{s^2-2s-3}, \\ \mathscr{L}[y_l^{-}(t,\alpha)] = \frac{(1+2\alpha)+(s-2)(2+2\alpha)}{s^2-2s-3}, \\ \mathscr{L}[y_r^{-}(t,\alpha)] = \frac{(5-2\alpha)+(s-2)(6-2\alpha)}{s^2-2s-3}. \end{cases} \tag{12}$$

Now, After performing partition of fractions and by using inverse classical Laplace transform, $y_l^+(t,\alpha)$, $y_r^+(t,\alpha)$, $y_l^-(t,\alpha)$ and $y_r^-(t,\alpha)$ are computed as follows

$$\begin{cases} y_l^+(t,\alpha) = (\frac{2\alpha+7}{4})\mathscr{L}^{-1}[\frac{1}{s+1}] + (\frac{5+2\alpha}{4})\mathscr{L}^{-1}[\frac{1}{s-3}], \\ y_r^+(t,\alpha) = (\frac{-2\alpha+11}{4})\mathscr{L}^{-1}[\frac{1}{s+1}] + (\frac{-2\alpha+9}{4})\mathscr{L}^{-1}[\frac{1}{s-3}], \\ y_l^-(t,\alpha) = (\frac{4\alpha+5}{4})\mathscr{L}^{-1}[\frac{1}{s+1}] + (\frac{4\alpha+3}{4})\mathscr{L}^{-1}[\frac{1}{s-3}], \\ y_r^-(t,\alpha) = (\frac{-4\alpha+13}{4})\mathscr{L}^{-1}[\frac{1}{s+1}] + (\frac{-4\alpha+11}{4})\mathscr{L}^{-1}[\frac{1}{s-3}]. \end{cases} \tag{13}$$

Thus,

$$\begin{cases} y_l^+(t,\alpha) = (\frac{2\alpha+7}{4})e^{-t} + (\frac{5+2\alpha}{4})e^{3t}, \\ y_r^+(t,\alpha) = (\frac{-2\alpha+11}{4})e^{-t} + (\frac{-2\alpha+9}{4})e^{3t}, \\ y_l^-(t,\alpha) = (\frac{4\alpha+5}{4})e^{-t} + (\frac{4\alpha+3}{4})e^{3t}, \\ y_r^-(t,\alpha) = (\frac{-4\alpha+13}{4})e^{-t} + (\frac{-4\alpha+11}{4})e^{3t}. \end{cases} \tag{14}$$

We remark that $y_l^+(t,\alpha) \leq y_r^+(t,\alpha)$; $y_l^-(t,\alpha) \leq y_r^-(t,\alpha)$ also the functions $y_l^+(t,\alpha)$, $y_l^-(t,\alpha)$ are increasing with respect to α and the functions $y_r^+(t,\alpha)$, $y_r^-(t,\alpha)$ are decreasing with respect to α.

So, this we shown that $\left(y_l^+(t,\alpha), y_r^+(t,\alpha), y_l^-(t,\alpha), y_r^-(t,\alpha)\right)$ is the parametric form of the solution of the problem (8).

References

1. N. Ahmad, M. Mamat, J.K. Kumar, N.S. Amir, Hamzah, Solving fuzzy Duffing's equation by the laplace transform decomposition. Appl. Math. Sci. **6**(59), 2935–2944 (2012)
2. T. Allahviranloo, M. Barkhordari Ahmadi, Fuzzy laplace transforms. Soft Comput. **14**(3), 235–243 (2010)
3. T.M. Apostol, *Mathematical Analysis*, 2nd edn. (Addison-Welesy, Boston, 1974)
4. K. Atanassov, Intuitionistic fuzzy sets. Fuzzy Sets Syst. **20**, 87–96 (1986)
5. K. Atanassov, *Intuitionistic Fuzzy Sets: Theory and Applications* (Physica-Verlag, Berlin, 1999)
6. R. Ettoussi, S. Melliani, L.S. Chadli, Differential equation with intuitionistic fuzzy parameters. Notes Intuitionistic Fuzzy Sets **23**(4), 46–61 (2017)
7. R. Ettoussi, S. Melliani, M. Elomari, L.S. Chadli, Solution of intuitionistic fuzzy differential equations by successive approximations method. Notes Intuitionistic Fuzzy Sets **21**(2), 51–62 (2015)
8. M. Keyanpour, T. Akbarian, Solving intuitionistic fuzzy nonlinear equations. J. Fuzzy Set Valued Anal. **2014**, 1–6 (2014)
9. S. Melliani, M. Elomari, L.S. Chadli, R. Ettoussi, Intuitionistic Fuzzy metric spaces. Notes Intuitionistic Fuzzy Sets **21**(1), 43–53 (2015)

10. S. Melliani, L.S. Chadli, Introduction to intuitionistic fuzzy partial differential equations. Notes Intuitionistic Fuzzy Sets **7**(3), 39–42 (2001)
11. S. Melliani, L.S. Chadli, Introduction to intuitionistic fuzzy differential equations. Notes on Intuitionistic Fuzzy Sets **6**(2), 31–41 (2000)
12. S.J. Ramazannia Tolouti, M. Barkhordary Ahmadi, Fuzzy laplace transform on two order derivative and solving fuzzy two order differential equation. Int. J. Ind. Math. **2**(4), 279–293 (2010)
13. H.L. Royden, *Real Analysis*, 2nd edn. (Macmillan, New York, 1968)
14. L.A. Zadeh, Fuzzy Sets. Inf. Control **8**(3), 338–353 (1965)

Approximate Solution of Intuitionistic Fuzzy Differential Equations by Using Picard's Method

R. Ettoussi, Said Melliani and L. S. Chadli

Abstract Our main result in this paper is to find the power series solution of an intuitionistic fuzzy differential equation $x'(t) = f(t, x(t)), x(t_0) = x_0$ by using successive approximation method and we prove that the approximate solution converge uniformly in t to the exact solution. Finally, we illustrate this result with a numerical example.

1 Introduction

The concept of intuitionistic fuzzy is introduced by Attanasov (1984) [1, 2]. This concept is a generalization of fuzzy theory introduced by Zadeh [3]. Intutionistic fuzzy differential equation is very rare. Melliani and Chadli [4] solve partial differential equation with intutionistic fuzzy number. By the metric space defined in [5] we have something that makes sense to study this problem in intuitionistic fuzzy theory. In [6] the authors discussed the existence and uniqueness of solution of the intuitionistic fuzzy differential equation, using the method of successive approximation.

In this paper we considered Picard's approximation methods for finding approximate solution of an intuitionistic fuzzy differential equation

$$\begin{cases} x'(t) = f(t, x(t)) \\ x(t_0) = x_0 \end{cases} \tag{1}$$

where x_0 is an intuitionistic fuzzy quantity and $f : I \times \mathrm{IF}_1 \to \mathrm{IF}_1$ is levelwise continuous.

R. Ettoussi · S. Melliani (✉) · L. S. Chadli
Sultan Moulay Slimane University, BP 523 Beni Mellal, Morocco
e-mail: s.melliani@usms.ma; saidmelliani@gmail.com

R. Ettoussi
e-mail: razika.imi@gmail.com

L. S. Chadli
e-mail: sa.chadli@yahoo.fr

S. Melliani and O. Castillo (eds.), *Recent Advances in Intuitionistic Fuzzy Logic Systems*, Studies in Fuzziness and Soft Computing 372,
https://doi.org/10.1007/978-3-030-02155-9_14

At first, in Sect. 2, we give some definitions and properties regarding the concept of an intuitionistic fuzzy sets. The main results of this work is discussed in Sect. 3. Finally, we illustrated our theorem numerically by considering an example.

2 Preliminaries

Let us $T = [c, d] \subset \mathbb{R}$ be a compact interval. We denote by

$$\mathrm{IF}_1 = \mathrm{IF}(\mathbb{R}) = \left\{ \langle u, v \rangle \ : \ \mathbb{R} \to [0, 1]^2 \ , |\forall\, x \in \mathbb{R} 0 \leq u(x) + v(x) \leq 1 \right\}$$

An element $\langle u, v \rangle$ of IF_1 is said an intuitionistic fuzzy number if it satisfies the following conditions

(i) $\langle u, v \rangle$ is normal i.e there exists $x_0, x_1 \in \mathbb{R}$ such that $u(x_0) = 1$ and $v(x_1) = 1$.
(ii) u is fuzzy convex and v is fuzzy concave.
(iii) u is upper semi-continuous and v is lower semi-continuous.
(iv) $supp \, \langle u, v \rangle = \mathrm{cl}\{x \in \mathbb{R} \ : | \ v(x) < 1\}$ is bounded.

so we denote the collection of all intuitionistic fuzzy number by IF_1.

Definition 1 [5] An intuitionistic fuzzy number $\langle u, v \rangle$ in parametric form is a pair $\langle u, v \rangle = \left((\underline{\langle u, v \rangle}^+, \overline{\langle u, v \rangle}^+), (\underline{\langle u, v \rangle}^-, \overline{\langle u, v \rangle}^-) \right)$ of functions $\underline{\langle u, v \rangle}^-(\alpha)$, $\overline{\langle u, v \rangle}^-(\alpha)$, $\underline{\langle u, v \rangle}^+(\alpha)$ and $\overline{\langle u, v \rangle}^+(\alpha)$, which satisfies the following requirements:

1. $\underline{\langle u, v \rangle}^+(\alpha)$ is a bounded monotonic increasing continuous function,
2. $\overline{\langle u, v \rangle}^+(\alpha)$ is a bounded monotonic decreasing continuous function,
3. $\underline{\langle u, v \rangle}^-(\alpha)$ is a bounded monotonic increasing continuous function,
4. $\overline{\langle u, v \rangle}^-(\alpha)$ is a bounded monotonic decreasing continuous function,
5. $\underline{\langle u, v \rangle}^-(\alpha) \leq \overline{\langle u, v \rangle}^-(\alpha)$ and $\underline{\langle u, v \rangle}^+(\alpha) \leq \overline{\langle u, v \rangle}^+(\alpha)$, for all $0 \leq \alpha \leq 1$.

Example
A Triangular Intuitionistic Fuzzy Number (TIFN) $\langle u, v \rangle$ is an intuitionistic fuzzy set in $\mathbb{R}$ with the following membership function u and non-membership function v :

$$u(x) = \begin{cases} \dfrac{x - a_1}{a_2 - a_1} & \text{if } a_1 \leq x \leq a_2 \\ \dfrac{a_3 - x}{a_3 - a_2} & \text{if } a_2 \leq x \leq a_3, \\ 0 & \text{otherwise} \end{cases}$$

$$v(x) = \begin{cases} \frac{a_2 - x}{a_2 - a_1'} & \text{if } a_1' \le x \le a_2 \\ \frac{x - a_2}{a_3' - a_2} & \text{if } a_2 \le x \le a_3', \\ 1 & otherwise. \end{cases}$$

where $a_1' \le a_1 \le a_2 \le a_3 \le a_3'$.
This TIFN is denoted by $\langle u, v\rangle = \langle a_1, a_2, a_3; a_1', a_2, a_3'\rangle$.
Its parametric form is

$$\underline{\langle u, v\rangle}^+(\alpha) = a_1 + \alpha(a_2 - a_1), \quad \overline{\langle u, v\rangle}^+(\alpha) = a_3 - \alpha(a_3 - a_2)$$

$$\underline{\langle u, v\rangle}^-(\alpha) = a_1' + \alpha(a_2 - a_1'), \quad \overline{\langle u, v\rangle}^-(\alpha) = a_3' - \alpha(a_3' - a_2)$$

For $\alpha \in [0, 1]$ and $\langle u, v\rangle \in IF_1$, the upper and lower α-cuts of $\langle u, v\rangle$ are defined by

$$[\langle u, v\rangle]^\alpha = \{x \in \mathbb{R} : v(x) \le 1 - \alpha\}$$

and

$$[\langle u, v\rangle]_\alpha = \{x \in \mathbb{R} : u(x) \ge \alpha\}$$

Remark 1 If $\langle u, v\rangle \in IF_1$, so we can see $[\langle u, v\rangle]_\alpha$ as $[u]^\alpha$ and $[\langle u, v\rangle]^\alpha$ as $[1 - v]^\alpha$ in the fuzzy case.

We define $0_{\langle 1,0\rangle} \in IF_1$ as

$$0_{\langle 1,0\rangle}(t) = \begin{cases} \langle 1, 0\rangle & t = 0 \\ \langle 0, 1\rangle & t \ne 0 \end{cases}$$

For $\langle u, v\rangle, \langle z, w\rangle \in IF_1$ and $\lambda \in \mathbb{R}$, the addition and scaler-multiplication are defined as follows

$$\Big[\langle u, v\rangle \oplus \langle z, w\rangle\Big]^\alpha = \Big[\langle u, v\rangle\Big]^\alpha + \Big[\langle z, w\rangle\Big]^\alpha, \quad \Big[\lambda\langle z, w\rangle\Big]^\alpha = \lambda\Big[\langle z, w\rangle\Big]^\alpha$$
$$\Big[\langle u, v\rangle \oplus \langle z, w\rangle\Big]_\alpha = \Big[\langle u, v\rangle\Big]_\alpha + \Big[\langle z, w\rangle\Big]_\alpha, \quad \Big[\lambda\langle z, w\rangle\Big]_\alpha = \lambda\Big[\langle z, w\rangle\Big]_\alpha$$

Definition 2 Let $\langle u, v\rangle$ an element of IF_1 and $\alpha \in [0, 1]$, we define the following sets:

$$\Big[\langle u, v\rangle\Big]_l^+(\alpha) = \inf\{x \in \mathbb{R} \mid u(x) \ge \alpha\}, \quad \Big[\langle u, v\rangle\Big]_r^+(\alpha) = \sup\{x \in \mathbb{R} \mid u(x) \ge \alpha\}$$
$$\Big[\langle u, v\rangle\Big]_l^-(\alpha) = \inf\{x \in \mathbb{R} \mid v(x) \le 1 - \alpha\}, \quad \Big[\langle u, v\rangle\Big]_r^-(\alpha) = \sup\{x \in \mathbb{R} \mid v(x) \le 1 - \alpha\}$$

Remark 2

$$\Big[\langle u,v\rangle\Big]_\alpha = \Big[\Big[\langle u,v\rangle\Big]_l^+(\alpha), \Big[\langle u,v\rangle\Big]_r^+(\alpha)\Big]$$
$$\Big[\langle u,v\rangle\Big]^\alpha = \Big[\Big[\langle u,v\rangle\Big]_l^-(\alpha), \Big[\langle u,v\rangle\Big]_r^-(\alpha)\Big]$$

On the space IF_1 we will consider the following metric,

$$\begin{aligned} d_\infty\Big(\langle u,v\rangle, \langle z,w\rangle\Big) &= \frac{1}{4}\sup_{0<\alpha\leq 1}\Big|\Big[\langle u,v\rangle\Big]_r^+(\alpha) - \Big[\langle z,w\rangle\Big]_r^+(\alpha)\Big| \\ &+ \frac{1}{4}\sup_{0<\alpha\leq 1}\Big|\Big[\langle u,v\rangle\Big]_l^+(\alpha) - \Big[\langle z,w\rangle\Big]_l^+(\alpha)\Big| \\ &+ \frac{1}{4}\sup_{0<\alpha\leq 1}\Big|\Big[\langle u,v\rangle\Big]_r^-(\alpha) - \Big[\langle z,w\rangle\Big]_r^-(\alpha)\Big| \\ &+ \frac{1}{4}\sup_{0<\alpha\leq 1}\Big|\Big[\langle u,v\rangle\Big]_l^-(\alpha) - \Big[\langle z,w\rangle\Big]_l^-(\alpha)\Big| \end{aligned}$$

Theorem 1 *([7]) The metric space (IF_1, d_∞) is complete.*

Definition 3 A mapping $F : T \to \mathrm{IF}_1$ is called levelwise continuous at $t_0 \in T$ if the set-valued mappings $F_\alpha(t) = [F(t)]_\alpha$ and $F^\alpha(t) = [F(t)]^\alpha$ are continuous at $t = t_0$ with respect to the Hausdorff metric d_H for all $\alpha \in [0, 1]$.

Definition 4 A mapping $F : T \to \mathrm{IF}_1$ is said to be differentiable at t_0 if there exist $F'(t_0) \in \mathrm{IF}_1$ such that limits:

$$\lim_{\Delta t\to 0^+}\frac{F(t_0+\Delta t)\ominus F(t_0)}{\Delta t} \quad and \quad \lim_{\Delta t\to 0^+}\frac{F(t_0)\ominus F(t_0-\Delta t)}{\Delta t}$$

exist and they are equal to $F'(t_0) = \big\langle u'(t_0), v'(t_0)\big\rangle$.

Here the limit is taken in the metric space $(\mathrm{IF}_1, d_\infty)$. At the end points of T we consider only the one-sided derivatives.

If $F : T \to \mathrm{IF}_1$ is differentiable at $t_0 \in T$, then we say that $F'(t_0)$ is the intuitionistic fuzzy derivative of $F(t)$ at the point t_0.

Theorem 2 *[6] Let $F : T \to IF_1$ be levelwise continuous. Then for every $t \in T$ the integral $G(t) = \int_a^t F(s)ds$ is differentiable and $G'(t) = F(t)$.*

Theorem 3 *[6] Let $F : T \to IF_1$ be differentiable and assume that the derivative F' is integrable over T. Then, for each $s \in T$, we have*

$$F(s) = F(a) \oplus \int_a^s F'(t)dt. \tag{2}$$

3 Main Result

Assume that $f : I \times \mathrm{IF}_1 \to \mathrm{IF}_1$ is levelwise continuous, where the interval $I = \{t : |t - t_0| \leq a\}$.
Consider the intuitionistic fuzzy differential equation

$$\begin{cases} x'(t) = f(t, x(t)) \\ x(t_0) = x_0 \end{cases} \tag{3}$$

where $x_0 \in \mathrm{IF}_1$.
We denote $J_0 = I \times B(x_0, b)$ where $a > 0,\ b > 0\ ,\ x_0 \in \mathrm{IF}_1$

$$B(x_0, b) = \left\{x \in \mathrm{IF}_1 | d_\infty(x, x_0) \leq b\right\}$$

with the parametric representation of $x(t)$, $f(t, x(t))$ and x_0 are respectively

$$\Big((\underline{x}^+(\alpha, t), \overline{x}^+(\alpha, t)), (\underline{x}^-(\alpha, t), \overline{x}^-(\alpha, t))\Big),$$

$$\Big((\underline{f}^+(\alpha, t, x(\alpha, t)), \overline{f}^+(\alpha, t, x(\alpha, t)), (\underline{f}^-(\alpha, t, x(\alpha, t)), \overline{f}^-(\alpha, t, x(\alpha, t))\Big)$$

and

$$\Big((\underline{x}^+(\alpha, 0), \overline{x}^+(\alpha, 0)), (\underline{x}^-(\alpha, 0), \overline{x}^-(\alpha, 0))\Big).$$

Definition 5 A mapping $x : I \to \mathrm{IF}_1$ is a solution to the problem (3) if it is levelwise continuous and satisfies the integral equation

$$x(t) = x_0 \oplus \int_{t_0}^{t} f(s, x(s))ds. \ for\ all\ t \in I \tag{4}$$

i.e

$$\underline{x}^+(\alpha, t) = \underline{x}^+(\alpha, t_0) + \int_{t_0}^{t} \underline{f}^+(\alpha, s, x(\alpha, s))ds. \ for\ all\ t \in I$$

$$\overline{x}^+(\alpha, t) = \overline{x}^+(\alpha, t_0) + \int_{t_0}^{t} \overline{f}^+(\alpha, s, x(\alpha, s))ds. \ for\ all\ t \in I$$

$$\underline{x}^-(\alpha, t) = \underline{x}^-(\alpha, t_0) + \int_{t_0}^{t} \underline{f}^-(\alpha, s, x(\alpha, s))ds. \ for\ all\ t \in I$$

$$\overline{x}^-(\alpha, t) = \overline{x}^-(\alpha, t_0) + \int_{t_0}^{t} \overline{f}^-(\alpha, s, x(\alpha, s))ds. \ for\ all\ t \in I$$

Remark 3

- If we have a mapping $f : I \to \mathrm{IF}_1$ is continuous when IF_1 is endowed with the topology generated by the metric d_∞ implies that lower and upper α-cuts of f are continuous with respect to the Hausdorff metric d_H which means f is levelwise continuous. In fact, Let $\varepsilon > 0$ and $t_0 \in I$, by continuity of f there exists a $\delta > 0$ such that
$$d_\infty(F(t), F(t_0)) < \varepsilon \text{ whenever } \mid t - t_0 \mid < \delta$$
$$d_\infty(F(t), F(t_0)) < \varepsilon \Rightarrow$$
$$|[(F(t))]_r^+(\alpha) - [F(t_0)]_r^+(\alpha)| < \varepsilon \text{ and } |[F(t)]_l^+(\alpha) - [F(t_0)]_l^+(\alpha)| < \varepsilon$$
hence
$$\max\Big(|[(F(t))]_r^+(\alpha) - [F(t_0)]_r^+(\alpha)|; |[F(t)]_l^+(\alpha) - [F(t_0)]_l^+(\alpha)|\Big) = d_H\Big(F_\alpha(t), F_\alpha(t_0)\Big) < \varepsilon$$
whenever $\mid t - t_0 \mid < \delta$. So $F_\alpha(t)$ is continuous with respect to the Hausdorff metric. The same for $F^\alpha(t)$.
- The Theorems 2 and 3 prove the equivalence between Cauchy problem (3) and integral equation (4).

Theorem 4 *[8]*
Assume that f is levelwise continuous and there exists a constant $k > 0$ such that

$$|[f(s, x(s))]_r^+(\alpha) - [f(s, y(s))]_r^+(\alpha)| \le k|[x(s)]_r^+(\alpha) - [y(s)]_r^+(\alpha)|$$
$$|[f(s, x(s))]_l^+(\alpha) - [f(s, y(s))]_l^+(\alpha)| \le k|[x(s)]_l^+(\alpha) - [y(s)]_l^+(\alpha)|$$
$$|[f(s, x(s))]_r^-(\alpha) - [f(s, y(s))]_r^-(\alpha)| \le k|[x(s)]_r^-(\alpha) - [y(s)]_r^-(\alpha)|$$
$$|[f(s, x(s))]_l^-(\alpha) - [f(s, y(s))]_l^-(\alpha)| \le k|[x(s)]_l^-(\alpha) - [y(s)]_l^-(\alpha)|$$

with $k(T - t_0) < 1$, for all $s \in I$, $x, y \in IF_1$. Then the initial value problem (3) has an unique solution on I.

3.1 Picard's approximation Method

Let f be levelwise continuous in a region J_0 containing the point (t_0, x_0) By integrating both sides of the differential equation (3) with respect to t, we get

$$\underline{x}^+(\alpha, t) = \underline{c}^+ + \int_{t_0}^{t} \underline{f}^+(\alpha, s, x(\alpha, s))ds. \ for \ all \ t \in I$$

$$\overline{x}^+(\alpha, t) = \overline{c}^+ + \int_{t_0}^{t} \overline{f}^+(\alpha, s, x(\alpha, s))ds. \ for \ all \ t \in I$$

$$\underline{x}^{-}(\alpha,t)=\underline{c}^{-}+\int_{t_0}^{t}\underline{f}^{-}(\alpha,s,x(\alpha,s))ds.\ for\ all\ t\in I$$

$$\overline{x}^{-}(\alpha,t)=\overline{c}^{-}+\int_{t_0}^{t}\overline{f}^{-}(\alpha,s,x(\alpha,s))ds.\ for\ all\ t\in I$$

where $\underline{c}^{+}$, $\overline{c}^{+}$, $\underline{c}^{-}$ and $\overline{c}^{-}$ are constants.
We replace t by t_0, we obtain that $\underline{c}^{+}=\underline{x}^{+}(\alpha,t_0)$, $\overline{c}^{+}=\overline{x}^{+}(\alpha,t_0)$,$\underline{c}^{-}=\underline{x}^{-}(\alpha,t_0)$ and $\overline{c}^{-}=\overline{x}^{-}(\alpha,t_0)$. Thus,

$$\underline{x}^{+}(\alpha,t)=\underline{x}^{+}(\alpha,t_0)+\int_{t_0}^{t}\underline{f}^{+}(\alpha,s,x(\alpha,s))ds. \quad (3.2a)$$

$$\overline{x}^{+}(\alpha,t)=\overline{x}^{+}(\alpha,t_0)+\int_{t_0}^{t}\overline{f}^{+}(\alpha,s,x(\alpha,s))ds. \quad (3.2b)$$

$$\underline{x}^{-}(\alpha,t)=\underline{x}^{-}(\alpha,t_0)+\int_{t_0}^{t}\underline{f}^{-}(\alpha,s,x(\alpha,s))ds. \quad (3.2c)$$

$$\overline{x}^{-}(\alpha,t)=\overline{x}^{-}(\alpha,t_0)+\int_{t_0}^{t}\overline{f}^{-}(\alpha,s,x(\alpha,s))ds. \quad (3.2d)\ for\ all\ t\in I$$

Now to get power series solution of (3.2) we will use Picard's successive approximation method.

Suppose $x_0(t)$ be an arbitrary levelwise continuous function that represents an approximation to the solution of (3.2). Since $f(t,x_0(t))$ is a known function depending solely on x it can be integrated with $x(t)$ replaced by $x_0(t)$. The right hand side of (3.2) defines another function, which we write as

$$\underline{x}_1^{+}(\alpha,t)=\underline{x}^{+}(\alpha,t_0)+\int_{t_0}^{t}\underline{f}^{+}(\alpha,s,x_0(\alpha,s))ds.$$

$$\overline{x}_1^{+}(\alpha,t)=\overline{x}^{+}(\alpha,t_0)+\int_{t_0}^{t}\overline{f}^{+}(\alpha,s,x_0(\alpha,s))ds.$$

$$\underline{x}_1^{-}(\alpha,t)=\underline{x}^{-}(\alpha,t_0)+\int_{t_0}^{t}\underline{f}^{-}(\alpha,s,x_0(\alpha,s))ds.$$

$$\overline{x}_1^{-}(\alpha,t)=\overline{x}^{-}(\alpha,t_0)+\int_{t_0}^{t}\overline{f}^{-}(\alpha,s,x_0(\alpha,s))ds.$$

for all $t\in I$.

When we repeat the procedure, we obtain a sequence of functions $x_1(t), x_2(t), x_3(t)$, ...whose n-th term is defined by the relation

$$\underline{x}_n^+(\alpha, t) = \underline{x}^+(\alpha, t_0) + \int_{t_0}^{t} \underline{f}^+(\alpha, s, x_{n-1}(\alpha, s))ds.$$

$$\overline{x}_n^+(\alpha, t) = \overline{x}^+(\alpha, t_0) + \int_{t_0}^{t} \overline{f}^+(\alpha, s, x_{n-1}(\alpha, s))ds.$$

$$\underline{x}_n^-(\alpha, t) = \underline{x}^-(\alpha, t_0) + \int_{t_0}^{t} \underline{f}^-(\alpha, s, x_{n-1}(\alpha, s))ds.$$

$$\overline{x}_n^-(\alpha, t) = \overline{x}^-(\alpha, t_0) + \int_{t_0}^{t} \overline{f}^-(\alpha, s, x_{n-1}(\alpha, s))ds.$$

for all $t \in I$ $n = 1, 2, 3, \ldots$

Theorem 5 *Let the exact solution* $\left\{(\underline{X}^+(\alpha, t), \overline{X^+}(\alpha, t)), (\underline{X}^-(\alpha, t), \overline{X^-}(\alpha, t))\right\}$ *be approximated by* $\left\{(\underline{x}^+(\alpha, t), \overline{x^+}(\alpha, t)), (\underline{x}^-(\alpha, t), \overline{x^-}(\alpha, t))\right\}$. *For arbitrarily fixed* $0 < \alpha \leq 1$, *the solution of (3.1) using Picard's approximation method Sect. 3.1 converge to the exact solutions* $\underline{X}^+(\alpha, t)$, $\overline{X^+}(\alpha, t)$ $\underline{X}^-(\alpha, t)$, $\overline{X^-}(\alpha, t)$ *uniformly in t.*

Proof It is sufficient to show that

$$\underline{X}^+(\alpha, t) = \lim_{n\to\infty} \underline{x}_n^+(\alpha, t)$$

$$\overline{X}^+(\alpha, t) = \lim_{n\to\infty} \overline{x}_n^+(\alpha, t)$$

$$\underline{X}^-(\alpha, t) = \lim_{n\to\infty} \underline{x}_n^-(\alpha, t)$$

and

$$\overline{X}^-(\alpha, t) = \lim_{n\to\infty} \overline{x}_n^-(\alpha, t),$$

for all $0 \leq \alpha \leq 1$
Let the exact solution of (3.1) is

$$X(\alpha, t) = X(\alpha, 0) + \int_0^t f(\alpha, s, X(s))ds \tag{5}$$

and the solution of (3.1) using Picard's approximation method is

$$x_n(\alpha, t) = x(\alpha, 0) + \int_0^t f(\alpha, s, x_{n-1}(s))ds \tag{6}$$

Now, we denote

$$l_n^+ = \underline{X}^+(\alpha, t) - \underline{x}_n^+(\alpha, t)$$

$$u_n^+ = \overline{X}^+(\alpha, t) - \overline{x}_n^+(\alpha, t)$$

$$l_n^- = \underline{X}^-(\alpha, t) - \underline{x}_n^-(\alpha, t)$$

$$u_n^- = \overline{X}^-(\alpha, t) - \overline{x}_n^-(\alpha, t)$$

Using (5) and (6), we have

$$\begin{aligned} l_n^+ &= \underline{X}^+(\alpha, 0) + \int_0^t \underline{f}^+(\alpha, s, \underline{X}^+(s))ds - \underline{x}^+(\alpha, 0) - \int_0^t \underline{f}^+(\alpha, s, \underline{x}_{n-1}^+(s))ds \\ &= \underline{X}^+(\alpha, 0) - \underline{x}^+(\alpha, 0) + \int_0^t \Big(\underline{f}^+(\alpha, s, \underline{X}^+(s)) - \underline{f}^+(\alpha, s, \underline{x}_{n-1}^+(s))\Big)ds \\ &\leq l_0^+ + k\int_0^t \Big(\underline{X}^+(\alpha, s)) - \underline{x}_{n-1}^+(\alpha, s))\Big)ds \end{aligned}$$

Since $l_0^+ = \underline{X}^+(\alpha, 0) - \underline{x}^+(\alpha, 0) = 0$, $\underline{X}^+(\alpha, s)) - \underline{x}_0^+(\alpha, s) \leq Mt$ where $M = \underline{f}^+(\alpha, s, \underline{X}^+(s)) - \underline{0}^+_{\langle 1,0\rangle}(\alpha, s)$. We have $l_1^+ \leq KMt$ and $l_2^+ \leq \frac{MK^2t^2}{2}$.

So, we can assume $l_n^+ \leq \frac{MK^nt^n}{n!}$ and this can be easily proved by induction.

Therefore, $\lim\limits_{n\to\infty} l_n^+ \leq \lim\limits_{n\to\infty} \frac{MK^nt^n}{n!} = 0$ i.e $\lim\limits_{n\to\infty} l_n^+ = \lim\limits_{n\to\infty} \underline{X}^+(\alpha, t) - \underline{x}_n^+(\alpha, t) = 0$.

Thus, $\lim\limits_{n\to\infty} \underline{x}_n^+(\alpha, t) = \underline{X}^+(\alpha, t)$.

Similarly we can prove $\lim\limits_{n\to\infty} \overline{x}_n^+(\alpha, t) = \overline{X}^+(\alpha, t)$, $\lim\limits_{n\to\infty} \underline{x}_n^-(\alpha, t) = \underline{X}^-(\alpha, t)$ and $\lim\limits_{n\to\infty} \overline{x}_n^-(\alpha, t) = \overline{X}^-(\alpha, t)$.

4 Numerical Example

Consider the intuitionistic fuzzy initial value problem

$$\begin{cases} x'(t) = x(t), & t \in I = [0, 1] \\ x(0) = \Big((0.75 + 0.25\alpha, 1.125 - 0.125\alpha); (0.5 + 0.5\alpha, 2 - \alpha)\Big) \end{cases} \tag{6.1}$$

To solve this IFIVP first we transform the problem into crisp system of ordinary initial value problems

$$\begin{cases} \underline{x}'^{+}(\alpha,t) = \underline{x}^{+}(\alpha,t) & ; \ \underline{x}^{+}(\alpha,0) = 0.75 + 0.25\alpha \\ \overline{x}'^{+}(\alpha,t) = \overline{x}^{+}(\alpha,t), & ; \ \overline{x}^{+}(\alpha,0) = 0.25\alpha \\ \underline{x}'^{-}(\alpha,t) = \underline{x}^{-}(\alpha,t), & ; \ \underline{x}^{-}(\alpha,0) = 0.5 + 0.5\alpha \\ \overline{x}'^{-}(\alpha,t) = \overline{x}^{-}(\alpha,t), & ; \ \overline{x}^{-}(\alpha,0) = 2 - \alpha \end{cases}$$

the parametric form of the exact solution is

$$\begin{cases} \underline{X}^{+}(\alpha,t) = & \underline{x}^{+}(\alpha,0)\exp(t), \\ \overline{X}^{+}(\alpha,t) = & \overline{x}^{+}(\alpha,0)\exp(t), \\ \underline{X}^{-}(\alpha,t) = & \underline{x}^{-}(\alpha,0)\exp(t), \\ \overline{X}^{-}(\alpha,t) = & \overline{x}^{-}(\alpha,0)\exp(t) \end{cases}$$

for all $0 < \alpha \leq 1$.

The approximate solution using Picard's method for n-times is

$$\begin{cases} \underline{x}^{+}(\alpha,t) = & \underline{x}^{+}(\alpha,0)\left(1 + t + \frac{t^2}{2} + \frac{t^3}{3!} + ...\right), \\ \overline{x}^{+}(\alpha,t) = & \overline{x}^{+}(\alpha,0)\left(1 + t + \frac{t^2}{2} + \frac{t^3}{3!} + ...\right), \\ \underline{x}^{-}(\alpha,t) = & \underline{x}^{-}(\alpha,0)\left(1 + t + \frac{t^2}{2} + \frac{t^3}{3!} + ...\right), \\ \overline{x}^{-}(\alpha,t) = & \overline{x}^{-}(\alpha,0)\left(1 + t + \frac{t^2}{2} + \frac{t^3}{3!} + ...\right) \end{cases}$$

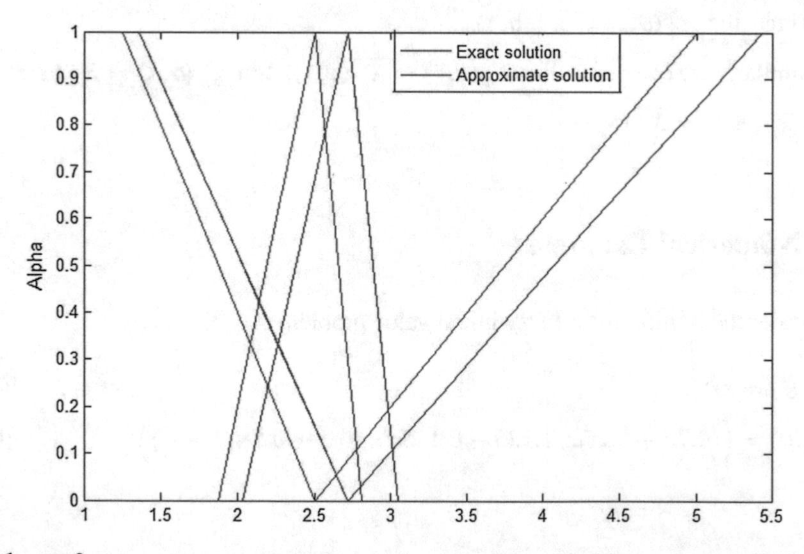

Fig. 1 $n = 2$

for all $0 < \alpha \leq 1$.
The exact and approximate solutions of IFIVP (5.1) for $n = 2, n = 3$ and $n = 4$ are compared and plotted at $t = 1$ shown in Figs. 1, 2 and 3.

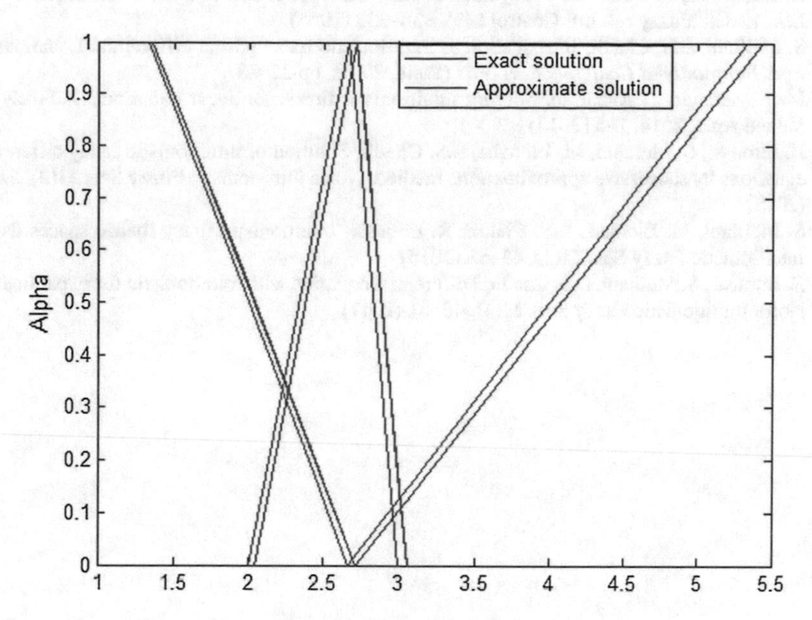

Fig. 2 $n = 3$

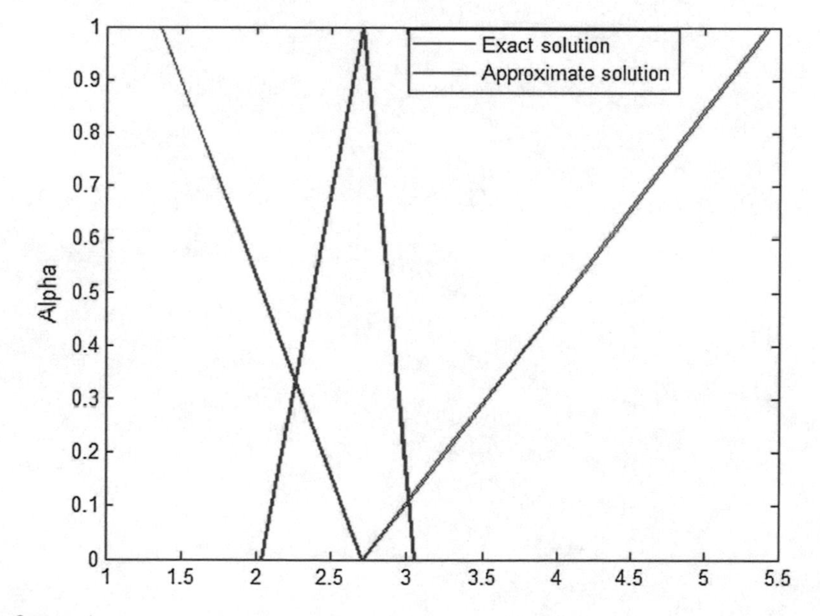

Fig. 3 $n = 4$

References

1. K. Atanassov, Intuitionistic fuzzy sets. Fuzzy Sets Syst. **20**, 87–96 (1986)
2. Atanassov K., Intuitionistic Fuzzy Sets. Theory and Applications, Physica-Verlag (1999)
3. L.A. Zadeh, Fuzzy set. Inf. Control **8**(3), 338–353 (1965)
4. S. Melliani, L.S. Chadli, Introduction to intuitionistic fuzzy partial differential Equations. in *Fifth International Conference on IFSs* (Sofia, 2001), pp 22–23
5. M. Keyanpour, T. Akbarian, Solving intuitionistic fuzzy nonlinear equations. J. Fuzzy Set Valued Anal. **2014**, 1–6 (2014)
6. R. Ettoussi, S. Melliani, M. Elomari, L.S. Chadli, Solution of intuitionistic fuzzy differential equations by successive approximations method. Notes Intuitionistic Fuzzy Sets **21**(2), 51–62 (2015)
7. S. Melliani, M. Elomari, L.S. Chadli, R. Ettoussi, Intuitionistic fuzzy metric spaces. Notes Intuitionistic Fuzzy Sets **21**(1), 43–53 (2015)
8. R. Ettoussi, S. Melliani, L.S. Chadli, Differential equation with intuitionistic fuzzy parameters. Notes Intuitionistic Fuzzy Sets **23**(4), 46–61 (2017)

Fuzzy Cross Language Plagiarism Detection Approach Based on Semantic Similarity and Hadoop MapReduce

H. Ezzikouri, M. Oukessou, M. Erritali and Y. Madani

Abstract Ranging from modifying texts into semantically equivalent up to translation and adopting ideas, without proper referencing to its originator, Cross Language Plagiarism can be of many different natures. Among the most common problems in any data processing system is reliable large-scale text comparison, especially in a fuzzy semantic based similarity due to the complexity of natural languages in particular Arabic, and the increasing number of publications which raise the rate of suspicious documents sources of plagiarism. CLPD is more complicated than monolingual plagiarism, it goes beyond copy+translate and paste, consequently the detecting process exposes the need for vague concept and fuzzy sets techniques in a big data environment to reveal dishonest practices of hidden plagiarism in Arabic documents translated from English or French sources. In this paper, we propose a fuzzy-semantic similarity for CLPD using WordNet taxonomy and three semantic approaches Wu and Palmer, Lin and Leacock-Chodorow for Arabic documents; the work has been parallelized using Apache Hadoop with HDFS file system and MapReduce programming model.

1 Introduction

Thanks to the rise of W3 and development of information technologies, information is within everyone's reach, people are able to generate more than you can imagine of data and information which lead to an explosive growth in the amount of data and the increase of originality and plagiarism issues. Big data is having a continuous exponential progress, intensified by the broadcast of digital information technologies, owing to this an already hard fuzzy task become even harder, Cross-Language Plagiarism detection (CLPD) consist of detecting plagiarism in documents from

H. Ezzikouri · M. Oukessou (✉) · M. Erritali · Y. Madani
Sultan Moulay Slimane University, BP 523 Beni Mellal, Morocco
e-mail: ouk_mohamed@yahoo.fr

H. Ezzikouri
e-mail: ezzikourihanane@gmail.com

S. Melliani and O. Castillo (eds.), *Recent Advances in Intuitionistic Fuzzy Logic Systems*, Studies in Fuzziness and Soft Computing 372,
https://doi.org/10.1007/978-3-030-02155-9_15

less-related languages such as English Arabic and French, in a massive amount of data seems to be impossible from the first sight even after applying some information retrieval systems that may generate thousands of candidate documents.

Cross-Language Plagiarism refers to the unacknowledged reuse of a text involving its translation from one natural language to another without proper referencing to the original source, its a sort of plagiarism idea, because texts are totally changed but ideas in the original texts remain unchanged; Such a change in the syntax and semantics of texts requires a deep and concentrated processing, then we confront two major factors, the management of large mass of data in all candidate documents and the number of operations required for this kind of plagiarism detection process. The nature of Cross-Language Plagiarism practices could be more complicated than simple copy translate and paste, in CLPD the languages from source and suspicious documents differ, thus the process exposes the need for a vague concept and fuzzy sets techniques to reveal dishonest practices in Arabic documents.

In this paper, we propose an detailed fuzzy semantic based similarity approach for analyzing and comparing texts in CLP cases using Big Data and WordNet lexical database (Miller, 1995) [1], to detect multilingual plagiarism in documents translated from English and French to Arabic. Arabic is known as one of the richest human languages in terms of words constructions and meanings diversity. We focus in our work on obfuscated plagiarism cases where texts are translated and rephrased from one language to another with no reference to the original source.

It's obvious that documents published in every field increases explosively, thus detecting plagiarism need a deep important treatment, therefore we need a large storage volume for storing all this data and also it is the problem of the required time to get results. To remedy this problem our proposal consists of parallelizing our method by working in a Big Data system with the Apache Hadoop using the HDFS (Hadoop Distributed File System) and the Hadoop MapReduce. Preliminary operations and text preprocessing are done to the documents such as tokenization, part-of-speech (POS) tagging, lemmatization and stop words removal, text segmentation (word 3-gram). The fuzzy semantic-based approach is based on the fact that words from two translated compared texts have in general, Strong fuzzy similarity words of the meaning from the second language

2 Related Work

The amount of generated data the frequency generated by on the web is incredible, produced the 'big data' term, defined by Gartner (Beyer and Douglas 2012) [2] as high volume, velocity and variety of data that demand cost-effective, innovative forms of processing for enhanced insight and decision making and the recently added Veracity and Value based on the fact that accurate analysis could be affected by the quality of captured data. Most research works have used big data in the information retrieval phase [3, 4]. Zhang et al. [5] Presented a sequence-based method to detect the partial similarity of web pages using MapReduce based on sentence level near-duplicate detection and sequence matching. Dwivedi et al. [6] introduced a SCAM (Standard Copy Analysis Mechanism) plagiarism detection algorithm, the proposed detection process is based on natural language processing by comparing documents and a modified Map-Reduce based SCAM algorithm for processing big data using Hadoop and detect plagiarism in big data.

3 Big Data a Solution for Fuzzy CLPD

Fuzzy set theory introduced by Lofti Zadeh in 1965 based on his mathematical theory of fuzzy sets, fuzzy set theory is a generalization of the theory of classical sets, and it permits the gradual assessment of the membership of elements in a set using a membership function valued in the real unit interval [0, 1]. Fuzzy set theory could be used in a wide range of domains especially for handling uncertain and imprecise data that linked with CLPD. In a cross language semantic based similarity detection process where words borders are not clear and the intersection of meanings of words are fuzzy, the fuzzy set theory seems to be the right way to treat such case. The huge masse of information and data generated from a voluminous corpuses, an efficient solution is the use of Big Data technologies to parallelize the fuzzy CLPD process in [7], the idea is to distribute and share it between several cluster, using Apache Hadoop framework with Hadoop distributed file system HDFS for distributing the storage of documents and also for storing results, and MapReduce programming model for the parallelisation and the development of our proposal.

CLP is a fuzzy complex process. Each word is associated with a fuzzy set that contains words with the same meaning with a similarity between 0 (for totally different) and 1 (for identic) (Fig. 1); thus fuzzy sets theory in CLPD looks to be an obvious way to solve the problem. Fuzzy set theory and CLPD turn up to be the perfect couple, however, the important number of operations and the running time implies to search for a solution to improve performance and results: is Big Data technologies.

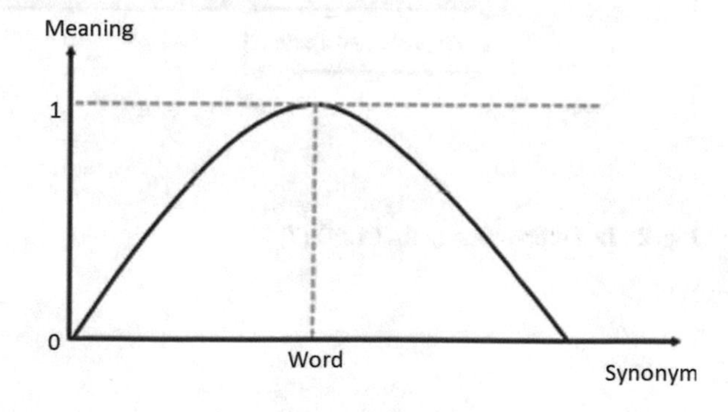

Fig. 1 A words fuzzy set synonyms

4 The Proposed Method

1. **Preprocessing and segmentation**

This paper is an extension and improvement of previous work by Ezzikouri et al. [7], where we present an intelligent multilingual plagiarism detection using Fuzzy-Semantic Similarity based methods (Wu and Palmer, Lin and Leacock-Chodorow) and Big data technologies (Hadoop, HDFS and MapReduce).

Input text and the collection corpus are from three distant languages Arabic-English and French, the creation of a suitable target data of each document is elementary, various text preprocessing methods based on NLP techniques are implemented (Fig. 2) and described in details in our previous work [7].

Semantic Similarity is defined here as the similarity between two concepts in a taxonomy(e.g. WordNet (Miller, 1995) [1]), where synonymous words are joined together to form synonyms sets called also synsets; Several similarity measures have been proposed in the last few years, LCH (Leacock and Chodorow, 1998), WuP (Wu and Palmer, 1994), RES (Resnik, 1995), LIN (Lin, 1998), LESK (Banerjee and Pedersen, 2003), and HSO (Hirst and St Onge, 1998) [8–12].

The proposed algorithm in this paper, is based on three semantic similarity approaches (Wu and Palmer, Lin and Leacock-Chodorow), that use WordNet to automatically evaluate semantic relations between words.

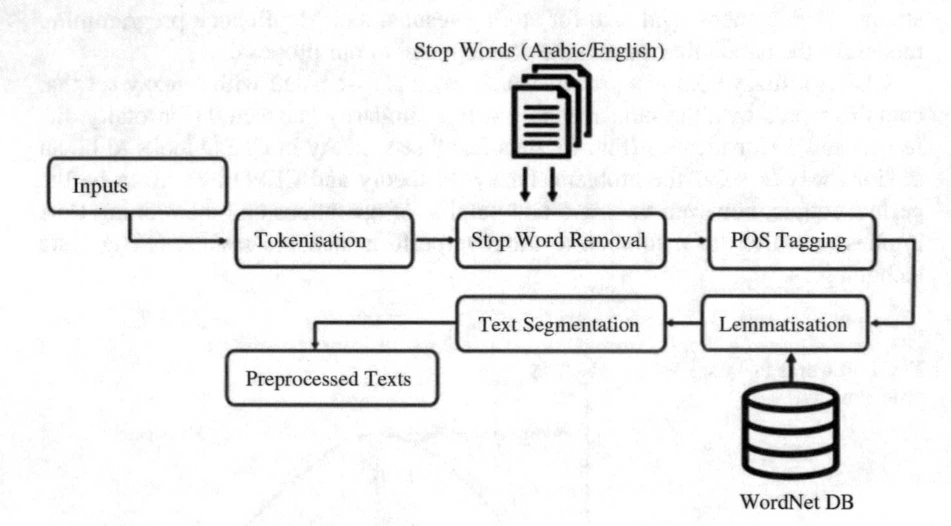

Fig. 2 Text preprocessing for CLPD [7]

2. **Complexity**

Leacock-Chodorow and Wu-Palmer are based on path length. They are based on counting the number of relation links between nodes in a taxonomy. Path length based measures are independent of corpus statistics and uninfluenced by sparse data. To compute the distance between two nodes in Wordnet, first we have to compute the shortest path between two nodes (length). This calculation process is equivalent to shortest path in a graph, i.e. the complexity could be estimated by $O(|V| + |E|)$ using a standard BFS approach where $|V|$ is the number of vertices and $|E|$ is the number of edges in the graph [13].

3. **Fuzzification**

Commonly two translated compared texts form a fuzzy similarity sets of words (differs in languages) sharing the same meaning class, based on that a Fuzzy semantic-based approach is obtained [4]. Many researches concentrate on text preprocessing methods especially Part-Of-Speech (POS) and its integration with fuzzy based methods for an efficient identification of similar documents [14].

Fuzzification is a fundamental operation in every fuzzy inference system, where relationship between inputs and linguistic variables is defined by a fuzzy membership function. In this paper we propose two semantic similarity approaches (Fuzzy-Lin and Fuzzy-Wup proposed in [7]) as fuzzy memberships function to fuzzify the relationship of word pairs (from text pairs), and compare results with Leacock-Chodorow (LCH).

The Lin and Wup fuzzy membership functions (expressed bellow Eq. 3) :

$$\mu_{1\,a_i\,b_j} = Lin(a_i, b_j) \quad (1)$$

$$\mu_{2\,a_i\,b_j} = Wup(a_i, b_j) \quad (2)$$

The membership degrees indicate to what extent an element belongs to a fuzzy set [15], i.e. for synonyms the value is 1, and 0 for dissimilar words.

To evaluate the similarity of two texts, a fuzzy inference system is needed. Fuzzy PROD operator is used for evaluating word relationship in first text with words in the second :

$$\begin{aligned} \mu_{a_1\,B} &= 1 - \prod_{\substack{b_j \in B \\ j \in [1,m]}} \left(1 - Wup(a_1, b_j)\right) \\ \mu_{a_n\,B} &= 1 - \prod_{\substack{b_j \in B \\ j \in [1,m]}} \left(1 - Wup(a_n, b_j)\right) \end{aligned} \quad (3)$$

The average sum is calculated by :

$$\mu_{A,B} = \left(\sum_{i=1}^{n} \mu_{a_i,B}\right) / n \quad (4)$$

4. Research methodology and Algorithm

Our main contribution in this article is fuzzifying cross language plagiarism detection using fuzzy semantic similarity approaches (fuzzy-WuP and fuzzy-Lin and LCH) in a parallel manner using Apache Hadoop (HDFS + MapReduce).

The idea is to store the inputs (the candidate documents (English, French) and the query document (Arabic text)) also results will be stored in HDFS for distributing the storage between several machines (Hadoop Cluster). Moreover, for the development of our proposal, MapReduce programming model was used (a parallel plagiarism's detection).

Inputs are from three different languages, an Arabic text and a corpus of potential candidates source of plagiarism in English and French. Some necessary steps must be done first like, text preprocessing, which contain several NLP processes (tokenization, stop words removal, post-tagging [16]) and word 3-grams (W3G) segmentation, this step is pivotal since Arabic has a complex morphology and one of the most difficult languages to treat. After that, then we prepare the inputs is the storage of them in HDFS (distributing the storage).

The resulting text (Arabic text) and every single text from the corpus (the English/French texts one by one) are used as inputs for the fuzzy inference system, then WuP Lin and LCH semantic similarity measures are modelled as membership functions. The output is a similarity score between the Arabic text and each input text from the corpus. Figure 3 shows the different steps of our work.

The MapReduce algorithm followed :

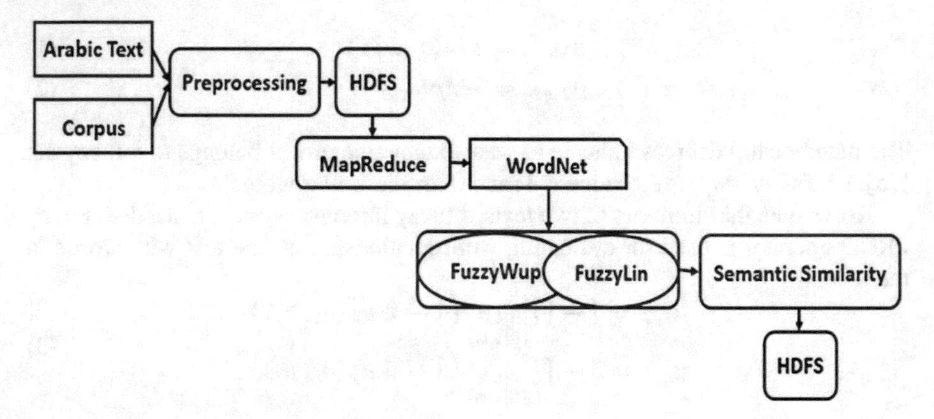

Fig. 3 Parallel Fuzzy CLPD system

Algorithm : MapReduce Programming Model for our proposed method

```
Inputs : Arabic Text , English text from the corpus
Require : Semantic Similarity between inputs
  S_Wup    0;
  S_Lin    0;
  S_LCH    0;
  C    0;
  AR      Text Preprocessing(Arabic Text);
  EN      Text Preprocessing(English Text);
  FR     Text Preprocessing(French Text);
  Segmentation1[]     W3G(AR);
  Segmentation2[]     W3G(EN);
  Segmentation3[]     W3G(FR);
  For all word In Segmentation1
     For all term In Segmentation2 and Segmentation3
        If word In WordNet
          Then word      WordNet(word);
        Else
          word     Translate(word);
        End If
        Fuzzy-Wup     1 - Wup(word, term);
        Fuzzy-Lin     1 - Lin(word, term);
        LCH     LCH(word, term);
        S_Wup     S_Wup + Fuzzy-Wup;
        S_Lin     S_Lin + Fuzzy-Lin;
        S_LCH     S_LCH + LCH;
        C     C + 1;
     End For
  End For
  Sim_Wup=S_Wup/C;
  Sim_Lin=S_Lin/C;
Write(Arabic Text || English (French) text, Sim_Wup /Sim_Lin/LCH)
```

The inputs storage distributing and plagiarism detection parallelising is done by constructing a Hadoop cluster that contains five machines Hadoop Nodes, this cluster has a master machine and four slave machines. Each node is an Ubuntu 16.04 machine.

5 Experimental Results and Discussion

Testing corpus is built up from 600 English/French and Arabic documents from different sources (news, articles, tweets, and academic works). To detect cross language plagiarism, 200 are translated from English/French to Arabic (machine based) with no change, and 400 documents are translated and modified with a high percentage of obfuscated plagiarism (paraphrasing, back-translation, etc.).

Fuzzy semantic metrics (Fuzzy WuP, Lin and LCH) are implemented and results are compared. Results presented in Figs. 4 and 5 are some of the experimental tests that demonstrate that the Fuzzy Wu and Palmer have high performance than Fuzzy Lin.

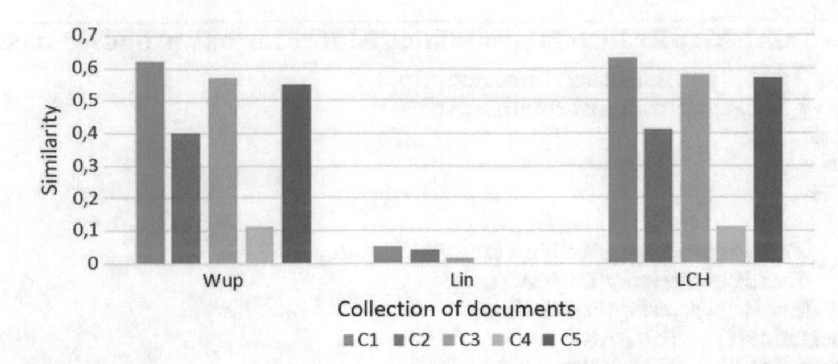

Fig. 4 Comparison of similarity for the proposed similarity measures

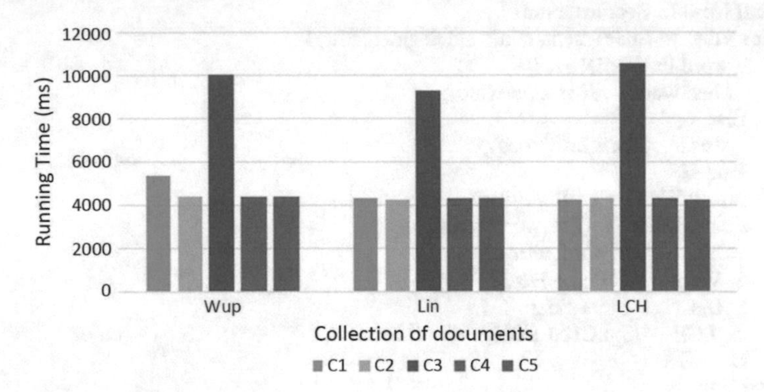

Fig. 5 Comparison of running time for the proposed similarity measures

6 Discussion

Based on the evaluation results and previous results in [8, 17, 18] and other ongoing works, Fuzzy-Lin gives good results in execution time but poor results for multilingual similarity and also it is not effective in detecting plagiarism in a big mass of data. Therefore Fuzzy-Lin is not suitable for detection multilingual plagiarism in high volumes of information. Fuzzy-WuP and LCH gives similar results in terms of similarity and execution time. For the execution time we notice a change in results of LCH for collections where documents are a little bit large. As said before our testing corpus contains several form of plagiarism ranging from simple translation to making serious changing in the text, which makes in addition to getting good results in detecting CLP, running time is also an important factor. For that we examined the two algorithms in terms of time with documents of different sizes (Fig. 6).

Fuzzy-WuP shows up to be the best semantic similarity measure for detection obfuscated plagiarism in a big data environment. Hence, no matter how the translated

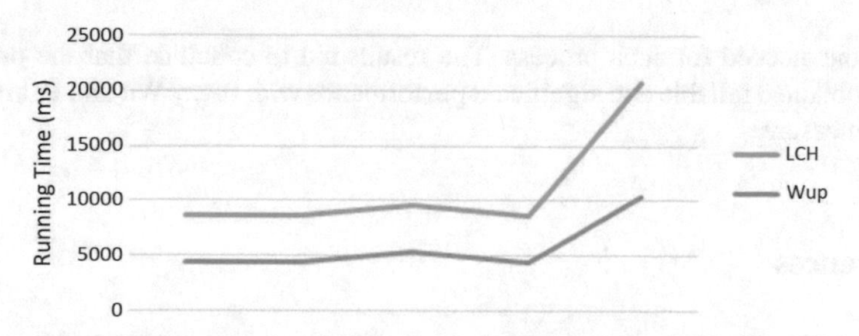

Fig. 6 Comparison of execution time by document size for WuP and LCH

plagiarized text is modified and the structure changed, such serious changing is the principal reason to involve fuzzy theory in plagiarism detection then similarity can still be detected.

The Parallel fuzzy-based Cross Language Plagiarism Detection using the lexical taxonomies WordNet in a big data environment presented in this paper achieves good results in comparison with some existing models and approaches namely fuzzy IR method in [19] such as word correlation factors obtained from large corpora that require allocation of disk space to save the word-to-word correlation factor tables. The processing time required to search for words and retrieve their correlation value is one of the main problems of former models and becomes a major issue for extending the use of parallel fuzzy-based CLPD approach, which have widely been reduced in our proposed model.

7 Conclusion

In this paper we have presented our Parallel fuzzy-based Cross Language Plagiarism Detection using WordNet and semantic similarity measures in a big data environment. Currently, most plagiarism detection tools are not suitable for detection a serious kind of plagiarism, where plagiarists are pushing to a high level using translation (human or machine based), paraphrasing, back-translation and a lots of manipulation to avoid to be caught with PD systems, obfuscated semantic plagiarism is a substantial issue and concern especially in academic works.

Different pre-processing methods based on NLP techniques were used (lemmatization, stop word removal and POS tagging), texts were segmented to 3-gram. Fuzzy semantic measures (Wu and Palmer, Lin and Leacock-Chodorow) were evaluated to judge the similarity in compared texts. Using a testing corpus of 600 handmade (rewording, paraphrasing, back-translation, idea adoption ...etc.) and artificial plagiarism cases, the fuzzy plagiarism detecting method using two fuzzy semantic similarity approaches (fuzzy WuP and fuzzy Lin) in a parallel manner using Apache Hadoop (HDFS+MapReduce), hadoop was used for performance enhancement and time reducing, data were distributed across the cluster of machines (master and four slaves); processing time doesnt increase in comparison of the important number of operations needed for such process. The results aid to conclude that the proposed model obtained reliable and significant performance with fuzzy Wu and Palmer similarity measure.

References

1. G.A. Miller, WordNet: a lexical database for English. Commun. ACM **38**(11), 39–41 (1995)
2. M.A. Beyer, D. Laney, The importance of big data : a definition, Stamford CT Gart., 2014–2018 (2012)
3. M.M. Najafabadi, F. Villanustre, T.M. Khoshgoftaar, N. Seliya, R. Wald, Muharemagic E., Deep learning applications and challenges in big data analytics. J. Big Data, **2** (2015)
4. B. Parhami, A Highly Parallel Computing System for Information Retrieval, in *Proceedings of the December 5–7, 1972, Fall Joint Computer Conference*, Part II (New York, NY, USA, 1972) pp. 681–690
5. Q. Zhang, Y. Zhang, H. Yu, X. Huang, Efficient partial-duplicate detection based on sequence matching, in *Proceedings of the 33rd International ACM SIGIR Conference on Research and Development in Information Retrieval*, pp. 675–682 (2010)
6. J. Dwivedi, A. Tiwary, Plagiarism detection on bigdata using modified map-reduced based SCAM algorithm. Int. Conference on Innovative Mechanisms Ind. Appl. (ICIMIA) **2017**, 608–610 (2017)
7. H. Ezzikouri, M. Erritali, M. Oukessou, Fuzzy-semantic similarity for automatic multilingual plagiarism detection. Int. J. Adv. Comput. Sci. Appl. **8**(9), 86–90 (2017)
8. C. Leacock, M. Chodorow, Combining local context and WordNet similarity for word sense identification. WordNet Electron. Lex. Database **49**(2), 265–283 (1998)
9. Z. Wu, M. Palmer, Verbs semantics and lexical selection, in *Proceedings of the 32nd Annual Meeting on Association for Computational Linguistics*, pp. 133–138 (1994)
10. P. Rensik, Using information content to evaluate semantic similarity, in *Proceedings of the 14th International Joint Conference on Artificial Intelligence*, pp. 448–453 (1995)
11. D. Lin, An information-theoretic definition of similarity, in Icml, **98**, 296–304 (1998)
12. G. Hirst, D. St-Onge, Lexical chains as representations of context for the detection and correction of malapropisms. WordNet Electron. Lex. Database **305**, 305–332 (1998)
13. T.H. Cormen, C.E. Leiserson, R.L. Rivest, C. Stein, Introduction to Algorithms-Secund Edition. (McGraw-Hill, 2001)
14. D. Gupta, K. Vani, C.K. Singh, Using Natural Language Processing techniques and fuzzy-semantic similarity for automatic external plagiarism detection, in *International Conference on Advances in Computing, Communications and Informatics* (ICACCI, 2014) pp. 2694–2699
15. N. Werro, *Fuzzy Classification of Online Customers* (Thesis University of Fribourg (Switzerland), Fuzzy Management Methods, 2008)
16. C. Manning, M. Surdeanu, J. Bauer, J. Finkel, S. Bethard, D. McClosky, The Stanford CoreNLP natural language processing toolkit, in *Proceedings of 52nd annual meeting of the association for computational linguistics: system demonstrations*, pp. 55–60 (2014)
17. H. Ezzikouri, M. Erritali, M. Oukessou, Semantic similarity/relatedness for cross language plagiarism detection. Indones. J. Electr. Eng. Comput. Sci. **1**(2), 371–374 (2016)
18. S. Alzahrani, N. Salim, Fuzzy semantic-based string similarity for extrinsic plagiarism detection. Braschler Harman **1176**, 1–8 (2010)
19. R. Yerra, Y.-K. Ng, A sentence-based copy detection approach for web documents, in *International Conference on Fuzzy Systems and Knowledge Discovery*. vol. 2005, pp. 557–570 (2005)

Existence Results for an Impulsive Fractional Integro-Differential Equations with a Non-compact Semigroup

K. Hilal, K. Guida, L. Ibnelazyz and M. Oukessou

Abstract In this paper we study a fractional differential equations problem with not instantaneous impulses involving a non-compact semigroup. We present some concepts and facts about the strongly continuous semigroup and the measure of noncompactness. After that we give an existence theorem of our problem using a condensing operator and the measure of noncompactness.

1 Introduction

The concept of fractional differential equations has become more popular among mathematicians, and is studied extensively in the recent years for its many applications, for more details about fractional differential equations we refer the readers to [1–7].

Furthermore, impulsive differential equations have known rapid growth because they play a main role in describing modern problems in fields such as physics, biology, economics and population dynamics; for more details the reader can see [8–10].

A lot of models about fractional impulsive differential equations were studied recently, for more details we give the references [11–13] and the references therein.

In [14] P. Chen, X. Zhang and Y. Li studied the existence of mild solutions for the initial value problem:

All authors contributed equally to the writing of this paper. All authors read and approved the final manuscript.

K. Hilal · K. Guida · L. Ibnelazyz · M. Oukessou (✉)
Sultan Moulay Slimane University, BP 523 Beni Mellal, Morocco
e-mail: ouk.mohamed@yahoo.fr

K. Guida
e-mail: guida.karim@gmail.com

L. Ibnelazyz
e-mail: ibnelazyzlahcen@gmail.com

S. Melliani and O. Castillo (eds.), *Recent Advances in Intuitionistic Fuzzy Logic Systems*, Studies in Fuzziness and Soft Computing 372,
https://doi.org/10.1007/978-3-030-02155-9_16

$$\begin{cases} x'(t) + Ax(t) = f(t, x(t)), \ t \in (s_i, t_{i+1}], i = 0, 1, 2, \ldots, m, \\ x(t) = g_i(t, x(t)), \ t \in (t_i, s_i], i = 1, 2, \ldots, m, \\ x(0) = x_0. \end{cases} \tag{1}$$

where $A : D(A) \subset E \longrightarrow E$ is a closed linear operator, $-A$ is the infinitesimal generator of a strongly continuous semigroup $(T(t))_{t\geq 0}$ in E, here the semigroup $(T(t))_{t\geq 0}$ is non-compact.

Motivated by this work, in this article we study the impulsive fractional evolution equation

$$\begin{cases} {}^c\mathcal{D}^{\alpha} x(t) = Ax(t) + f(t, x(t), Bx(t)) + C(t)u(t), \ t \in (s_i, t_{i+1}], \\ \qquad\qquad i = 0, 1, 2, \ldots, m, \ u \in U_{ad}, \\ {}^c\mathcal{D}^{\beta} x(t) = g_i(t, x(t)), \ t \in (t_i, s_i], i = 1, 2, \ldots, m, \\ x(0) = x_0, \end{cases} \tag{2}$$

involving ${}^c\mathcal{D}^{\alpha}$ and ${}^c\mathcal{D}^{\beta}$ which are the Caputo fractional derivatives of order $\alpha \in (0, 1)$ and $\beta \in (0, 1)$ respectively with the lower limit zero, $A : D(A) \subset X \longrightarrow X$ is the generator of a non-compact C_0-semigroup of bounded operators $(T(t))_{t\geq 0}$ on a Banach space $(X, \|\,.\,\|)$, $x_0 \in X$, $0 = t_0 = s_0 < t_1 < s_1 < t_2 < s_2 < \cdots < t_m \leq s_m < t_{m+1} = T$ are fixed numbers, $g_i \in C(J \times X, X)$, and $Bx(t) = \int_0^t B(t, s)x(s)ds$, $B \in C(D_1, \mathbb{R}^+)$, with $D_1 = \{(t, s) \in \mathbb{R}^2 : 0 \leq s \leq t \leq \omega\}$, $K_0 = \max\limits_{(t,s)\in D_1} B(t, s)$ and U_{ad} is a set that will be defined later.

The rest of the paper is organized as follows. In Sect. 2 we present the notations, definitions and preliminary results needed in the following sections. In Sect. 3, a suitable concept of PC-mild solutions for our problems is introduced. Section 4 is concerned with the existence results of our problems.

2 Preliminaries and Notations

Let us set $J = [0, T], J_0 = [0, t_1], J_1 = (t_1, t_2], \ldots, J_{m-1} = (t_{m-1}, t_m], J_m = (t_m, t_{m+1}]$ and introduce the space $PC(J, X) := \{x : J \to X : x \in C((t_i, t_{i+1}], X)$, $i = 0, 1, \ldots, m$ *and there exist* $x(t_i^-)$ *and* $x(t_i^+)$, $i = 1, \ldots, m$ *with* $x(t_i^-) = x(t_i^+)\}$. It is clear that $PC(J, X)$ is a Banach space with the norm $\|u\|_{PC} = sup\,\{\|u(t)\| : t \in J\}$. Let Y be a separable reflexive Banach space where controls u takes values, and $P_f(Y)$ is a class of nonempty closed and convex subsets of Y. We suppose that the multi-valued map $w : [0, a] \longrightarrow P_f(Y)$ is measurable, $w(.) \subset E$, where E is bounded set of Y, and the admissible control set $U_{ad} = \Big\{u \in L^p(E) : u(t) \in w(t), a.e\Big\}$, $p > \frac{1}{\tau}$, ($\tau \in (0, \alpha)$), for more details about admissible control set, we refer the readers to [15].

Let us recall the following well-known definitions.

Definition 1 [6] The Riemann-Liouville fractional integral of order q with lower limit zero for a function f is defined as

$$I^q f(t) = \frac{1}{\Gamma(q)} \int_0^t (t-s)^{q-1} f(s)ds, \ q > 0,$$

provided the integral exists, where $\Gamma(.)$ is the gamma function.

Definition 2 [6] The Riemann-Liouville derivative of order q with the lower limit zero for a function $f : [0, \infty) \to \mathbb{R}$ can be written as

$$^{L}D^q f(t) = \frac{1}{\Gamma(n-q)} \frac{d^n}{dt^n} \int_0^t (t-s)^{n-q-1} f(s)ds, \ n-1 < q < n, t > 0.$$

Definition 3 [6] The Caputo derivative of order q for a function $f : [0, \infty) \longrightarrow \mathbb{R}$ can be written as

$$^{C}D^q f(t) = {}^{L}D^q \left(f(t) - \sum_{k=0}^{k} \frac{t^k}{k!} f^k(0) \right), n-1 < q < n, t > 0.$$

Definition 4 [13] A function $x \in C(J, X)$ is said to be a mild solution of the following problem:

$$\begin{cases} {}^{C}D^{\alpha} x(t) = Ax(t) + y(t), \ t \in (0, T], \\ x(0) = x_0, \end{cases}$$

if it satisfies the integral equation

$$x(t) = P_\alpha(t)x_0 + \int_0^t (t-s)^{\alpha-1} Q_\alpha(t-s)y(s)ds,$$

Here

$$P_\alpha(t) = \int_0^\infty \xi_\alpha(\theta)T(t^\alpha\theta)d\theta, \ Q_\alpha(t) = \alpha \int_0^\infty \theta\xi_\alpha(\theta)T(t^\alpha\theta)d\theta,$$

$\xi_\alpha(\theta) = \frac{1}{\alpha}\theta^{-1-\frac{1}{\alpha}}\overline{\omega_\alpha}(\theta^{-\frac{1}{\alpha}}) \geq 0,$

$\overline{\omega_\alpha}(\theta) = \frac{1}{\pi}\sum_{n=1}^{\infty}(-1)^{n-1}\theta^{-n\alpha-1}\frac{\Gamma(n\alpha+1)}{n!}sin(n\pi\alpha), \theta \in (0, \infty)$, and $\xi_\alpha(\theta) \geq 0, \theta \in (0, \infty)$, $\int_0^\infty \xi_\alpha(\theta)d\theta = 1$.

It is easy to verify that $\int_0^\infty \theta\xi_\alpha(\theta)d\theta = \frac{1}{\Gamma(1+\alpha)}$.

Theorem 1 (Darbo-Sadovskii [16]) *if $D \subset X$ is bounded, closed and convex, the continuous map $S : D \to D$ is β- condensing, then S has a fixed point in D.*

Lemma 1 [17–19] *The operators P_α and Q_α have the following properties:*

(1) For any fixed $t \geq 0$, $P_\alpha(t)$ and $Q_\alpha(t)$ are linear and bounded operators, and for any $x \in X$,

$$\| P_\alpha(t)x \| \leq M_A \| x \|, \quad \| Q_\alpha(t)x \| \leq \frac{\alpha M_A}{\Gamma(1+\alpha)} \| x \|,$$

(2) $\{P_\alpha(t), t \geq 0\}$ and $\{Q_\alpha(t), t \geq 0\}$ are strongly continuous,

(3) if $(T(t))_{t\geq 0}$ is an equicontinuous semigroup, then $P_\alpha(t)$ and $Q_\alpha(t)$ are continuous in $(0, \infty)$ by the norm, which means that for $0 < t' < t'' < T$ we have:
$\| P_\alpha(t'') - P_\alpha(t') \| \longrightarrow 0$ *and* $\| Q_\alpha(t'') - Q_\alpha(t') \| \longrightarrow 0$ *as* $t'' \longrightarrow t'$.

Definition 5 [16] Let X be a Banach space and Ω_X be the bounded subsets of X. The Kuratowski measure of noncompactness is the map $\beta : \Omega_X \longrightarrow [0, \infty)$ defined by :

$$\beta(B) = \inf\left\{ \varepsilon > 0 : B = \cup_{i=1}^{n} B_i \text{ and } diam(B_i) \leq \varepsilon \text{ for } i = 1, 2, \ldots, n\right\}$$

with $diamB_i = \sup\{|x-y| : x, y \in B_i\}$ and $B \in \Omega_X$

Remark 1 It is clear that $\beta(B) \leq diam(B)$.

Next, we are going to look back on some properties of the measure of noncompactness that will be used in the proof of our main results.

Lemma 2 [16] *Let A and B be bounded sets of X and λ be a real number. Then the measure of noncompactness has the following properties:*

(1) $\beta(A) = 0$ if and only if A is a relatively compact set,
(2) $A \subset B$ implies that $\beta(A) \leq \beta(B)$,
(3) $\beta(\overline{A}) = \beta(A)$,
(4) $\beta(A \cup B) = max\{\beta(A), \beta(B)\}$,
(5) $\beta(\lambda A) = |\lambda|\beta(A)$,
(6) $\beta(A + B) \leq \beta(A) + \beta(B)$,
(7) $\beta(\overline{co}A) = \beta(A)$.

Where $\overline{co}$ means the closure of the convex hull.

Lemma 3 [20] *Let X be a Banach space, $W \subset X$ be bounded. Then there exists a countable set $W_1 \subset W$ such that*

$$\beta(W) \leq 2\beta(W_1).$$

Lemma 4 [16] *Let X be a Banach space, $W \subset C(J, X)$ be bounded and equicontinuous. Then $\beta(W(t))$ is continuous on J and*

$$\beta(W) = \max_{t\in J} \beta(W(t)) = \beta(W(J)).$$

Lemma 5 [21] *Let X be a Banach space, $W = \{u_n\} \in C(J, X)$ be bounded and countable set. Then $\beta(W(t))$ is a Lebesgue integral on J and*

$$\beta\left(\left\{\int_J u_n(t)dt : n \in \mathbb{N}\right\}\right) \leq 2\int_J \beta(W(t))dt.$$

3 The Construction of Mild Solutions

Let $x \in PC(J, X)$. We first consider the following fractional impulsive problem:

$$\begin{cases} {}^cD^{\alpha}x(t) = Ax(t) + f(t, x(t), Bx(t)) + C(t)u(t), \ t \in (s_i, t_{i+1}], \\ \qquad i = 0, 1, 2, \ldots, m, u \in U_{ad}, \\ {}^cD^{\beta}x(t) = g_i(t, x(t)), \ t \in (t_i, s_i], i = 1, 2, \ldots, m, \\ x(0) = x_0. \end{cases}$$

From the property of the Caputo derivative, a general solution can be written as

$$x(t) = \begin{cases} x_0 + \dfrac{1}{\Gamma(\alpha)}\displaystyle\int_0^t (t-s)^{\alpha-1}\left[Ax(s) + f(s, x(s), Bx(s)) + C(s)u(s)\right]ds, \\ \qquad t \in (0, t_1], \\ d_{1x} + \dfrac{1}{\Gamma(\beta)}\int_0^t (t-s)^{\beta-1} g_1(s, x(s))ds, \ t \in (t_1, s_1], \\ K_{1x} + \dfrac{1}{\Gamma(\alpha)}\displaystyle\int_0^t (t-s)^{\alpha-1}\left[Ax(s) + f(s, x(s), Bx(s)) + C(s)u(s)\right]ds, \\ \qquad t \in (s_1, t_2], \\ \vdots \\ d_{ix} + \dfrac{1}{\Gamma(\beta)}\int_0^t (t-s)^{\beta-1} g_i(s, x(s))ds, \ t \in (t_i, s_i], \\ K_{ix} + \dfrac{1}{\Gamma(\alpha)}\displaystyle\int_0^t (t-s)^{\alpha-1}\left[Ax(s) + f(s, x(s), Bx(s)) + C(s)u(s)\right]ds, \\ \qquad t \in (s_i, t_{i+1}], \end{cases}$$

where d_{ix} and K_{ix}, $i = 1, 2, \ldots, m$, are elements of X.

We obtain

$$x(t) = \begin{cases} d_{ix} + \dfrac{1}{\Gamma(\beta)}\int_0^t (t-s)^{\beta-1} g_i(s, x(s))ds, \ t \in (t_i, s_i], 1 \leq i \leq m, \\ P_\alpha(t-s_i)K_{ix} + \int_0^t (t-s)^{\alpha-1} Q_\alpha(t-s)\left[f(s, x(s), Bx(s)) + C(s)u(s)\right]ds, \\ \qquad t \in (s_i, t_{i+1}], 0 \leq i \leq m, \\ K_{0x} = x_0. \end{cases}$$

And using the fact that x is continuous at the points t_i, we get:

$$\begin{aligned}x(t_i) &= P_\alpha(t_i - s_{i-1})K_{(i-1)x} + \int_0^{t_i} (t_i - s)^{\alpha-1} Q_\alpha(t_i - s)\,[f(s, x(s), Bx(s)) \\ &\quad + C(s)u(s)]\,ds \\ &= d_{ix} + \frac{1}{\Gamma(\beta)} \int_0^{t_i} (t_i - s)^{\beta-1} g_i(s, x(s))ds.\end{aligned}$$

Which implies that:

$$\begin{aligned}d_{ix} &= P_\alpha(t_i - s_{i-1})K_{(i-1)x} + \int_0^{t_i} (t_i - s)^{\alpha-1} Q_\alpha(t_i - s)\,[f(s, x(s), Bx(s)) \\ &\quad + C(s)u(s)]\,ds - \frac{1}{\Gamma(\beta)} \int_0^{t_i} (t_i - s)^{\beta-1} g_i(s, x(s))ds.\end{aligned}$$

Using the fact that x is continuous at the points s_i, we get:

$$\begin{aligned}x(s_i) &= d_{ix} + \frac{1}{\Gamma(\beta)} \int_0^{s_i} (s_i - s)^{\beta-1} g_i(s, x(s))ds \\ &= K_{ix} + \int_0^{s_i} (s_i - s)^{\alpha-1} Q_\alpha(s_i - s)\,[f(s, x(s), Bx(s)) + C(s)u(s)]\,ds.\end{aligned}$$

Which implies that:

$$\begin{aligned}K_{ix} &= d_{ix} + \frac{1}{\Gamma(\beta)} \int_0^{s_i} (s_i - s)^{\beta-1} g_i(s, x(s))ds \\ &\quad - \int_0^{s_i} (s_i - s)^{\alpha-1} Q_\alpha(s_i - s)\,[f(s, x(s), Bx(s)) + C(s)u(s)]\,ds.\end{aligned}$$

Therefore, a mild solution of problem (1.2) is given by

$$x(t) = \begin{cases} P_\alpha(t)x_0 + \displaystyle\int_0^t (t-s)^{\alpha-1} Q_\alpha(t-s)\,[f(s, x(s), Bx(s)) + C(s)u(s)]\,ds, \\ \qquad\qquad t \in (0, t_1], \\ d_1 + \dfrac{1}{\Gamma(\beta)} \int_0^t (t-s)^{\beta-1} g_1(s, x(s))ds, \ t \in (t_1, s_1], \\ P_\alpha(t - s_1)K_1 + \displaystyle\int_0^t (t-s)^{\alpha-1} Q_\alpha(t-s)\,[f(s, x(s), Bx(s)) + C(s)u(s)]\,ds, \\ \qquad\qquad t \in (s_1, t_2], \\ \vdots \\ d_{ix} + \dfrac{1}{\Gamma(\beta)} \int_0^t (t-s)^{\beta-1} g_i(s, x(s))ds, \ t \in (t_i, s_i], \, 1 \le i \le m, \\ P_\alpha(t - s_i)K_{ix} + \int_0^t (t-s)^{\alpha-1} Q_\alpha(t-s)\,[f(s, x(s), Bx(s)) + C(s)u(s)]\,ds, \\ t \in (s_i, t_{i+1}], \, 1 \le i \le m, \end{cases}$$

where

$$
\begin{aligned}
K_{0x} &= x_0,\\
d_{ix} &= P_\alpha(t_i - s_{i-1})K_{(i-1)x} + \int_0^{t_i} (t_i - s)^{\alpha-1} Q_\alpha(t_i - s)\,[f(s, x(s), Bx(s))\\
&\quad + C(s)u(s)]\,ds - \frac{1}{\Gamma(\beta)}\int_0^{t_i} (t_i - s)^{\beta-1} g_i(s, x(s))ds,\\
K_{ix} &= d_{ix} - \int_0^{s_i} (s_i - s)^{\alpha-1} Q_\alpha(s_i - s)\,[f(s, x(s), Bx(s)) + C(s)u(s)]\,ds\\
&\quad + \frac{1}{\Gamma(\beta)}\int_0^{s_i} (s_i - s)^{\beta-1} g_i(s, x(s))ds.
\end{aligned}
$$

Definition 6 A function $x \in PC(J, X)$ is said to be a mild solution of problem (1.2) if it satisfies the following relation:

$$
x(t) = \begin{cases}
P_\alpha(t)K_{0x} + \int_0^t (t - s)^{\alpha-1} Q_\alpha(t - s)\,[f(s, x(s), Bx(s)) + C(s)u(s)]\,ds, \\
\qquad\qquad t \in (0, t_1],\ u \in U_{ad},\\
d_{ix} + \frac{1}{\Gamma(\beta)} \int_0^t (t - s)^{\beta-1} g_i(s, x(s))ds,\ t \in (t_i, s_i],\ 1 \le i \le m,\\
P_\alpha(t - s_i)K_{ix} + \int_0^t (t - s)^{\alpha-1} Q_\alpha(t - s)\,[f(s, x(s), Bx(s)) + C(s)u(s)]\,ds,\\
\qquad\qquad t \in (s_i, t_{i+1}],\ 1 \le i \le m.
\end{cases}
$$

where

$$
\begin{aligned}
K_{0x} &= x_0,\\
d_{ix} &= P_\alpha(t_i - s_{i-1})K_{(i-1)x} + \int_0^{t_i} (t_i - s)^{\alpha-1} Q_\alpha(t_i - s)\,[f(s, x(s), Bx(s))\\
&\quad + C(s)u(s)]\,ds - \frac{1}{\Gamma(\beta)}\int_0^{t_i} (t_i - s)^{\beta-1} g_i(s, x(s))ds,\\
K_{ix} &= d_{ix} - \int_0^{s_i} (s_i - s)^{\alpha-1} Q_\alpha(s_i - s)\,[f(s, x(s), Bx(s)) + C(s)u(s)]\,ds\\
&\quad + \frac{1}{\Gamma(\beta)}\int_0^{s_i} (s_i - s)^{\beta-1} g_i(s, x(s))ds.
\end{aligned}
$$

4 Existence Results

This section deals with the existence results of the problem (1.2).

To prove our first existence result we introduce the following assumptions.

(H_1) A generates an equicontinuous and uniformly bounded strongly continuous semigroup $T(t)_{t\geq 0}$ on a Banach space X such that $\| T(t) \| \leq M_A$ for all $t \in J$,

(H_2) $C : [0, T] \longrightarrow L(Y, X)$ is essentially bounded, i.e. $C \in L^\infty([0, T], L(Y, X))$,

(H_3) The function $f \in C(J \times X \times X, X)$,

(H_4)there exists $\tau \in (0, \alpha)$ and a positive function $m \in L^{\frac{1}{\tau}}(J, \mathbb{R})$ such that $\| f(t, u, v) \| \leq m(t)$, for $u, v \in X$ and $t \in J$,

(H_5) For $i = 1, 2, \ldots, m$, the function $g_i \in C([t_i, s_i] \times X; X)$ and there exists constants $K_i > 0$ such that:

$\| g_i(t, u) - g_i(t, v) \| \leq K_i \| u - v \|$ for all $u, v \in X$, and $K = \max\limits_{1 \leq i \leq m} \{K_i\}$

(H_5') For $i = 1, 2, \ldots, m$, $g_i \in C(J \times X; X)$ is completely continuous, and there exists constants $b, d > 0$ such that:

$\| g_i(t, u) \| \leq b_i \| u \| + d_i$, for all $u \in B_r$. B_r is a set that will be defined later, and $b = \max\limits_{1 \leq i \leq m} \{b_i\}$, $d = \max\limits_{1 \leq i \leq m} \{d_i\}$,

(H_6) There exists constants $L_1, L_2 > 0$ such that:

$\beta(f(t, D_1, D_2) \leq L_1 \beta(D_1) + L_2 \beta(D_2)$ for all $t \in J$ and D_1, D_2 bounded and countable sets in X.

Theorem 2 *Assume that* $(H_1) - (H_6)$ *hold. In addition, let's suppose that the following property is verified:*

$$max\{A, B\} < 1,$$

with

$$A = \left[\frac{(s_m^\beta + t_m^\beta) + M_A(s_{m-1}^\beta + t_{m-1}^\beta) + \cdots + M_A^{m-1}(s_1^\beta + t_1^\beta)}{\Gamma(\beta + 1)} \right] K$$
$$+ 4M_A(L_1 + \omega L_2 K_0) \left[\frac{t_m^\alpha + M_A(t_{m-1}^\alpha + s_{m-1}^\alpha) + \cdots + M_A^{m-1}(t_1^\alpha + s_1^\alpha)}{\Gamma(\alpha + 1)} \right]$$

and

$$B = \left[\frac{M_A(s_m^\beta + t_m^\beta) + M_A^2(s_{m-1}^\beta + t_{m-1}^\beta) + \cdots + M_A^m(s_1^\beta + t_1^\beta)}{\Gamma(\beta + 1)} \right] K$$
$$+ 4M_A(L_1 + \omega L_2 K_0) \left[\frac{t_{m+1}^\alpha + M_A(t_m^\alpha + s_m^\alpha) + \cdots + M_A^m(t_1^\alpha + s_1^\alpha)}{\Gamma(\alpha + 1)} \right]$$

Then the problem (1.2) *has at least one mild solution.*

Proof We introduce the composition $Q = Q_1 + Q_2$ where:

$$Q_1 x(t) = \begin{cases} P_\alpha(t)x_0 + \int_0^t (t - s)^{\alpha - 1} Q_\alpha(t - s) C(s) u(s) ds, \ t \in [0, t_1], \\ d_{i1x} + \dfrac{1}{\Gamma(\beta)} \int_0^t (t - s)^{\beta - 1} g_i(s, x(s)) ds, \ t \in (t_i, s_i], i = 1, 2, \ldots, m, \\ P_\alpha(t - s_i) K_{i1x} + \int_0^t (t - s)^{\alpha - 1} Q_\alpha(t - s) C(s) u(s) ds, \\ \qquad\qquad t \in (s_i, t_{i+1}], i = 1, 2, \ldots, m, \end{cases}$$

$$Q_2x(t) = \begin{cases} \int_0^t (t-s)^{\alpha-1} Q_\alpha(t-s) f(s, x(s), Bx(s))ds, \ t \in [0, t_1], \\ d_{i2x}, \ t \in (t_i, s_i], i = 1, 2, \ldots, m, \\ P_\alpha(t-s_i)K_{i2x} + \int_0^t (t-s)^{\alpha-1} Q_\alpha(t-s) f(s, x(s), Bx(s))ds, \\ \qquad\qquad\qquad\qquad t \in (s_i, t_{i+1}], i = 1, 2, \ldots, m, \end{cases}$$

with

$$\begin{cases} d_{i1x} = P_\alpha(t_i - s_{i-1})K_{(i-1)1x} + \int_0^{t_i} (t_i - s)^{\alpha-1} Q_\alpha(t_i - s)C(s)u(s)ds \\ \quad -\frac{1}{\Gamma(\beta)} \int_0^{t_i} (t_i - s)^{\beta-1} g_i(s, x(s))ds, \qquad i = 1, 2, \ldots, m, \\ K_{i1x} = d_{i1x} + \frac{1}{\Gamma(\beta)} \int_0^{s_i} (s_i - s)^{\beta-1} g_i(s, x(s))ds \\ \quad - \int_0^{s_i} (s_i - s)^{\alpha-1} Q_\alpha(s_i - s)C(s)u(s)ds, i = 1, 2, \ldots, m, \\ K_{01x} = x_0, \end{cases}$$

and

$$\begin{cases} d_{i2x} = P_\alpha(t_i - s_{i-1})K_{(i-1)2x} + \int_0^{t_i} (t_i - s)^{\alpha-1} Q_\alpha(t_i - s) f(s, x(s), Bx(s))ds, \\ \qquad i = 1, 2, \ldots, m, \\ K_{i2x} = d_{i2x} - \int_0^{s_i} (s_i - s)^{\alpha-1} Q_\alpha(s_i - s) f(s, x(x), Bx(s))ds, i = 1, 2, \ldots, m, \\ K_{02x} = 0. \end{cases}$$

Our proof will be divided into several steps.
Step 1: We show that $QB_r(J) \subset B_r(J)$
where $B_r = \{x \in PC(J, X); \| x \| \leq r\}$ the ball with radius $r > 0$;

$$K_{\alpha,\tau} = \left(\frac{1-\tau}{\alpha-\tau}\right)^{1-\tau} \| Cu \|_{L^{1/\tau}} \text{ and } S_{\alpha,\tau} = \left(\frac{1-\tau}{\alpha-\tau}\right)^{1-\tau} \| m \|_{L^{1/\tau}},$$

$$\gamma_1 = M_A^{m+1} \| x_0 \| \frac{\Gamma(\beta+1)}{\Gamma(\beta+1) - M_A K(t_1^\beta + s_1^\beta)},$$

$$\gamma_2 = \frac{(M_A t_{m+1}^{\alpha-\tau} + M_A^2(t_m^{\alpha-\tau} + s_m^{\alpha-\tau}) + \cdots + M_A^{m+1}(t_1^{\alpha-\tau} + s_1^{\alpha-\tau}))\Gamma(\beta+1)}{(\Gamma(\beta+1) - M_A K(t_1^\beta + s_1^\beta))\Gamma(\alpha)} (K_{\alpha,\tau} + S_{\alpha,\tau}),$$

$$\gamma_3 = \frac{M_A^2(t_{m-1}^\beta + s_{m-1}^\beta) + \cdots + M_A^m(t_1^\beta + s_1^\beta)}{(\Gamma(\beta+1) - M_A K(t_1^\beta + s_1^\beta))} \| g_m(t, 0) \|,$$

here $\gamma_1 + \gamma_1 + \gamma_1 \leq r$.
For any $x \in B_r$, we have:

Case 1. For $t \in [0, t_1]$

$$\| Qx(t) \| \leq \| P_\alpha(t)K_{0x} \| + \| \int_0^t (t-s)^{\alpha-1} Q_\alpha(t-s)\,[f(s, x(s), Bx(s)) + C(s)u(s)]\,ds \| \leq M_A \| x_0 \| + \frac{M_A t_1^{\alpha-\tau}}{\Gamma(\alpha)} \left(K_{\alpha,\tau} + S_{\alpha,\tau}\right).$$

Case 2. For $t \in (t_i, s_i]$, $i = 1, 2, \ldots, m$.
For $t \in (t_1, s_1]$

$$\begin{aligned}
\| Qx(t) \| &\leq \| d_{1x} \| + \| \frac{1}{\Gamma(\beta)} \int_0^t (t-s)^{\beta-1} g_1(s, x(s))ds \| \\
&\leq \| P_\alpha(t_1)K_{0x} \\
&+ \int_0^{t_1} (t_1-s)^{\alpha-1} Q_\alpha(t_1-s)\,[f(s, x(s), Bx(s)) + C(s)u(s)]\,ds \| \\
&+ \frac{1}{\Gamma(\beta)} \| \int_0^{t_1} (t_1-s)^{\beta-1} g_1(s, x(s))ds \| \\
&+ \| \frac{1}{\Gamma(\beta)} \int_0^t (t-s)^{\beta-1} g_1(s, x(s)ds \| \\
&\leq M_A \| x_0 \| + \frac{M_A t_1^{\alpha-\tau}}{\Gamma(\alpha)} \left(K_{\alpha,\tau} + S_{\alpha,\tau}\right) \\
&+ \frac{t_1^\beta + s_1^\beta}{\Gamma(\beta+1)} (K \| x \| + \| g_1(t, 0) \|) \\
&\leq r.
\end{aligned}$$

For $t \in (s_1, t_2]$

$$\begin{aligned}
\| Qx(t) \| &\leq \| P_\alpha(t-s_1)K_{1x} \| + \| \int_0^t (t-s)^{\alpha-1} Q_\alpha(t-s)\,[f(s, x(s), Bx(s)) \\
&+ C(s)u(s)]\,ds \| \leq \| P_\alpha(t-s_1)d_{1x} \| \\
&+ \| P_\alpha(t-s_1)(\int_0^{s_1} (s_1-s)^{\alpha-1} Q_\alpha(t-s)\,[f(s, x(s), Bx(s)) \\
&+ C(s)u(s)]\,ds + \frac{1}{\Gamma(\beta)} \int_0^{s_1} (s_1-s)^{\beta-1} g_1(s, x(s))ds) \| \\
&+ \| \int_0^t (t-s)^{\alpha-1} Q_\alpha(t-s)\,[f(s, x(s), Bx(s)) + C(s)u(s)]\,ds \| \\
&\leq M_A^2 \| x_0 \| \\
&+ \frac{M_A^2 (t_1^{\alpha-\tau} + s_1^{\alpha-\tau}) + M_A t_2^{\alpha-\tau}}{\Gamma(\alpha)} (K_{\alpha,\tau} + S_{\alpha,\tau})
\end{aligned}$$

$$+\frac{M_A(t_1^\beta+s_1^\beta)}{\Gamma(\beta+1)}(K\parallel x\parallel+\parallel g_1(t,0)\parallel)$$
$$\le r.$$

We suppose that: for $1\le j\le i$.
For $t\in(t_j,s_j]$

$$\begin{aligned}\parallel Qx(t)\parallel &\le M_A^j\parallel x_0\parallel\\ &+\frac{M_At_j^{\alpha-\tau}+M_A^2(t_{j-1}^{\alpha-\tau}+s_{j-1}^{\alpha-\tau})+\cdots+M_A^j(t_1^{\alpha-\tau}+s_1^{\alpha-\tau})}{\Gamma(\alpha)}(K_{\alpha,\tau}+S_{\alpha,\tau})\\ &+\frac{(t_j^\beta+s_j^\beta)+M_A(t_{j-1}^\beta+s_{j-1}^\beta)+\cdots+M_A^{j-1}(t_1^\beta+s_1^\beta)}{\Gamma(\beta+1)}\big(K\parallel x\parallel+\parallel g_j(t,0)\parallel\big)\\ &\le r.\end{aligned}$$

For $t\in(s_j,t_{j+1}]$

$$\begin{aligned}\parallel Qx(t)\parallel &\le M_A^{j+1}\parallel x_0\parallel\\ &+\frac{M_At_{j+1}^{\alpha-\tau}+M_A^2(t_j^{\alpha-\tau}+s_j^{\alpha-\tau})+\cdots+M_A^{j+1}(t_1^{\alpha-\tau}+s_1^{\alpha-\tau})}{\Gamma(\alpha)}(K_{\alpha,\tau}+S_{\alpha,\tau})\\ &+\frac{M_A(t_j^\beta+s_j^\beta)+M_A^2(t_{j-1}^\beta+s_{j-1}^\beta)+\cdots+M_A^j(t_1^\beta+s_1^\beta)}{\Gamma(\beta+1)}\big(K\parallel x\parallel+\parallel g_j(t,0)\parallel\big)\\ &\le r.\end{aligned}$$

And we prove the relations for $j=i+1$.
For $t\in(t_{i+1},s_{i+1}]$

$$\begin{aligned}\parallel Qx(t)\parallel &\le\parallel d_{(i+1)x}\parallel+\parallel\frac{1}{\Gamma(\beta)}\int_0^t(t-s)^{\beta-1}g_{i+1}(s,x(s)ds\parallel\\ &\le\parallel P_\alpha(t_{i+1}-s_i))K_{ix}+\int_0^{t_{i+1}}(t_{i+1}-s)^{\alpha-1}Q_\alpha(t_{i+1}-s)\,[f(s,x(s),Bx(s))\\ &+C(s)u(s)]\,ds\parallel+\parallel\frac{1}{\Gamma(\beta)}\int_0^{t_{i+1}}(t_{i+1}-s)^{\beta-1}g_{i+1}(s,x(s))ds\\ &+\parallel\frac{1}{\Gamma(\beta)}\int_0^t(t-s)^{\beta-1}g_{i+1}(s,x(s)ds\parallel\\ &\le M_A^{i+1}\parallel x_0\parallel+\frac{M_At_{i+1}^{\alpha-\tau}+M_A^2(t_i^{\alpha-\tau}+s_i^{\alpha-\tau})+\cdots+M_A^{i+1}(t_1^{\alpha-\tau}+s_1^{\alpha-\tau})}{\Gamma(\alpha)}(K_{\alpha,\tau}+S_{\alpha,\tau})\\ &+\frac{(t_{i+1}^\beta+s_{i+1}^\beta)+M_A(t_i^\beta+s_i^\beta)+\cdots+M_A^i(t_1^\beta+s_1^\beta)}{\Gamma(\beta+1)}(K\parallel x\parallel+\parallel g_{i+1}(t,0)\parallel)\\ &\le r.\end{aligned}$$

For $t \in (s_{i+1}, t_{i+2}]$

$$\begin{aligned}
&\| Qx(t) \| \leq M_A^{i+2} \| x_0 \| \\
&+\frac{M_A t_{i+2}^{\alpha-\tau} + M_A^2(t_{i+1}^{\alpha-\tau} + s_{i+1}^{\alpha-\tau}) + \cdots + M_A^{i+2}(t_1^{\alpha-\tau} + s_1^{\alpha-\tau})}{\Gamma(\alpha)}(K_{\alpha,\tau} + S_{\alpha,\tau}) \\
&+\frac{M_A(t_{i+1}^{\beta} + s_{i+1}^{\beta}) + M_A^2(t_i^{\beta} + s_i^{\beta}) + \cdots + M_A^{i+1}(t_1^{\beta} + s_1^{\beta})}{\Gamma(\beta+1)}\big(K \| x \| + \| g_{i+1}(t,0) \|\big) \\
&\leq r.
\end{aligned}$$

We proved that $QB_r(J) \subset B_r(J)$.
Step 2: Q_1 is lipschitz. Let $x, y \in PC(J, X)$,
Case 1. For $t \in [0, t_1]$, we have:

$$\| Q_1x(t) - Q_1y(t) \| = 0.$$

Similar to the proof on Step1, we prove that:
Case 2. For $t \in [t_i, s_i]$, $1 \leq i \leq m$,

$$\begin{aligned}
&\| Q_1x(t) - Q_1y(t) \| \\
&\leq \left[\frac{(s_i^{\beta} + t_i^{\beta}) + M_A(s_{i-1}^{\beta} + t_{i-1}^{\beta}) + \cdots + M_A^{i-1}(s_1^{\beta} + t_1^{\beta})}{\Gamma(\beta+1)} \right] K \| x - y \|_{PC} .
\end{aligned}$$

For $t \in [s_i, t_{i+1}]$, $1 \leq i \leq m$,

$$\begin{aligned}
&\| Q_1x(t) - Q_1y(t) \| \\
&\leq \left[\frac{M_A(s_i^{\beta} + t_i^{\beta}) + M_A^2(s_{i-1}^{\beta} + t_{i-1}^{\beta}) + \cdots + M_A^{i}(s_1^{\beta} + t_1^{\beta})}{\Gamma(\beta+1)} \right] K \| x - y \|_{PC} .
\end{aligned}$$

This implies that Q_1 is Lipschitz.
Step 3: Q_2 is continuous.
Let $(x_n)_{n\geq 0}$ be a sequence such that $\lim\limits_{x\to\infty} \| x_n - x \|_{PC} = 0$, we have:
Case 1. For $t \in [0, t_1]$

$$\begin{aligned}
&\| Q_2x_n(t) - Q_2x(t) \| \leq \| \textstyle\int_0^t (t-s)^{\alpha-1} Q_\alpha(t-s)\,(f(s, x_n(s), Bx_n(s)) \\
&\quad - f(s, x(s), Bx(s)))\, ds \| \leq \frac{M_A t_1^{\alpha}}{\Gamma(\alpha+1)} \| f(., x_n(.), Bx_n(.)) - f(., x(.), Bx(.)) \|_{PC} .
\end{aligned}$$

Similar to the proof we did in Step1, we prove that:

Case 2. For $t \in (t_i, s_i]$, $i = 1, 2, \ldots, m$,

$$\| Q_2x_n(t) - Q_2x(t) \| \leq \left[\frac{M_A t_i^\alpha + M_A^2(t_{i-1}^\alpha + s_{i-1}^\alpha) + \cdots + M_A^i(t_1^\alpha + s_1^\alpha)}{\Gamma(\alpha+1)} \right]$$
$$\| f(.,x_n(.),Bx_n(.)) - f(.,x(.),Bx(.)) \|_{PC}.$$

Case 3. For $t \in (s_i, t_{i+1})$, $i = 1, 2, \ldots, m$,

$$\| Q_2x_n(t) - Q_2x(t) \| \leq \left[\frac{M_A t_{i+1}^\alpha + M_A^2(t_i^\alpha + s_i^\alpha) + \cdots + M_A^{i+1}(t_1^\alpha + s_1^\alpha)}{\Gamma(\alpha+1)} \right]$$
$$\| f(.,x_n(.),Bx_n(.)) - f(.,x(.),Bx(.)) \|_{PC}.$$

Step 4: Q_2 is equicontinuous, which means $\| Q_2x(t_2) - Q_2x(t_1) \| \to 0$ as $t_2 \to t_1$.

For $0 \leq t' < t'' \leq t_1$, we have:

$$\begin{aligned}\| Q_2x(t'') - Q_2x(t') \| &\leq \| \int_0^{t''} (t''-s)^{\alpha-1} Q_\alpha(t''-s) f(s,x(s),Bx(s))ds \\ &\quad - \int_0^{t'} (t'-s)^{\alpha-1} Q_\alpha(t''-s) f(s,x(s),Bx(s))ds \| \\ &\leq I_1 + I_2 + I_3,\end{aligned}$$

where $I_1 = \| \int_{t'}^{t''} (t''-s)^{\alpha-1} Q_\alpha(t''-s) f(s,x(s),Bx(s))ds \|$,

$I_2 = \| \int_0^{t'} (t'-s)^{\alpha-1} \left[Q_\alpha(t''-s) - Q_\alpha(t'-s)\right] f(s,x(s),Bx(s))ds \|$,

$I_3 = \| \int_0^{t'} \left[(t''-s)^{\alpha-1} - (t'-s)^{\alpha-1}\right] Q_\alpha(t''-s) f(s,x(s),Bx(s))ds \|$,

$$\begin{aligned}I_1 &\leq \frac{\alpha M_A}{\Gamma(\alpha+1)} \int_{t'}^{t''} \| (t''-s)^{\alpha-1} f(s,x(s),Bx(s)) \| ds \\ &\leq \frac{M_A S_{\alpha,\tau}}{\Gamma(\alpha)} (t''-t')^{\alpha-\tau} \longrightarrow 0 \text{ as } t''-t' \longrightarrow 0,\end{aligned}$$

$I_1 \longrightarrow 0$ as $t'' - t' \longrightarrow 0$.

For $t' = 0$, $0 < t'' < t_1$, it is easy to see that $I_2 = 0$.

For $t' > 0$ and $\varepsilon > 0$ small enough, we have

$$\begin{aligned}I_2 &\leq \| \int_0^{t'-\varepsilon} (t'-s)^{\alpha-1} \left[Q_\alpha(t''-s) - Q_\alpha(t'-s)\right] f(s,x(s),Bx(s))ds \| \\ &\quad + \| \int_{t'-\varepsilon}^{t'} (t'-s)^{\alpha-1} \left[Q_\alpha(t''-s) - Q_\alpha(t'-s)\right] f(s,x(s),Bx(s))ds \| \\ &\leq \sup_{s\in[0,t'-\varepsilon]} \| Q_\alpha(t''-s) - Q_\alpha(t'-s) \| \int_0^{t'-\varepsilon} \| (t'-s)^{\alpha-1} f(s,x(s),Bx(s)) \| ds\end{aligned}$$

$$+\frac{2M_A}{\Gamma(\alpha)}\int_{t'-\varepsilon}^{t'} \| (t'-s)^{\alpha-1} f(s,x(s),Bx(s)) \| ds$$
$$\leq S_{\alpha,\tau}\left(t'^{\frac{\alpha-\tau}{1-\tau}} - \varepsilon^{\frac{\alpha-\tau}{1-\tau}}\right)^{1-\tau} \sup_{s\in[0,t'-\varepsilon]} \| Q_\alpha(t''-s) - Q_\alpha(t'-s) \| + \frac{2M_A}{\Gamma(\alpha)} S_{\alpha,\tau}\varepsilon^{\alpha-\tau}$$

$I_2 \longrightarrow 0$ as $t''-t' \longrightarrow 0$ and $\varepsilon \longrightarrow 0$.

$$\begin{aligned} I_3 &\leq \int_0^{t'} \| \left[(t''-s)^{\alpha-1} - (t'-s)^{\alpha-1}\right] Q_\alpha(t''-s) f(s,x(s),Bx(s)) \| ds \\ &\leq \frac{M_A S_{\alpha,\tau}}{\Gamma(\alpha)}\left((t''-t')^{\frac{\alpha-\tau}{1-\tau}} + t'^{\frac{\alpha-\tau}{1-\tau}} + t''^{\frac{\alpha-\tau}{1-\tau}}\right)^{1-\tau} \\ &\leq \frac{M_A S_{\alpha,\tau}}{\Gamma(\alpha)}(t''-t')^{\alpha-\tau} \longrightarrow 0;\ t''-t' \longrightarrow 0 \end{aligned}$$

$I_3 \longrightarrow 0$ as $t''-t' \longrightarrow 0$.

Case 1. For $t_i \leq t' < t'' \leq s_i$,

$$\| Q_2x(t'') - Q_2x(t') \| = 0.$$

Case 2. For $s_i \leq t' < t'' \leq t_{i+1}$,

$$\| Q_2x(t'') - Q_2x(t') \| \leq I_1 + I_2 + I_3 + \| \left(P_\alpha(t''-s_i) - P_\alpha(t'-s_i)\right) K_{i2x} \| . \quad (3)$$

The right-hand side of (4.1) tends to 0 independently of $x \in B_r$ as $t'' \to t'$.

Case 3. For $t_i \leq t' < s_i < t'' \leq t_{i+1}$,

$$\begin{aligned} \| Q_2x(t'') - Q_2x(t') \| &\leq \| P_\alpha(t''-s_i)K_{i2x} \\ &+ \int_0^{t''} (t''-s)^{\alpha-1} Q_\alpha(t''-s) f(s,x(s),Bx(s))ds - d_{i2x} \| \longrightarrow 0 \end{aligned}$$

independently of $x \in B_r$, as $t'' \longrightarrow t'$ we have $(t'' \longrightarrow s_i)$.

In conclusion, $\| Q_2x(t'') - Q_2x(t') \| \longrightarrow 0$, as $t''-t' \longrightarrow 0$, which implies that $Q_2(B_r(J))$ is equicontinuous.

We have $Q_2B_r \subseteq B_r$, where $Q_2B_r(t) = \{Q_2x(t); x \in B_r\}$ for $t \in J$.

Step 5: Q is β-condensing in B_r.

For any $W \subset B_r$, $Q_2(W)$ is bounded and equicontinuous. Hence, by Lemma (2.10), there exists a countable set $W_1 = \{u_n\}_{n=1}^{\infty} \subset W$ such that $\beta(Q_2(W)) \leq 2\beta(Q_2(W_1))$.

Since $Q_2(W_1) \subset B_r$ is equicontinuous, Lemma (2.11) implies $Q_2(W_1) = max\beta Q_2(W_1(t))$.

We have: $\beta\left(\left\{\int_0^t B(t,s)u(s)ds \big/ u \in B_r, t \in J\right\}_{n=1}^{\infty}\right) \leq \omega K_0 \beta\left(\{u(t)/u \in B_r, t \in J\}_{n=1}^{\infty}\right)$.

Case 1. For $t \in [0, t_1]$, we have:

$$\begin{aligned}
\beta(Q_2(W_1(t))) &= \beta\left(\left\{\int_0^t (t-s)^{\alpha-1} Q_\alpha(t-s) f(s, u_n(s), Bu_u(s))ds\right\}_{n=1}^{\infty}\right) \\
&\leq \frac{2M_A}{\Gamma(\alpha)} \int_0^t (t-s)^{\alpha-1} \beta\left(\{f(s, u_n(s), Bu_n(s))\}_{n=1}^{\infty}\right) ds \\
&\leq \frac{2M_A}{\Gamma(\alpha)} \int_0^t (t-s)^{\alpha-1} \left(L_1 \beta(W_1(s)) + L_2 \beta(B(W_1)(s))\right) ds \\
&\leq \frac{2M_A L_1}{\Gamma(\alpha)} \int_0^t (t-s)^{\alpha-1} \beta(W_1(s))ds \\
&\quad + \frac{2M_A L_2 \omega K_0}{\Gamma(\alpha)} \int_0^t (t-s)^{\alpha-1} \beta(W_1(s))ds \\
&\leq \frac{2M_A(L_1 + \omega L_2 K_0) t_1^{\alpha}}{\Gamma(\alpha+1)} \beta(W).
\end{aligned}$$

Since $Q_2(W_1)$ is bounded and equicontinuous, by Lemma (2.11)

$$\begin{aligned}
\beta(Q_2(W)) \leq 2\beta(Q_2(W_1)) &= 2 \max_{t \in J} \beta(Q_2(W_1(t))) \\
&\leq \frac{4M_A(L_1 + \omega L_2 K_0) t_1^{\alpha}}{\Gamma(\alpha+1)} \beta(W) \\
&< \beta(W).
\end{aligned}$$

In the other hand we have:
$\| Q_1 x(t) - Q_1 y(t) \| = 0$ which implies that $\beta(Q_1(W)) = 0$.
Then

$$\begin{aligned}
\beta(Q(W)) &\leq \beta(Q_1(W)) + \beta(Q_2(W)) \\
&\leq \frac{4M_A(L_1 + \omega L_2 K_0) t_1^{\alpha}}{\Gamma(\alpha+1)} \beta(W) < \beta(W).
\end{aligned}$$

Case 2. For $t \in (t_i, s_i]$, $i = 1, 2, \ldots, m$, we have:

$$\begin{aligned}
&\| Q_1 x(t) - Q_1 y(t) \| \\
&\leq \left[\frac{(s_i^{\beta} + t_i^{\beta}) + M_A(s_{i-1}^{\beta} + t_{i-1}^{\beta}) + \cdots + M_A^{i-1}(s_1^{\beta} + t_1^{\beta})}{\Gamma(\beta+1)}\right] K \| x - y \|_{PC}.
\end{aligned}$$

Hence, by Definition (2.7) we get:

$$\beta(Q_1(W)) \leq \left[\frac{(s_i^\beta + t_i^\beta) + M_A(s_{i-1}^\beta + t_{i-1}^\beta) + \cdots + M_A^{i-1}(s_1^\beta + t_1^\beta)}{\Gamma(\beta+1)}\right] K\beta(W).$$

On the other hand:

$$\beta(Q_2(W)) \leq 4M_A(L_1 + \omega L_2 K_0)\left[\frac{t_i^\alpha + M_A(t_{i-1}^\alpha + s_{i-1}^\alpha + \cdots + M_A^{i-1}(t_1^\alpha + s_1^\alpha)}{\Gamma(\alpha+1)}\right]\beta(W).$$

Then

$$\begin{aligned}
\beta(Q(W)) &\leq \beta(Q_1(W)) + \beta(Q_2(W)) \\
&\leq \left[\frac{(s_i^\beta + t_i^\beta) + M_A(s_{i-1}^\beta + t_{i-1}^\beta) + \cdots + M_A^{i-1}(s_1^\beta + t_1^\beta)}{\Gamma(\beta+1)}\right] K\beta(W) \\
&+ 4M_A(L_1 + \omega L_2 K_0)\left[\frac{t_i^\alpha + M_A(t_{i-1}^\alpha + s_{i-1}^\alpha) + \cdots + M_A^{i-1}(t_1^\alpha + s_1^\alpha)}{\Gamma(\alpha+1)}\right]\beta(W) \\
&< \beta(W).
\end{aligned}$$

Case 3. For $t \in [s_i, t_{i+1}], i = 1, 2, \ldots, m$ we have:

$$\begin{aligned}
&\| Q_1x(t) - Q_1y(t) \| \\
&\leq \left[\frac{M_A(s_i^\beta + t_i^\beta) + M_A^2(s_{i-1}^\beta + t_{i-1}^\beta) + \cdots + M_A^i(s_1^\beta + t_1^\beta)}{\Gamma(\beta+1)}\right] K \| x - y \|_{PC}.
\end{aligned}$$

Hence, by Definition (2.7) we get:

$$\beta(Q_1(W)) \leq \left[\frac{M_A(s_i^\beta + t_i^\beta) + M_A^2(s_{i-1}^\beta + t_{i-1}^\beta) + \cdots + M_A^i(s_1^\beta + t_1^\beta)}{\Gamma(\beta+1)}\right] K\beta(W).$$

On the other hand:

$$\beta(Q_2(W)) \leq 4M_A(L_1 + \omega L_2 K_0)\left[\frac{t_{i+1}^\alpha + M_A(t_i^\alpha + s_i^\alpha) + \cdots + M_A^i(t_1^\alpha + s_1^\alpha}{\Gamma(\alpha+1)}\right]\beta(W).$$

Then

$$
\begin{aligned}
\beta(Q(W)) &\leq \beta(Q_1(W)) + \beta(Q_2(W)) \\
&\leq \left[\frac{M_A(s_i^\beta + t_i^\beta) + M_A^2(s_{i-1}^\beta + t_{i-1}^\beta) + \cdots + M_A^i(s_1^\beta + t_1^\beta)}{\Gamma(\beta+1)}\right] K\beta(W) \\
&+ 4M_A(L_1 + \omega L_2 K_0)\left[\frac{t_{i+1}^\alpha + M_A(t_i^\alpha + s_i^\alpha + \cdots + M_A^i(t_1^\alpha + s_1^\alpha)}{\Gamma(\alpha+1)}\right]\beta(W) \\
&< \beta(W).
\end{aligned}
$$

Conclusion: in all cases we have:

$$
\beta(Q(W)) \leq \beta(Q_1(W)) + \beta(Q_2(W)) \leq C\beta(W) < \beta(W) \text{ with } C > 0.
$$

Since the operator Q is continuous and β-condensing. According to Darbo-Sadovskii's fixed point theorem, Q has a fixed point in B_r. Therefore, the problem (1.2) has at least one mild solution in B_r. This completes the proof.

Theorem 3 *Assume that* $(H_1) - (H_4)$,*and* $(H_5') - (H_6)$ *hold. In addition, let's suppose that the following property is verified:*

$$
max\{C, D\} < 1
$$

where $C = 4M_A(L_1 + \omega L_2 K_0)\left[\frac{t_m^\alpha + M_A(t_{m-1}^\alpha + s_{m-1}^\alpha) + \cdots + M_A^{m-1}(t_1^\alpha + s_1^\alpha)}{\Gamma(\alpha+1)}\right]$,
and $D = 4M_A(L_1 + \omega L_2 K_0)\left[\frac{t_{m+1}^\alpha + M_A(t_m^\alpha + s_m^\alpha) + \cdots + M_A^m(t_1^\alpha + s_1^\alpha)}{\Gamma(\alpha+1)}\right]$.

Then the problem (1.2) *has at least one mild solution.*

Proof We introduce the composition $Q = Q_1 + Q_2$ where:

$$
Q_1x(t) = \begin{cases}
P_\alpha(t)x_0 + \int_0^t (t-s)^{\alpha-1} Q_\alpha(t-s)C(s)u(s)ds, \; t \in [0, t_1], \\
d_{i1x} + \dfrac{1}{\Gamma(\beta)} \int_0^t (t-s)^{\beta-1} g_i(s, x(s))ds, \;\; t \in (t_i, s_i], \; i = 1, 2, \ldots, m, \\
P_\alpha(t-s_i)K_{i1x} + \int_0^t (t-s)^{\alpha-1} Q_\alpha(t-s)C(s)u(s)ds, \\
\qquad\qquad t \in (s_i, t_{i+1}], \; i = 1, 2, \ldots, m,
\end{cases}
$$

$$Q_2x(t) = \begin{cases} \int_0^t (t-s)^{\alpha-1} Q_\alpha(t-s) f(s,x(s),Bx(s))ds, \ t \in [0,t_1], \\ d_{i2x}, \ t \in (t_i, s_i], \ i = 1,2,\ldots,m, \\ P_\alpha(t-s_i)K_{i2x} + \int_0^t (t-s)^{\alpha-1} Q_\alpha(t-s) f(s,x(s),Bx(s))ds, \\ \qquad\qquad t \in (s_i, t_{i+1}], \ i = 1,2,\ldots,m, \end{cases}$$

with

$$\begin{cases} d_{i1x} = P_\alpha(t_i - s_{i-1})K_{(i-1)1x} + \int_0^{t_i} (t_i - s)^{\alpha-1} Q_\alpha(t_i - s)C(s)u(s)ds \\ \qquad -\frac{1}{\Gamma(\beta)} \int_0^{t_i} (t_i - s)^{\beta-1} g_i(s,x(s))ds, \ i = 1,2,\ldots,m, \\ K_{i1x} = d_{i1x} + \frac{1}{\Gamma(\beta)} \int_0^{s_i} (s_i - s)^{\beta-1} g_i(s,x(s))ds \\ \qquad - \int_0^{s_i} (s_i - s)^{\alpha-1} Q_\alpha(s_i - s)C(s)u(s)ds, \ i = 1,2,\ldots,m, \\ K_{01x} = x_0, \end{cases}$$

and

$$\begin{cases} d_{i2x} = P_\alpha(t_i - s_{i-1})K_{(i-1)2x} + \int_0^{t_i} (t_i - s)^{\alpha-1} Q_\alpha(t_i - s) f(s,x(s),Bx(s))ds, \\ \qquad i = 1,2,\ldots,m, \\ K_{i2x} = d_{i2x} - \int_0^{s_i} (s_i - s)^{\alpha-1} Q_\alpha(s_i - s) f(s,x(x),Bx(s))ds, \ i = 1,2,\ldots,m, \\ K_{02x} = 0. \end{cases}$$

Our proof will be divided into several steps.
Step 1: We show that $QB_r(J) \subset B_r(J)$,
where $B_r = \{x \in PC(J,X); \parallel x \parallel \leq r\}$ the ball with radius $r > 0$;

$$K_{\alpha,\tau} = \alpha \left(\frac{1-\tau}{\alpha-\tau}\right)^{1-\tau} \parallel Cu \parallel_{L^{1/\tau}} \text{ and } S_{\alpha,\tau} = \alpha \left(\frac{1-\tau}{\alpha-\tau}\right)^{1-\tau} \parallel m \parallel_{L^{1/\tau}},$$

$$\gamma_1 = M_A^{m+1} \parallel x_0 \parallel \frac{\Gamma(\beta+1)}{\Gamma(\beta+1) - M_A K(t_1^\beta + s_1^\beta)},$$

$$\gamma_2 = \frac{(M_A t_{m+1}^{\alpha-\tau} + M_A^2(t_m^{\alpha-\tau} + s_m^{\alpha-\tau}) + \cdots + M_A^{m+1}(t_1^{\alpha-\tau} + s_1^{\alpha-\tau}))\Gamma(\beta+1)}{(\Gamma(\beta+1) - M_A K(t_1^\beta + s_1^\beta))\Gamma(\alpha)}(K_{\alpha,\tau} + S_{\alpha,\tau}),$$

$$\gamma_3 = \frac{M_A^2(t_{m-1}^\beta + s_{m-1}^\beta) + \cdots + M_A^m(t_1^\beta + s_1^\beta)}{(\Gamma(\beta+1) - M_A K(t_1^\beta + s_1^\beta))} d.$$

Here $\gamma_1 + \gamma_1 + \gamma_1 \leq r$.
For any $x \in B_r$, we have:

Case 1. For $t \in [0, t_1]$

$$\| Qx(t) \| \leq \| P_\alpha(t)K_{0x} \| + \| \int_0^t (t-s)^{\alpha-1} Q_\alpha(t-s)\,[f(s, x(s), Bx(s)) + C(s)u(s)]\,ds \| \leq M_A \| x_0 \| + \frac{M_A t_1^{\alpha-\tau}}{\Gamma(\alpha)} \left(K_{\alpha,\tau} + S_{\alpha,\tau}\right).$$

Similar to the proof of the previous theorem we show that:
for $t \in (t_i, s_i]$

$$\begin{aligned}\| Qx(t) \| &\leq M_A^i \| x_0 \| \\ &+ \frac{M_A t_i^{\alpha-\tau} + M_A^2(t_{i-1}^{\alpha-\tau} + s_{i-1}^{\alpha-\tau}) + \cdots + M_A^i(t_1^{\alpha-\tau} + s_1^{\alpha-\tau})}{\Gamma(\alpha)}(K_{\alpha,\tau} + S_{\alpha,\tau}) \\ &+ \frac{(t_i^\beta + s_i^\beta) + M_A(t_{i-1}^\beta + s_{i-1}^\beta) + \cdots + M_A^{i-1}(t_1^\beta + s_1^\beta)}{\Gamma(\beta+1)}(b \| x \| + d) \\ &\leq r.\end{aligned}$$

For $t \in (s_i, t_{i+1}]$

$$\begin{aligned}\| Qx(t) \| &\leq M_A^{i+1} \| x_0 \| \\ &+ \frac{M_A t_{i+1}^{\alpha-\tau} + M_A^2(t_i^{\alpha-\tau} + s_i^{\alpha-\tau}) + \cdots + M_A^{i+1}(t_1^{\alpha-\tau} + s_1^{\alpha-\tau})}{\Gamma(\alpha)}(K_{\alpha,\tau} + S_{\alpha,\tau}) \\ &+ \frac{M_A(t_i^\beta + s_i^\beta) + M_A^2(t_{i-1}^\beta + s_{i-1}^\beta) + \cdots + M_A^i(t_1^\beta + s_1^\beta)}{\Gamma(\beta+1)}(b \| x \| + d) \\ &\leq r.\end{aligned}$$

We proved that $QB_r(J) \subset B_r(J)$.

Step 2: Q_2 is continuous.
Let $(x_n)_{n\geq 0}$ be a sequence such that $\lim_{n\to\infty} \| x_n - x \|_{PC} = 0$, we have by (H_5') : $g_i(t, x_n(t)) \longrightarrow g_i(t, x(t))$.

Case 1. For $t \in [0, t_1]$, we have:
$\| Q_1 x_n(t) - Q_1 x(t) \| = 0$
Case 2. For $t \in [t_i, s_i]$, $1 \leq i \leq m$,

$$\| Q_1x(t) - Q_1y(t) \| \leq \left[\frac{(s_i^\beta + t_i^\beta) + M_A(s_{i-1}^\beta + t_{i-1}^\beta) + \cdots + M_A^{i-1}(s_1^\beta + t_1^\beta)}{\Gamma(\beta+1)} \right] \| g_i(t, x_n(t)) - g_i(t, x(t)) \|_{PC}.$$

For $t \in [s_i, t_{i+1}]$, $1 \leq i \leq m$,

$$\| Q_1x(t) - Q_1y(t) \| \leq \left[\frac{M_A(s_i^\beta + t_i^\beta) + M_A^2(s_{i-1}^\beta + t_{i-1}^\beta) + \cdots + M_A^i(s_1^\beta + t_1^\beta)}{\Gamma(\beta+1)} \right] \| g_i(t, x_n(t)) - g_i(t, x(t)) \|_{PC} .$$

Thus we get $\| Q_1x_n(t) - Q_1x(t) \|_{PC} \longrightarrow 0$ as $n \longrightarrow \infty$.
Then we can say that Q_1 is continuous.
We already have that Q_2 is continuous. Finally Q is continuous.
Step 3: Q is β- condensing in B_r.
Case 1. For $t \in [0, t_1]$, we have:
Considering the condition (H_5') and using the same method in the previous theorem we get:

$$\begin{aligned}\beta(Q(W)) &\leq \beta(Q_2(W)) \\ &\leq \frac{4M_A(L_1 + \omega L_2 K_0)t_1^\alpha}{\Gamma(\alpha+1)}\beta(W) \\ &< \beta(W).\end{aligned}$$

Case 2. For $t \in (t_i, s_i]$,$i = 1, 2, \ldots, m$, we have:

$$\begin{aligned}&\beta(Q(W)) \leq \beta(Q_2(W)) \\ &\leq 4M_A(L_1 + \omega L_2 K_0)\left[\frac{t_i^\alpha + M_A(t_{i-1}^\alpha + s_{i-1}^\alpha) + \cdots + M_A^{i-1}(t_1^\alpha + s_1^\alpha)}{\Gamma(\alpha+1)} \right]\beta(W).\end{aligned}$$

Case 3: For $t \in [s_i, t_{i+1}]$, $i = 1, 2, \ldots, m$, we have:

$$\begin{aligned}&\beta(Q(W)) \leq \beta(Q_2(W)) \\ &\leq 4M_A(L_1 + \omega L_2 K_0)\left[\frac{t_{i+1}^\alpha + M_A(t_i^\alpha + s_i^\alpha) + \cdots + M_A^{i}(t_1^\alpha + s_1^\alpha)}{\Gamma(\alpha+1)} \right]\beta(W).\end{aligned}$$

Conclusion: in all cases we have:

$$\beta(Q(W)) \leq \beta(Q_2(W)) \leq C\beta(W) < \beta(W) \text{ with } C > 0.$$

Since the operator Q is continuous and β-condensing. According to Darbo-Sadovskii's fixed point theorem, Q has a fixed point in B_r. Therefore, the problem (1.2) has at least one mild solution in B_r. This completes the proof.

References

1. R. Hilfer, *Apllications of Fractional Calculus in Physics* (World Scientific, Singapore, 2000)
2. A.A. Kilbas, H.M. Srivastava, J.J. Trujillo, *Theory and Applications of Fractional Differential Equations*, vol. 204, North-Holland Mathematics Studies (Elsevier, Amsterdam, 2006)
3. V. Lakshmikantham, Theory of fractional functional differential equations. Nonlinear Anal. (2007). https://doi.org/10.1016/j.na.2007.09.025
4. V. Lakshmikantham, A.S. Vatsala, Basic theory of fractional differential equations. Nonlinear Anal. (2007). https://doi.org/10.1016/j.na.2007.09.025
5. K.S. Miller, B. Ross, *An Introduction to the Fractional Calulus andd Fractional Differential Equations* (Wiley, New York, 1993)
6. I. Podlubny, *Fractional Differential Equations* (Academic Press, New York, 1993)
7. Z. Yong, *Basic Theory of Fractional Differential Equations* (Xiangtan University, China, 2014)
8. M. Benchohra, J. Henderson, S.K. Ntouyas, *Impulsive Differential Equations and Inclusions* (Hindawi Publishing, New York, 2006)
9. M. Gou, X. Xue, R. Li, Controllability of impulsive evolution inclusions with nonlocal conditions. J. Optim. Theory Appl. **120**, 255–374 (2004)
10. V. Lakshmikantham, D.D. Bainov, P.S. Simeonov, *Theory of impulsive Differential Equations* (World Scientific, Singapore, 1989)
11. P. Kumar, D. Pandey, D. Bahuguna, On a new class of abstract impulsive functional differential equations of fractional order. J. Nonlinear Sci. Appl. **7**, 102–114 (2014)
12. J. Wang, M. Feckan, Y. Zhou, On the new concept of solutions and existence results for impulsive fractional evolution equations. Dyn. Partial Differ. Equ. **8**, 345–361 (2011)
13. X. Fu, X. Liu, B. Lu, On a new class of impulsive fractional evolution equations. Adv. Differ. Equ. **2015**, article 227 (2015)
14. P. Chen, X. Zhang, Y. Li, Existence of mild solutions to partial differential equations with non-instantenous impulses. Electron. J. Differ. Equ. **2016**(241), 1–11 (2016)
15. P.L. Falb, Infinite dimensional control problems: on the closure of the set of attainable states for linear systems. Math. Anal. Appl. **9**, 12–22 (1964)
16. J. Banas, K. Goebel, *Measure of Noncompactness in Banach Space*, vol. 60, Lecture notes in Pure and Applied Mathematics (Marcel Dekker, New York, 1980)
17. M.M. El-Borai, Some probability densities and fundamental solutions of fractional evolution equations. Chaos Solitons Fractals **14**, 433–440 (2002)
18. M.M. El-Borai, Semigroups and some nonlinear fractional differential equations. Appl. Math. Comput. **149**, 823–831 (2004)
19. J. Wang, Y. Zhou, A class of fractional evolution equations and optimal controls. Nonlinear Anal. RWA **12**, 262–272 (2011)
20. P. Chen, Y. Li, Monotone iterative technique for a class of semilinear evolution equations with nonlocal conditions. Results Math. **63**, 731–744 (2013)
21. H.P. Heinz, On the behaviour of measure of noncompactness with respect to differentiation and integration of vector-valued functions. Nonlinear Anal. **7**, 1351–1371 (1983)

Existence Results of Hybrid Fractional Integro-Differential Equations

Said Melliani, K. Hilal and M. Hannabou

Abstract We study in this paper, the existence results for initial value problems for hybrid fractional integro-differential equations. By using fixed point theorems for the sum of three operators are used for proving the main results.An example is also given to demonstrate the applications of our main results.

1 Introduction

Fractional differential equations arise in the mathematical modeling of systems and processes occurring in many engineering and scientific disciplines such as physics, chemistry, aerodynamics, electrodynamics of complex medium, polymer rheology, economics, control theory, signal and image processing, biophysics, blood flow phenomena, etc. (see [1–4]). Compared with integer order models, the fractional order models describe the underlying processes in a more effective manner by taking into account their past history. This has led to a great interest and considerable attention in the ubject of fractional order differential equations.

For some recent developments on the topic, (see [5–8]), and the references therein. Hybrid fractional differential equations have also been studied by several researchers.

This class of equations involves the fractional derivative of an unknown function hybrid with the nonlinearity depending on it. Some recent results on hybrid differential equations can be found in a series of papers [9–16].

S. Melliani (✉) · K. Hilal · M. Hannabou
Sultan Moulay Slimane University, BP 523 Beni Mellal, Morocco
e-mail: s.melliani@usms.ma

K. Hilal
e-mail: Khalid.hilal.usms@gmail.com

M. Hannabou
e-mail: hnnabou@gmail.com

S. Melliani and O. Castillo (eds.), *Recent Advances in Intuitionistic Fuzzy Logic Systems*, Studies in Fuzziness and Soft Computing 372,
https://doi.org/10.1007/978-3-030-02155-9_17

Hybrid fractional differential equations have also been studied by several researchers. This class of equations involves the fractional derivative of an unknown function hybrid with the nonlinearity depending on it. In Sitho et al. [17] discussed the following existence results for hybrid fractional integro-differential equations

$$\begin{cases} D^{\alpha}\left(\frac{x(t)-\sum_{i=1}^{m} I^{\beta_i} h_i(t,x(t))}{f(t,x(t))}\right) = g(t,x(t)) \quad t \in J = [0,T], \quad 0 < \alpha \leq 1 \\ x(0) = 0 \end{cases}$$

where D^{α} denotes the Riemann-Liouville fractional derivative of order $\alpha, 0 < \alpha \leq 1$, I^{ϕ} is the Riemann-Liouville fractional integral of order $\phi > 0, \phi \in \{\beta_1, \beta_2, \ldots, \beta_m\}$, $f \in C(J \times \mathbb{R}, \mathbb{R} \setminus \{0\})$, $g \in C(J \times \mathbb{R}, \mathbb{R})$, with $h_i \in C(J \times \mathbb{R}, \mathbb{R})$ and $h_i(0,0) = 0$, $i = 1, 2, \ldots, m$.

In Hilal and Kajouni [18] considered boundary value problems for hybrid differential equations with fractional order (BVPHDEF of short) involving Caputo differential operators of order $0 < \alpha < 1$,

$$\begin{cases} D^{\alpha}\left(\frac{x(t)}{f(t,x(t))}\right) = g(t,x(t)) \quad a.e. \quad t \in J = [0,T] \\ a\frac{x(0)}{f(0,x(0))} + b\frac{x(T)}{f(T,x(T))} = c \end{cases}$$

where $f \in C(J \times \mathbb{R}, \mathbb{R}\backslash\{0\})$, $g \in \mathscr{C}(J \times \mathbb{R}, \mathbb{R})$ and a, b, c are real constants with $a + b \neq 0$.

Dhage and Lakshmikantham [15], discussed the following first order hybrid differential equation

$$\begin{cases} \frac{d}{dt}\left[\frac{x(t)}{f(t,x(t))}\right] = g(t,x(t)) \quad t \in J = [0,T] \\ x(t_0) = x_0 \in \mathbb{R} \end{cases}$$

where $f \in C(J \times \mathbb{R}, \mathbb{R}\backslash\{0\})$ and $g \in \mathscr{C}(J \times \mathbb{R}, \mathbb{R})$. They established the existence, uniqueness results and some fundamental differential inequalities for hybrid differential equations initiating the study of theory of such systems and proved utilizing the theory of inequalities, its existence of extremal solutions and a comparison results.

Zhao et al. [19], are discussed the following fractional hybrid differential equations involving Riemann-Liouville differential operators

$$\begin{cases} D^{q}\left[\frac{x(t)}{f(t,x(t))}\right] = g(t,x(t)) \quad t \in J = [0,T] \\ x(0) = 0 \end{cases}$$

where $f \in C(J \times \mathbb{R}, \mathbb{R}\backslash\{0\})$ and $g \in \mathscr{C}(J \times \mathbb{R}, \mathbb{R})$. They established the existence theorem for fractional hybrid differential equation, some fundamental differential inequalities are also established and the existence of extremal solutions.

Benchohra et al. [20], we study the following boundary value problems for differential equations with fractional order

$$\begin{cases} {}^cD^{\alpha}y(t) = f(t, y(t)), \quad for \quad each \quad t \in J = [0, T], \quad 0 < \alpha < 1 \\ ay(0) + by(T) = c \end{cases}$$

where ${}^cD^{\alpha}$ is the Caputo fractional derivative, $f : [0, T] \times \mathbb{R} \to \mathbb{R}$, is a continuous function, a, b, c are real constants with $a + b \neq 0$.

Motivated by some recent studies on hybrid fractional integro-differential equations see [17, 18], we consider the following value problem:

$$\begin{cases} D^{\alpha}\left(\frac{x(t)-I^{\beta}h(t,x(t),I^{\alpha_1}x(t),\ldots,I^{\alpha_n}x(t))}{f(t,x(t),I^{\alpha_1}x(t),\ldots,I^{\alpha_n}x(t))}\right) = g(t, x(t), I^{\beta_1}x(t), \ldots, I^{\beta_k}x(t)) \\ t \in J = [0, T], \quad 1 < \alpha \leq 2 \\ \frac{x(0)}{f(0,x(0),\underbrace{0, 0 \ldots, 0}_{n})} = x_0, \qquad \frac{x(T)}{f(T,x(T),I^{\alpha_1}x(T),\ldots,I^{\alpha_n}x(T))} = x_T, \end{cases} \tag{1}$$

where $\alpha_1, \ldots, \alpha_n > 0\,, \beta_1, \ldots, \beta_n > 0, x \in \mathbb{R}, D^{\alpha}$ denotes Caputo fractional derivative of order α. I^{β} is the Riemann-Liouville fractional integral of order $\beta > 0$. $f : J \times \mathbb{R}^n \longrightarrow \mathbb{R} \setminus \{0\}\,, h : J \times \mathbb{R}^n \longrightarrow \mathbb{R}$ is continuous with $h(0, x(0), \underbrace{0, 0 \ldots, 0}_{n}) = 0$, and $g \in \mathscr{C}(J \times \mathbb{R}^k, \mathbb{R})$ is a function via some properties.

The problem 1 considered here is general in the sense that it includes the following three well-known classes of initial value problems of fractional differential equations.

Case I: Let $f(t, x(t), I^{\alpha_1}x(t), \ldots, I^{\alpha_n}x(t)) = 1$ and $h(t, x(t), I^{\alpha_1}x(t), \ldots, I^{\alpha_n}x(t)) = 0$, for all $t \in J$ and $x \in \mathbb{R}$. Then the problem 1 reduces to standard initial value problem of fractional differential equation,

$$\begin{cases} D^{\alpha}(x(t)) = g(t, x(t), I^{\beta_1}x(t), \ldots, I^{\beta_k}x(t)) \quad t \in J = [0, T], \quad 1 < \alpha \leq 2 \\ x(0) = x_0, \qquad x(T) = x_T, \end{cases}$$

Case II: If $h(t, x(t), I^{\alpha_1}x(t), \ldots, I^{\alpha_n}x(t)) = 0$ for all $t \in J$ and $x \in \mathbb{R}$ in 1. We obtain the following quadratic fractional differential equation,

$$\begin{cases} D^{\alpha}\left(\frac{x(t)}{f(t,x(t),I^{\alpha_1}x(t),\ldots,I^{\alpha_n}x(t))}\right) = g(t, x(t), I^{\beta_1}x(t), \ldots, I^{\beta_k}x(t)) \quad t \in J = [0, T], \quad 1 < \alpha \leq 2 \\ \frac{x(0)}{f(0,x(0),\underbrace{0, 0 \ldots, 0}_{n})} = x_0, \qquad \frac{x(T)}{f(T,x(T),I^{\alpha_1}x(T),\ldots,I^{\alpha_n}x(T))} = x_T, \end{cases}$$

Case III: If $f(t, x(t), I^{\alpha_1}x(t), \ldots, I^{\alpha_n}x(t)) = 1$ for all $t \in J$ and $x \in \mathbb{R}$ in 1.We obtain the following interesting fractional differential equation,

$$\begin{cases} D^{\alpha}\Big(x(t) - I^{\beta}h(t, x(t), I^{\alpha_1}x(t), \ldots, I^{\alpha_n}x(t))\Big) = g(t, x(t), I^{\beta_1}x(t), \ldots, I^{\beta_k}x(t)) \quad t \in J = [0, T], \\ 1 < \alpha \leq 2 \\ x(0) = x_0, \qquad\qquad x(T) = x_T, \end{cases}$$

Therefore, the main result of this paper also includes the existence the results for the solutions of above mentioned initial value problems of fractional differential equations as special cases.

An existence result is obtained for the initial value problem 1. by using a hybrid fixed point theorem for three operators in a Banach algebra due to Dhage [21].

As a second problem we discuss in Sect. 4 an initial value problem for hybrid fractional sequential integro-differential equations,

$$\begin{cases} D^{\alpha}\left(\frac{D^{\omega}x(t) - I^{\beta}h(t, x(t), I^{\alpha_1}x(t), \ldots, I^{\alpha_n}x(t))}{f(t, x(t), I^{\alpha_1}x(t), \ldots, I^{\alpha_n}x(t))}\right) = g(t, x(t), I^{\beta_1}x(t), \ldots, I^{\beta_k}x(t)) \\ \quad t \in J = [0, T], \quad 0 < \alpha < 1 \\ x(0) = x_0, \quad D^{\omega}x(0) = 0, \end{cases} \tag{2}$$

where $0 < \alpha, \omega \leq 1, 1 < \alpha + \omega \leq 2$, functions $f \in C(J \times \mathbb{R}^n, \mathbb{R} \setminus \{0\}), h \in C(J \times \mathbb{R}^n, \mathbb{R})$ whith $h(0, x(0), \underbrace{0, 0 \ldots, 0}_{n}) = 0$ and $g \in \mathscr{C}(J \times \mathbb{R}^k, \mathbb{R})$. I^{β} is the Riemann-Liouville fractional integral of order β.

D^{α}, D^{ω} are denotes Caputo fractional derivative of order α, β respectively.

By using a useful generalization of Krasnoselskii's fixed point theorem due to Dhage , we prove an existence result for the initial value problem 2.

This paper is arranged as follows. In Sect. 2, we recall some concepts and some fractional calculation law and establish preparation results. In Sect. 3, we study the existence of the initial value problem 1, based on the Dhage fixed point theorem,while in Sect. 4 we deal with the initial value problem 2. In Sect. 5, we give an example to demonstrate the application of our main result.

2 Preliminaries

Next, we review some basic concepts, notations, and technical results that are necessary in our study.

By $E = C(J, \mathbb{R})$ we denote the Banach space of all continuous functions from $J = [0, T]$ into $\mathbb{R}$ with the norm

$$\|y\| = \sup\{|y(t)|, t \in J\}$$

and a multiplication in E by

$$(xy)(t) = x(t)y(t), \forall t \in J.$$

Clearly E is a Banach algebra with respect to above supremum norm and the multiplication in it. and let $\mathscr{C}(J \times \mathbb{R}^k, \mathbb{R})$ denote the class of functions $g : J \times \mathbb{R}^k \longrightarrow \mathbb{R}$ such that

(i) the map $s \longrightarrow g(s, x_1, x_2, \ldots, x_k)$ is mesurable for all $x_1, x_2, \ldots, x_k \in \mathbb{R}$.
(ii) $(x_1, x_2, \ldots, x_k) \longrightarrow g(s, x_1, x_2, \ldots, x_k)$ is continuous map for almost all $s \in J$.

The class $\mathscr{C}(J \times \mathbb{R}^k, \mathbb{R})$ is called the Carathéodory class of functions on $J \times \mathbb{R}^k$.

Also, a Carathéodory function $g : J \times \mathbb{R}^k \longrightarrow \mathbb{R}$ is called L^1-Carathéodory whenever for each $\rho > 0$ there exists $\phi_\rho \in L^1(J, \mathbb{R}^+)$ such that

$$\|g(s, x_1, x_2, \ldots, x_k)\| = \sup\{|v| : v \in g(s, x_1, x_2, \ldots, x_k)\} \leq \phi_\rho(s)$$

for all $|x_1|, |x_2|, \ldots, |x_k| \leq \rho$ and for almost all $s \in J$.

By $L^1(J, \mathbb{R})$ denote the space of Lebesgue integrable real-valued functions on J equipped with the norm $\|.\|_{L^1}$ defined by
$\|x\|_{L^1} = \int_0^T |x(s)|ds$

Definition 1 [1] The fractional integral of the function $h \in L^1([a, b], \mathbb{R}^+)$ of order $\alpha \in \mathbb{R}^+$ is defined by

$$I_a^\alpha h(t) = \int_a^t \frac{(t-s)^{\alpha-1}}{\Gamma(\alpha)} h(s)ds$$

where Γ is the gamma function.

Definition 2 [1] For a function h given on the interval $[a, b]$, the Riemann-Liouville fractional-order derivative of h, is defined by

$$({}^c D_{a^+}^\alpha h)(t) = \frac{1}{\Gamma(n-\alpha)} \left(\frac{d}{dt}\right)^n \int_a^t \frac{(t-s)^{n-\alpha-1}}{\Gamma(\alpha)} (s)ds$$

where $n = [\alpha] + 1$ and $[\alpha]$ denotes the integer part of α.

Definition 3 [1] For a function h given on the interval $[a, b]$, the Caputo fractional-order derivative of h, is defined by

$$({}^c D_{a^+}^\alpha h)(t) = \frac{1}{\Gamma(n-\alpha)} \int_a^t \frac{(t-s)^{n-\alpha-1}}{\Gamma(\alpha)} h^{(n)}(s)ds$$

where $n = [\alpha] + 1$ and $[\alpha]$ denotes the integer part of α.

Lemma 1 *[1] Let $\alpha > 0$ and $x \in C(0, T) \cap L(0, T)$. Then the fractional differential equation*

$$D^{\alpha}x(t) = 0$$

has a unique solution

$$x(t) = k_1 t^{\alpha-1} + k_2 t^{\alpha-2} + \ldots + k_n t^{\alpha-n},$$

where $k_i \in \mathbb{R}$, $i = 1, 2, \ldots, n$, and $n - 1 < \alpha < n$.

Lemma 2 *[1] Let $\alpha > 0$. Then for $x \in C(0, T) \cap L(0, T)$ we have*

$$I^{\alpha}D^{\alpha}x(t) = x(t) + c_0 + c_1 t + \ldots + c_{n-1}t^{n-1},$$

fore some $c_i \in \mathbb{R}$, $i = 1, 2, \ldots, n - 1$. Where $n = [\alpha] + 1$.

Lemma 3 *[1] For $\alpha, \beta > 0$ and f as a suitable function, we have*

(i) $I^{\alpha}I^{\beta}f(t) = I^{\alpha+\beta}f(t)$
(ii) $I^{\alpha}I^{\beta}f(t) = I^{\beta}I^{\alpha}f(t)$
(iii) $I^{\alpha}(f(t) + g(t)) = I^{\alpha}f(t) + I^{\alpha}g(t)$

3 Fractional Hybrid Differential Equation

In this section we prove the existence of a solution the initial value problem 1 by a fixed point theorem in the Banach algebra due to Dhage [21].

Lemma 4 *Let S be a nonempty, closed convex and bounded subset of a Banach algebra E and let $A, C : E \longrightarrow E$ and $B : S \longrightarrow E$ be three operators satisfying:*

(a_1) A and C are Lipschitzian with Lipschitz constants δ and ρ, respectively,
(b_1) B is compact and continuous,
(c_1) $x = AxBy + Cx \Longrightarrow x \in S$ for all $y \in S$,
(d_1) $\delta M + \rho < 1$, where $M = \|B(S)\|$.

Then the operator equation $x = AxBx + Cx$ has a solution.

For brevity let us take,

$$d = \frac{I^{\beta}h(T, x(T), I^{\alpha_1}x(T), \ldots, I^{\alpha_n}x(T))}{f(T, x(T), I^{\alpha_1}x(T), \ldots, I^{\alpha_n}x(T))}$$

Lemma 5 *Suppose that $1 < \alpha \leq 2$.*

Then, for any $k \in L^1(J, \mathbb{R})$, the function $x \in C(J, \mathbb{R})$ is a solution of the

$$\begin{cases} D^{\alpha}\left(\frac{x(t)-I^{\beta}h(t,x(t),I^{\alpha_1}x(t),\ldots,I^{\alpha_n}x(t))}{f(t,x(t),I^{\alpha_1}x(t),\ldots,I^{\alpha_n}x(t))}\right) = k(t) \quad t \in J = [0,T] \\ \frac{x(0)}{f(0,x(0),\underbrace{0,0\ldots,0}_{n})} = x_0, \qquad \frac{x(T)}{f(T,x(T),I^{\alpha_T}x(t),\ldots,I^{\alpha_n}x(T))} = x_T, \end{cases} \tag{3}$$

if and only if x satisfies the hybrid integral equation

$$\begin{aligned} x(t) = & \left(f(t,x(t),I^{\alpha_1}x(t),\ldots,I^{\alpha_n}x(t))\right)\Big(\frac{1}{\Gamma(\alpha)}\int_0^t (t-s)^{\alpha-1}k(s)ds \\ & +(1-\frac{t}{T})x_0 + \frac{t}{T}x_T - \frac{t}{T\Gamma(\alpha)}\int_0^T (T-s)^{\alpha-1}k(s)ds - \frac{td}{T}\Big) \\ & + \int_0^t \frac{(t-s)^{\beta-1}}{\Gamma(\beta)}h(s,x(s),I^{\beta_1}x(s),\ldots,I^{\alpha_n}x(s))ds, \quad t \in [0,T] \end{aligned} \tag{4}$$

Proof Assume that x is a solution of the problem 5. By definition, $\frac{x(t)}{f(t,x(t),I^{\alpha_1}x(t),\ldots,I^{\alpha_n}x(t))}$ is continuous. Applying the Caputo fractional operator of the order α, we obtain the first equation in 3. Again, substituting $t = 0$ and $t = T$ in 5 we have

$$\frac{x(0)}{f(0,x(0),\underbrace{0,0\ldots,0}_{n})} = x_0, \qquad \frac{x(T)}{f(T,x(T),I^{\alpha_1}x(t),\ldots,I^{\alpha_n}x(T))} = x_T,$$

Conversely, $D^{\alpha}\left(\frac{x(t)-I^{\beta}h(t,x(t),I^{\alpha_1}x(t),\ldots,I^{\alpha_n}x(t))}{f(t,x(t),I^{\alpha_1}x(t),\ldots,I^{\alpha_n}x(t))}\right) = k(t)$

so we get

$$\frac{x(t)-I^{\beta}h(t,x(t),I^{\alpha_1}x(t),\ldots,I^{\alpha_n}x(t))}{f(t,x(t),I^{\alpha_1}x(t),\ldots,I^{\alpha_n}x(t))} = I^{\alpha}k(t) - c_0 - c_1 t$$

$$\frac{x(t)}{f(t,x(t),I^{\alpha_1}x(t),\ldots,I^{\alpha_n}x(t))} = I^{\alpha}k(t) - c_0 - c_1 t + \frac{I^{\beta}h(t,x(t),I^{\alpha_1}x(t),\ldots,I^{\alpha_n}x(t))}{f(t,x(t),I^{\alpha_1}x(t),\ldots,I^{\alpha_n}x(t))}$$

Substituting $t = 0$ we have

$$c_0 = -\frac{x(0)}{f(0,x(0),\underbrace{0,0\ldots,0}_{n})} = -x_0$$

And substituting $t = T$ we have

$$\frac{x(T)}{f(T, x(T), I^{\alpha_1}x(T), \ldots, I^{\alpha_n}x(T))} = I^{\alpha}k(T) + x_0 - c_1T + d$$

Then

$$c_1 = \frac{1}{T}(x_0 + I^{\alpha}k(T) - x_T + d)$$

In consequence, we have

$$\begin{aligned} x(t) = &\Big(f(t, x(t), I^{\alpha_1}x(t), \ldots, I^{\alpha_n}x(t))\Big)\Big(\frac{1}{\Gamma(\alpha)}\int_0^t (t-s)^{\alpha-1}k(s)ds \\ &+(1-\frac{t}{T})x_0 + \frac{t}{T}x_T \\ &-\frac{t}{T\Gamma(\alpha)}\int_0^T (T-s)^{\alpha-1}k(s)ds - \frac{td}{T}\Big) \\ &+\int_0^t \frac{(t-s)^{\beta-1}}{\Gamma(\beta)}h(s, x(s), I^{\beta_1}x(s), \ldots, I^{\alpha_n}x(s))ds, t \in [0, T] \end{aligned}$$

In the forthcoming analysis, we need the following assumptions. Assume that :

(H_1) The functions $f : J \times \mathbb{R}^{n+1} \longrightarrow \times\mathbb{R} \setminus \{0\}$, $h : J \times \mathbb{R}^{n+1} \longrightarrow \mathbb{R}$, $g : J \times \mathbb{R}^{k+1} \longrightarrow \mathbb{R}$ are be a Carathéodory function, $h(0, x(0), \underbrace{0, 0 \ldots, 0}_{n})) = 0$ and there exist two positive functions

$p, m : J \longrightarrow (0, \infty)$ with bound $\|p\|$ and $\|m\|$ respectively, such that

$$|f(t, y_1, y_2, \ldots, y_{n+1}) - f(t, x_1, x_2, \ldots, x_{n+1})| \leq p(t)\sum_{i=1}^{n+1}|y_i - x_i|$$

and

$$|h(t, y_1, y_2, \ldots, y_{n+1}) - h(t, x_1, x_2, \ldots, x_{n+1})| \leq m(t)\sum_{i=1}^{n+1}|y_i - x_i|$$

for $t \in J$ and $(x_1, x_2, \ldots, x_{n+1}), (y_1, y_2, \ldots, y_{n+1}) \in \mathbb{R}^{n+1}$. (H_2) There exists a function $h \in L^1(J, \mathbb{R})$ such that .

$$|g(t, x_1, x_2, \ldots, x_k)| \leq h(t) \quad a.e \quad (t, x_1, x_2, \ldots, x_k) \in J \times \mathbb{R}^k$$

(H_3) There exists a real number $r > 0$ such that

$$r \geq \frac{F_0\Big(\frac{2\|h\|_{L^1}T^{\alpha}}{\Gamma(\alpha+1)} + |x_0| + |x_T| + |d|\Big) + \frac{k_0 T^{\beta}}{\Gamma(\beta+1)}}{1 - \Big(1 + \frac{T^{\alpha_1}}{\Gamma(\alpha_1+1)} + \ldots + \frac{T^{\alpha_n}}{\Gamma(\alpha_n+1)}\Big)\Big[\|p\|\Big(\frac{2\|h\|_{L^1}T^{\alpha}}{\Gamma(\alpha+1)} + |x_0| + |x_T| + |d|\Big) - \|m\|\frac{T^{\beta}}{\Gamma(\beta+1)}\Big]}$$

where $F_0 = \sup_{t\in J} |f(t, x(0), \underbrace{0, 0 \ldots, 0}_{n})|$ and $K_0 = \sup_{t\in J} |h(t, x(0), \underbrace{0, 0 \ldots, 0}_{n})|$

Theorem 1 *Assume that the conditions (H_1)–(H_3) hold. Then the initial value problem 1 has at least one solution on J provided that*

$$\Big(1+ \frac{T^{\alpha_1}}{\Gamma(\alpha_1+1)} + \ldots + \frac{T^{\alpha_n}}{\Gamma(\alpha_n+1)}\Big)\Big[\|p\|\Big(\frac{2\|h\|_{L^1}T^{\alpha}}{\Gamma(\alpha+1)} + |x_0| + |x_1| + |d|\Big) + \frac{\|m\|T^{\beta}}{\Gamma(\beta+1)}\Big] < 1.$$

Proof Set $E = C(J, \mathbb{R})$ and define a subset S of E as

$$S = \{x \in E : \|x\| \leq r\},$$

where r satisfies inequality 3.

Clearly S is closed, convex, and bounded subset of the Banach space E. By Lemma 5, problem 1 is equivalent to the integral Eq. 5. Now we define three operators,
$\mathscr{A} : E \longrightarrow E$ by

$$\mathscr{A}x(t) = f(t, x(t), I^{\alpha_1}x(t), \ldots, I^{\alpha_n}x(t)), \quad t \in J,$$

$\mathscr{B} : S \longrightarrow E$ by

$$\mathscr{B}x(t) = \int_0^t \frac{(t-s)^{\alpha-1}}{\Gamma(\alpha)} g(s, x(s), I^{\beta_1}x(s), \ldots, I^{\beta_k}x(s))ds + (1 - \frac{t}{T})x_0 + \frac{t}{T}x_T$$
$$- \frac{t}{T\Gamma(\alpha)} \int_0^T (T-s)^{\alpha-1} g(s, x(s), I^{\beta_1}x(s), \ldots, I^{\beta_k}x(s))ds - \frac{t}{T}d\Big], \quad t \in J,$$

and $\mathscr{C} : E \longrightarrow E$ by

$$\mathscr{C}x(t) = \int_0^t \frac{(t-s)^{\beta-1}}{\Gamma(\beta)} h(s, x(s), I^{\alpha_1}x(s), \ldots, I^{\alpha_n}x(s))ds, \quad t \in J$$

We shall show that the operators $\mathscr{A}$, $\mathscr{B}$, and $\mathscr{C}$ satisfy all the conditions of Lemma 4. The proof is constructed in several claims.
Claim 1. We will show that $\mathscr{A}$ and $\mathscr{C}$ are lipschitzian on E, that is, the assumption (a_1) of Lemma 4 holds.

Let $x, y \in E$. Then by (H_1), for $t \in J$ we have

$$\begin{aligned}|\mathscr{A}x(t) - \mathscr{A}y(t)| &= |f(t, x(t), I^{\alpha_1}x(t), \ldots, I^{\alpha_n}x(t)) - f(t, y(t), I^{\alpha_1}y(t), \ldots, I^{\alpha_n}y(t))|\\ &\leq \sup_{t\in J}(|p||x(t) - y(t)|)\left(1 + \frac{T^{\alpha_1}}{\Gamma(\alpha_1+1)} + \ldots + \frac{T^{\alpha_n}}{\Gamma(\alpha_n+1)}\right)\\ &\leq \|p\|\left(1 + \frac{T^{\alpha_1}}{\Gamma(\alpha_1+1)} + \ldots + \frac{T^{\alpha_n}}{\Gamma(\alpha_n+1)}\right)\|x-y\|\end{aligned}$$

for all $t \in J$.

Taking the supremum over the interval J, we obtain $\|\mathscr{A}x - \mathscr{A}y\| \leq \|p\|\left(1 + \frac{T^{\alpha_1}}{\Gamma(\alpha_1+1)} + \ldots + \frac{T^{\alpha_n}}{\Gamma(\alpha_n+1)}\right)\|x-y\|$ for all $x, y \in E$. So $\mathscr{A}$ is a Lipschitz on E with Lipschitz constant $\|p\|\left(1 + \frac{T^{\alpha_1}}{\Gamma(\alpha_1+1)} + \ldots + \frac{T^{\alpha_n}}{\Gamma(\alpha_n+1)}\right)$.

Analogously, for any $x, y \in E$, we have

$$\begin{aligned}|\mathscr{C}x(t) - \mathscr{C}y(t)| &= \Big|\int_0^t \frac{(t-s)^{\beta-1}}{\Gamma(\beta)}[h(s, x(s), I^{\alpha_1}x(s), \ldots, I^{\alpha_n}x(s))\\ &\quad -h(s, y(s), I^{\alpha_1}y(s), \ldots, I^{\alpha_n}y(s))]ds\Big|\\ &\leq \sup_{t\in J}(|m||x(t) - y(t)|)\left(1 + \frac{T^{\alpha_1}}{\Gamma(\alpha_1+1)} + \ldots + \frac{T^{\alpha_n}}{\Gamma(\alpha_n+1)}\right)\int_0^t \frac{(t-s)^{\beta-1}}{\Gamma(\beta)}ds\\ &\leq \|m\|\left(1 + \frac{T^{\alpha_1}}{\Gamma(\alpha_1+1)} + \ldots + \frac{T^{\alpha_n}}{\Gamma(\alpha_n+1)}\right)\frac{T^{\beta}}{\Gamma(\beta+1)}\|x-y\|\end{aligned}$$

for all $t \in J$.

Taking the supremum over the interval J, we obtain

$$\|\mathscr{C}x - \mathscr{C}y\| \leq \|m\|\left(1 + \frac{T^{\alpha_1}}{\Gamma(\alpha_1+1)} + \ldots + \frac{T^{\alpha_n}}{\Gamma(\alpha_n+1)}\right)\frac{T^{\beta}}{\Gamma(\beta+1)}\|x-y\|$$

So, $\mathscr{C}$ is a Lipschitzian on E with Lipschitz constant $\|m\|\left(1 + \frac{T^{\alpha_1}}{\Gamma(\alpha_1+1)} + \ldots + \frac{T^{\alpha_n}}{\Gamma(\alpha_n+1)}\right)\frac{T^{\beta}}{\Gamma(\beta+1)}$.

Claim 2. The operator $\mathscr{B}$ is completely continuous on S, that is, the assumption (b_1) of Lemma 4 holds.

We first show that the operator $\mathscr{B}$ is continuous on E.

Let $\{x_n\}$ be a sequence in S converging to a point $x \in S$. Then by the Lebesgue dominated convergence theorem, for all $t \in J$, we obtain

$$\begin{aligned}
&\lim_{n\to\infty} \int_0^t \frac{(t-s)^{\alpha-1}}{\Gamma(\alpha)} g(t, x_n(s), I^{\beta_1}x_n(s), \ldots, I^{\beta_k}x_n(s))ds \\
&= \int_0^t \frac{(t-s)^{\alpha-1}}{\Gamma(\alpha)} \lim_{n\to\infty} g(s, x_n(s), I^{\beta_1}x_n(s), \ldots, I^{\beta_k}x_n(s))ds \\
&= \int_0^t \frac{(t-s)^{\alpha-1}}{\Gamma(\alpha)} g(s, x(s), I^{\beta_1}x(s), \ldots, I^{\beta_k}x(s))ds
\end{aligned}$$

and

$$\begin{aligned}
&\lim_{n\to\infty} \Big[(1-\frac{t}{T})x_0 + \frac{t}{T}x_T \\
&\quad - \frac{t}{T\Gamma(\alpha)} \int_0^T (T-s)^{\alpha-1} g(s, x_n(s), I^{\beta_1}x_n(s), \ldots, I^{\beta_k}x_n(s))ds - \frac{t}{T}d \Big] \\
&= \lim_{n\to\infty} \Big[(1-\frac{t}{T})x_0 + \frac{t}{T}x_T \Big] \\
&\quad - \frac{t}{T\Gamma(\alpha)} \int_0^T (T-s)^{\alpha-1} \lim_{n\to\infty} g(t, x_n(t), I^{\beta_1}x_n(s), \ldots, I^{\beta_k}x_n(s))ds \\
&\quad - \frac{t}{T}d = (1-\frac{t}{T})x_0 + \frac{t}{T}x_T \\
&\quad - \frac{t}{T\Gamma(\alpha)} \int_0^T (T-s)^{\alpha-1} g(s, x(s), I^{\beta_1}x(s), \ldots, I^{\beta_k}x(s))ds - \frac{t}{T}d
\end{aligned}$$

In consequence, we have

$$\lim_{n\to\infty} \mathscr{B}x_n = \mathscr{B}x$$

This shows that $\mathscr{B}$ is continuous on S.

It is sufficient to show that the set $\mathscr{B}(S)$ is a uniformly bounded in S. For any $x \in S$, we have

$$\begin{aligned}
|\mathscr{B}x(t)| &= \Big| \int_0^t \frac{(t-s)^{\alpha-1}}{\Gamma(\alpha)} g(s, x(s), I^{\beta_1}x(s), \ldots, I^{\beta_k}x(s))ds + (1-\frac{t}{T})x_0 + \frac{t}{T}x_T \\
&\quad - \frac{t}{T\Gamma(\alpha)} \int_0^T (T-s)^{\alpha-1} g(t, x(t), I^{\beta_1}x(t), \ldots, I^{\beta_k}x(t))ds - \frac{t}{T}d \Big| \\
&\leq \Big(2\|h\|_{L^1} \frac{T^\alpha}{\Gamma(\alpha+1)} \Big) + |x_0| + |x_1| + |d| = K_1
\end{aligned}$$

for all $t \in J$. Taking supremum over the interval J, the above inequality becomes, $\|\mathscr{B}x\| \leq K_1$ for all $x \in S$. This shows that $\mathscr{B}(S)$ is uniformly bounded on S.

Next we show that $\mathscr{B}(S)$ is an equicontinuous set in E. We take, $\tau_1, \tau_2 \in J$ with $\tau_1 < \tau_2$ and $x \in S$.

Then we have

$$|\mathscr{B}x(\tau_2) - \mathscr{B}x(\tau_1)| = \Big| \int_0^{\tau_2} \frac{(\tau_2 - s)^{\alpha-1}}{\Gamma(\alpha)} g(s, x(s), I^{\beta_1}x(s), \ldots, I^{\beta_k}x(s))ds$$
$$- \int_0^{\tau_1} \frac{(\tau_1 - s)^{\alpha-1}}{\Gamma(\alpha)} g(s, x(s), I^{\beta_1}x(s), \ldots, I^{\beta_k}x(s))ds + \Big[(1 - \frac{\tau_2}{T}) - (1 - \frac{\tau_1}{T})\Big]x_0$$
$$+(\frac{\tau_1}{T} - \frac{\tau_2}{T})\Big(\int_0^{T} \frac{(T - s)^{\alpha-1}}{\Gamma(\alpha)} g(t, x(t), I^{\beta_1}x(t), \ldots, I^{\beta_k}x(t))ds - x_T + d\Big)\Big|$$
$$\leq \int_0^{\tau_1} \frac{|(\tau_2 - s)^{\alpha-1} - (\tau_1 - s)^{\alpha-1}|}{\Gamma(\alpha)} |g(t, x(t), I^{\beta_1}x(t), \ldots, I^{\beta_k}x(t))|ds$$
$$+ \int_{\tau_1}^{\tau_2} \frac{(\tau_2 - s)^{\alpha-1}}{\Gamma(\alpha)} |g(t, x(t), I^{\beta_1}x(t), \ldots, I^{\beta_k}x(t))|ds + (\frac{\tau_1 - \tau_2}{T})x_0$$
$$+\Big(\frac{\tau_1 - \tau_2}{T}\Big)\Big(x_T + \int_0^{T} \frac{(T - s)^{\alpha-1}}{\Gamma(\alpha)} |g(t, x(t), I^{\beta_1}x(t), \ldots, I^{\beta_k}x(t))|ds + d\Big)$$
$$\leq \int_0^{\tau_1} \frac{|(\tau_2 - s)^{\alpha-1} - (\tau_1 - s)^{\alpha-1}|}{\Gamma(\alpha)} \|h\|_{L^1}ds + \int_{\tau_1}^{\tau_2} \frac{(\tau_2 - s)^{\alpha-1}}{\Gamma(\alpha)} \|h\|_{L^1}ds$$
$$+(\frac{\tau_1 - \tau_2}{T})x_0 + \Big(\frac{\tau_1 - \tau_2}{T}\Big)\Big(x_T + \int_0^{T} \frac{(T - s)^{\alpha-1}}{\Gamma(\alpha)} \|h\|_{L^1}ds + d\Big)$$

Thus, we have that $|\mathscr{B}x(\tau_2) - \mathscr{B}x(\tau_1)| \longrightarrow 0$ as $\tau_2 \longrightarrow \tau_1$ which is independent of $x \in S$.

which is independent of $x \in S$. Thus, $\mathscr{B}(S)$ is equicontinuous. So $\mathscr{B}$ is relatively compact on S. Hence, by the Arzelá-Ascoli theorem, $\mathscr{B}$ is compact on S.

Claim 3. The hypothesis (c_1) of Lemma 4 is satisfied.

Let $x \in E$ and $y \in S$ be arbitrary elements such that $x = \mathscr{A}x\mathscr{B}y + \mathscr{C}x$. Then we have

$$|x(t)| \leq |\mathscr{A}x(t)||\mathscr{B}y(t)| + |\mathscr{C}x(t)|$$
$$\leq |f(t, x(t), I^{\alpha_1}x(t), \ldots, I^{\alpha_n}x(t))|$$
$$\Big[\int_0^t \frac{(t - s)^{\alpha-1}}{\Gamma(\alpha)}\Big|g(s, y(s), I^{\beta_1}y(s), \ldots, I^{\beta_k}y(s))\Big|ds$$
$$+ (1 - \frac{t}{T})x_0 + \frac{t}{T}x_T + \frac{t}{T\Gamma(\alpha)} \int_0^T (T - s)^{\alpha-1}\Big|g(s, y(s), I^{\beta_1}y(s), \ldots, I^{\beta_k}y(s))\Big|ds$$
$$+ \frac{t|d|}{T}\Big] + \int_0^t \frac{(t - s)^{\beta-1}}{\Gamma(\beta)}\Big|h(t, x(t), I^{\alpha_1}x(t), \ldots, I^{\alpha_n}x(t))\Big|ds$$
$$\leq (|f(t, x(t), I^{\alpha_1}x(t), \ldots, I^{\alpha_n}x(t)) - f(t, 0, \ldots, 0)| + |f(t, 0, \ldots, 0)|)$$
$$\Big[\int_0^t \frac{(t - s)^{\alpha-1}}{\Gamma(\alpha)}|h(s)|ds$$
$$+ (1 - \frac{t}{T})x_0 + \frac{t}{T}x_T + \frac{t}{T\Gamma(\alpha)} \int_0^T (T - s)^{\alpha-1}|h(s)|ds + |\frac{t}{T}||d|\Big]$$

$$
\begin{aligned}
&+\int_0^t \frac{(t-s)^{\beta-1}}{\Gamma(\beta)}(|h(t,x(t),I^{\alpha_1}x(t),\ldots,I^{\alpha_n}x(t))\\
&\quad -h(s,0,\ldots,0)+|h(s,0,\ldots,0)|)\Big|\\
&\leq \Big[r\|p\|\Big(1+\frac{T^{\alpha_1}}{\Gamma(\alpha_1+1)}+\ldots+\frac{T^{\alpha_n}}{\Gamma(\alpha_n+1)}\Big)+F_0\Big]\\
&\quad \Big(\frac{2\|h\|_{L^1}T^{\alpha}}{\Gamma(\alpha+1)}+|x_0|+|x_T|+|d|\Big)\\
&\quad +\frac{r\|m\|T^{\beta}}{\Gamma(\beta+1)}\Big(1+\frac{T^{\alpha_1}}{\Gamma(\alpha_1+1)}+\ldots+\frac{T^{\alpha_n}}{\Gamma(\alpha_n+1)}\Big)+\frac{T^{\beta}}{\Gamma(\beta+1)}k_0
\end{aligned}
$$

Taking supremum for $t \in J$, we obtain

$$
\begin{aligned}
\|x\| \leq &\Big[r\|p\|\Big(1+\frac{T^{\alpha_1}}{\Gamma(\alpha_1+1)}+\ldots+\frac{T^{\alpha_n}}{\Gamma(\alpha_n+1)}\Big)+F_0\Big]\Big(\frac{2\|h\|_{L^1}T^{\alpha}}{\Gamma(\alpha+1)}+|x_0|+|x_T|\\
&+|d|\Big)+\frac{r\|m\|T^{\beta}}{\Gamma(\beta+1)}\Big(1+\frac{T^{\alpha_1}}{\Gamma(\alpha_1+1)}+\ldots+\frac{T^{\alpha_n}}{\Gamma(\alpha_n+1)}\Big)+\frac{T^{\beta}}{\Gamma(\beta+1)}k_0
\end{aligned}
$$

that is, $x \in S$.
Claim 4. Finally we show that $\delta M+\rho<1$, that is, (d_1) of Lemma 4 holds.
Since

$$
\begin{aligned}
M=\|\mathscr{B}(S)\| &= \sup_{x\in S}\{\sup_{t\in J}|\mathscr{B}x(t)|\}\\
&\leq \frac{2\|h\|_{L^1}T^{\alpha}}{\Gamma(\alpha+1)}+|x_0|+|x_T|+|d|
\end{aligned}
$$

and by Theorem 1 we have

$$
\Big(1+\frac{T^{\alpha_1}}{\Gamma(\alpha_1+1)}+\ldots+\frac{T^{\alpha_n}}{\Gamma(\alpha_n+1)}\Big)\Big(\|p\|M+\frac{\|m\|T^{\beta}}{\Gamma(\beta+1)}\Big)<1
$$

with $\delta=\Big(1+\frac{T^{\alpha_1}}{\Gamma(\alpha_1+1)}+\ldots+\frac{T^{\alpha_n}}{\Gamma(\alpha_n+1)}\Big)\|p\|$ and $\rho=\frac{\|m\|T^{\beta}}{\Gamma(\beta+1)}\Big(1+\frac{T^{\alpha_1}}{\Gamma(\alpha_1+1)}+\ldots+\frac{T^{\alpha_n}}{\Gamma(\alpha_n+1)}\Big)$.
Thus all the conditions of Lemma 4 are satisfied and hence the operator equation $x=\mathscr{A}x\mathscr{B}x+\mathscr{C}x$ has a solution in S. In consequence, problem 3 has a solution on J. This completes the proof.

4 Hybrid Fractional Sequential Integro-Differential Equations

In this section we consider the initial value problem 2. An existence result will be proved by using the following fixed point theorem due to Dhage.

Lemma 6 *Let M be a nonempty, closed, convex and bounded subset of the Banach space X and let $A : X \longrightarrow X$ and $B : M \longrightarrow X$ be two operators such that*

(i) A is a contraction,
(ii) B is completely continuous, and
(iii) $x = Ax + By$ for all $y \in M \Longrightarrow x \in M$.

Then the operator equation $Ax + Bx = x$ has a solution.

Lemma 7 *Suppose that $0 < \alpha, \omega \leq 1$, $0 < \alpha + \omega \leq 1$, and the functions f, g, and h satisfy the problem 2. The function $x \in C(J, \mathbb{R})$ is a solution of the problem 2 if and only if x satisfies the hybrid integral equation,*

$$
\begin{aligned}
x(t) = & \int_0^t \Big(\frac{(t-s)^{\omega-1}}{\Gamma(\omega)} f(s, x(s), I^{\alpha_1}x(s), \ldots, I^{\alpha_n}x(s)) \\
& \int_0^s \frac{(s-\tau)^{\alpha-1}}{\Gamma(\alpha)} g(\tau, x(\tau), I^{\beta_1}x(\tau), \ldots, I^{\beta_k}x(\tau))) d\tau \Big) ds \\
& + \int_0^t \frac{(t-s)^{\beta+\omega-1}}{\Gamma(\beta+\omega)} h(s, x(s), I^{\alpha_1}x(s), \ldots, I^{\alpha_n}x(s)) ds + x_0, \quad t \in [0, T] \quad (5)
\end{aligned}
$$

Proof Assume that x is a solution of the problem 5. By definition, $\frac{x(t)}{f(t, x(t), I^{\alpha_1}x(t), \ldots, I^{\alpha_n}x(t))}$ is continuous. Applying the Caputo fractional operator of the order α, we obtain the first equation in 5 .

Again, substituting $t = 0$ in 5 we have
$x(0) = x_0, \qquad D^{\omega}x(0) = 0$
Conversely,
by lemma 2 we have

$$
\frac{D^{\omega}x(t) - I^{\beta}h(t, x(t), I^{\alpha_1}x(t), \ldots, I^{\alpha_n}x(t))}{f(t, x(t), I^{\alpha_1}x(t), \ldots, I^{\alpha_n}x(t))} = I^{\alpha}g(t, x(t), I^{\beta_1}x(t), \ldots, I^{\beta_k}x(t))) - c_0
$$

By condition $D^{\beta}x(0) = 0$, implies that $c_0 = 0$
Applying the semigroup property, i.e., $I^{\omega}I^{\beta}h = I^{\omega+\beta}h$ proposition 3, we obtain the,

$$
\begin{aligned}
x(t) = & I^{\omega}\Big[f(t, x(t), I^{\alpha_1}x(t), \ldots, I^{\alpha_n}x(t)) I^{\alpha} g(t, x(t), I^{\beta_1}x(t), \ldots, I^{\beta_k}x(t)) \Big] \\
& + I^{\omega+\beta}h(t, x(t), I^{\alpha_1}x(t), \ldots, I^{\alpha_n}x(t)) - c_1
\end{aligned}
$$

By condition $x(0) = x_0$, implies that $\qquad c_1 = -x_0$

Then,

$$x(t) = I^{\omega}\Big[f(t, x(t), I^{\alpha_1}x(t), \ldots, I^{\alpha_n}x(t))I^{\alpha}g(t, x(t), I^{\beta_1}x(t), \ldots, I^{\beta_k}x(t))\Big]$$
$$+ I^{\omega+\beta}h(t, x(t), I^{\alpha_1}x(t), \ldots, I^{\alpha_n}x(t)) + x_0$$

Consequently,

$$x(t) = \int_0^t \Big[\frac{(t-s)^{\omega-1}}{\Gamma(\omega)} f(s, x(s), I^{\alpha_1}x(s), \ldots, I^{\alpha_n}x(s))$$
$$\int_0^s \frac{(s-\tau)^{\alpha-1}}{\Gamma(\alpha)} g(\tau, x(\tau), I^{\beta_1}x(\tau), \ldots, I^{\beta_k}x(\tau))d\tau\Big]ds$$
$$+ \int_0^t \frac{(t-s)^{\beta+\omega-1}}{\Gamma(\beta+\omega)} h(s, x(s), I^{\alpha_1}x(s), \ldots, I^{\alpha_n}x(s))ds + x_0 \quad, t \in [0, T]$$

In the forthcoming analysis, we need the following assumptions. Assume that :

(A_1) The functions $f : J \times \mathbb{R}^{n+1} \longrightarrow \mathbb{R} \setminus \{0\}$ and $g : J \times \mathbb{R}^{k+1} \longrightarrow \mathbb{R}$, are continuous and there exist two positive functions ϕ, χ with bound $\|\phi\|$ and $\|\chi\|$, respectively, such that

$$|f(t, y_1, y_2, \ldots, y_{n+1}) - f(t, x_1, x_2, \ldots, x_{n+1})| < \phi(t)\sum_{i=1}^{n+1}|y_i - x_i|$$

for $t \in J$ and $(x_1, x_2, \ldots, x_{n+1}), (y_1, y_2, \ldots, y_{n+1}) \in \mathbb{R}^{n+1}$.
and

$$|g(t, y_1, y_2, \ldots, y_{k+1}) - g(t, x_1, x_2, \ldots, x_{k+1})| \le \chi(t)\sum_{i=1}^{k+1}|y_i - x_i|$$

for $t \in J$ and $(x_1, x_2, \ldots, x_{k+1}), (y_1, y_2, \ldots, y_{k+1}) \in \mathbb{R}^{k+1}$.

(A_2) $|f(t, x_1, x_2, \ldots, x_{n+1})| \le \mu(t), \quad \forall (t, x_1, x_2, \ldots, x_{n+1}) \in J \times \mathbb{R}^{n+1}, \quad \mu \in C(J, \mathbb{R}^+)$,
$|g(t, x_1, x_2, \ldots, x_{k+1})| \le \nu(t), \quad \forall (t, x_1, x_2, \ldots, x_{k+1}) \in J \times \mathbb{R}^{k+1}, \quad \nu \in C(J, \mathbb{R}^+)$ and
$|h(t, x_1, x_2, \ldots, x_{n+1})| \le \theta(t), \forall (t, x_1, x_2, \ldots, x_{n+1}) \in J \times \mathbb{R}^{n+1}, \quad \theta \in C(J, \mathbb{R}^+)$.

Theorem 2 *Assume that the conditions $(A_1) - (A_2)$ hold. Then the initial value problem 2 has at least one solution on J provided that*

$$\frac{T^{\alpha+\omega}}{\Gamma(\alpha+1)\Gamma(\omega+1)}\Big\{\Big(1+\frac{T^{\alpha_1}}{\Gamma(\alpha_1+1)}+\ldots+\frac{T^{\alpha_n}}{\Gamma(\alpha_n+1)}\Big)\|\nu\|\|\phi\| \\ +\|\mu\|\|\chi\|\Big(1+\frac{T^{\beta_1}}{\Gamma(\beta_1+1)}+\ldots+\frac{T^{\beta_k}}{\Gamma(\beta_k+1)}\Big)\Big\}<1$$

Proof Setting $\sup_{t\in J}|\mu(t)|=\|\mu\|$, $\sup_{t\in J}|\nu(t)|=\|\nu\|$, $\sup_{t\in J}|\theta(t)|=\|\theta\|$, and choosing

$$R\geq\frac{T^{\omega+\beta}}{\Gamma(\omega+\beta+1)}\|\theta\|+\frac{T^{\alpha+\beta}}{\Gamma(\alpha+\beta+1)}\|\mu\|\|\nu\|+|x_0|$$

We consider $B_R=\{x\in C(J,\mathbb{R}):\|x\|\leq R\}$. We define the operators $\mathscr{A}:E\longrightarrow E$ as in 3,
$\mathscr{D}:B_R\longrightarrow E$ by

$$\mathscr{D}x(t)=\int_0^t\frac{(t-s)^{\alpha-1}}{\Gamma(\alpha)}g(s,x(s),I^{\beta_1}x(s),\ldots,I^{\beta_k}x(s))ds,\quad t\in J$$

and

$$\mathscr{Q}x(t)=\int_0^t\frac{(t-s)^{\beta+\omega-1}}{\Gamma(\beta+\omega)}h(s,x(s),I^{\alpha_1}x(s),\ldots,I^{\alpha_n}x(s))ds,\quad t\in J$$

and

$$\mathscr{T}x(t)=\int_0^t\frac{(t-s)^{\omega-1}}{\Gamma(\omega)}\mathscr{A}x(t)\mathscr{D}x(s)ds+x_0,\quad t\in J$$

For any $y\in B_R$, we have

$$\begin{aligned}|x(t)|&=|\mathscr{Q}x(t)+\mathscr{T}y(t)|\\&\leq\int_0^t\frac{(t-s)^{\omega+\beta-1}}{\Gamma(\omega+\beta)}|h(s,x(s),I^{\alpha_1}x(s),\ldots,I^{\alpha_n}x(s))|ds\\&\quad+\int_0^t\frac{(t-s)^{\omega-1}}{\Gamma(\omega)}|\mathscr{A}y(s)||\mathscr{D}y(s)|ds+|x_0|\\&\leq\int_0^t\frac{(t-s)^{\omega+\beta-1}}{\Gamma(\omega+\beta)}|\theta(s)|ds+\int_0^t\frac{(t-s)^{\omega-1}}{\Gamma(\omega)}|\mu(t)|\int_0^t\frac{(t-s)^{\alpha-1}}{\Gamma(\alpha)}|\nu(s)|ds+|x_0|\\&\leq\frac{T^{\omega+\beta}}{\Gamma(\omega+\beta+1)}\|\theta\|+\frac{T^{\alpha+\beta}}{\Gamma(\alpha+\beta+1)}\|\mu\|\|\nu\|+|x_0|\end{aligned}$$

Taking supremum for $t\in J$, we obtain $\|x\|\leq R$, which means that $x\in B_R$. So, the condition *(iii)* of Lemma 6 holds.

Next we will show that $\mathscr{Q}$ satisfy the condition (*ii*) of Lemma 6. The operator $\mathscr{Q}$ is obviously continuous. Also, $\mathscr{Q}$ is uniformly bounded on B_R as

$$\|\mathscr{Q}x\| \leq \frac{T^{\omega+\beta}}{\Gamma(\omega+\beta+1)}\|\theta\|$$

Let $\tau_1, \tau_2 \in J$ with $\tau_1 < \tau_2$ and $(x_1, x_2, \ldots, x_{n+1}) \in B_R^{n+1}$. We define $\sup_{(t,x_1,x_2,\ldots,x_{n+1})\in J\times B_R^{n+1}} |h(t, x_1, x_2, \ldots, x_{n+1})| = \bar{h} < \infty.$
Then we have

$$\begin{aligned}
|\mathscr{Q}x(\tau_2) - \mathscr{Q}x(\tau_1)| &= \Big| \int_0^{\tau_2} \frac{(\tau_2 - s)^{\omega+\beta-1}}{\Gamma(\omega+\beta)} h(s, x(s), I^{\alpha_1}x(s), \ldots, I^{\alpha_n}x(s))ds \\
&\quad - \int_0^{\tau_1} \frac{(\tau_1 - s)^{\omega+\beta-1}}{\Gamma(\omega+\beta)} h(s, x(s), I^{\alpha_1}x(s), \ldots, I^{\alpha_n}x(s))ds \Big| \\
&\leq \frac{\bar{h}}{\Gamma(\omega+\beta)} \Big| \int_0^{\tau_1} [(\tau_2 - s)^{\omega+\beta-1} - (\tau_1 - s)^{\omega+\beta-1}]ds \\
&\quad + \int_{\tau_1}^{\tau_2} (\tau_2 - s)^{\omega+\beta-1}ds \Big| \\
&\leq \frac{\bar{h}}{\Gamma(\omega+\beta+1)} |\tau_2^{\omega+\beta} - \tau_1^{\omega+\beta}|
\end{aligned}$$

Thus, we have that $|\mathscr{Q}x(\tau_2) - \mathscr{Q}x(\tau_1)| \longrightarrow 0$ as $\tau_2 \longrightarrow \tau_1$

which is independent of $x \in S$. Thus, $\mathscr{Q}$ is equicontinuous. So $\mathscr{Q}$ is relatively compact on B_R. Hence, by the Arzelá-Ascoli theorem, $\mathscr{Q}$ is compact on B_R.

Now we show that $\mathscr{T}$ is a contraction mapping. Let $x, y \in B_R$, then for $t \in J$ we have

$$\begin{aligned}
&|\mathscr{T}x(t) - \mathscr{T}y(t)| \\
&= \Big| \int_0^t \frac{(t-s)^{\omega-1}}{\Gamma(\omega)} [\mathscr{A}x(s)\mathscr{D}x(s)ds - \mathscr{A}y(s)\mathscr{D}y(s)]ds \Big| \\
&= \Big| \int_0^t \frac{(t-s)^{\omega-1}}{\Gamma(\omega)} [\mathscr{A}x(s)\mathscr{D}x(s) - \mathscr{A}y(s)\mathscr{D}x(s) + \mathscr{A}y(s)\mathscr{D}x(s) - \mathscr{A}y(s)\mathscr{D}y(s)]ds \Big| \\
&\leq \int_0^t \frac{(t-s)^{\omega-1}}{\Gamma(\omega)} \Big\{ |\mathscr{D}x(s)||\mathscr{A}x(s) - \mathscr{A}y(s)| + |\mathscr{A}y(s)||\mathscr{D}x(s) - \mathscr{D}y(s)] \Big\} ds \\
&\leq \int_0^t \frac{(t-s)^{\omega-1}}{\Gamma(\omega)} \Big\{ \Big(1 + \frac{T^{\alpha_1}}{\Gamma(\alpha_1+1)} + \ldots + \frac{T^{\alpha_n}}{\Gamma(\alpha_n+1)}\Big) \frac{T^{\alpha}}{\Gamma(\alpha+1)} \|\nu\|\|\phi\|\|x-y\| \\
&\quad + \|\mu\|\|\chi\| \Big(1 + \frac{T^{\beta_1}}{\Gamma(\beta_1+1)} + \ldots + \frac{T^{\alpha_k}}{\Gamma(\beta_k+1)}\Big) \frac{T^{\alpha}}{\Gamma(\alpha+1)} \|x-y\| \Big\} ds \\
&\leq \frac{T^{\alpha}}{\Gamma(\alpha+1)} \frac{T^{\omega}}{\Gamma(\omega+1)} \Big\{ \Big(1 + \frac{T^{\alpha_1}}{\Gamma(\alpha_1+1)} + \ldots + \frac{T^{\alpha_n}}{\Gamma(\alpha_n+1)}\Big) \|\nu\|\|\phi\| \\
&\quad + \|\mu\|\|\chi\| \Big(1 + \frac{T^{\beta_1}}{\Gamma(\beta_1+1)} + \ldots + \frac{T^{\alpha_k}}{\Gamma(\beta_k+1)}\Big) \Big\} \|x-y\|
\end{aligned}$$

So, by Theorem 2, $\mathscr{T}$ is a contraction mapping, and thus the condition (i) of Lemma 2 is satisfied.

Thus all the assumptions of Lemma 6 are satisfied. Therefore, the conclusion of Lemma 6 implies that problem 2 has at least one solution on J.

5 Example

In this section we give an example to illustrate the usefulness of our main results. Let us consider the following fractional boundary value problem:

$$\begin{cases} D^{\frac{1}{2}}\left(\frac{x(t)-I^{\frac{1}{2}}\left[\frac{2te^{-3t}}{15(3+t)}\left(\sin x(t)+\frac{x(t)+9I^{\sqrt{2}}|x(t)|}{I^{\sqrt{2}}|x(t)|+5}\right)\right]}{\frac{(t+1)^2}{100}\left(\sin x(t)+\frac{|I^{\sqrt{2}}x(t)|}{1+|I^{\sqrt{2}}x(t)|}+3\right)}\right) = t^2\sin x(t)+\cos(I^{\frac{1}{4}}x(t))+1 \\ t \in J = [0,1] \\ \frac{x(0)}{f(0,x(0),0)} = \frac{\pi}{2}, \quad \frac{x(1)}{f(1,x(1),I^{\alpha_1}x(1))} = 0, \end{cases} \tag{6}$$

Put $\alpha = \frac{1}{2}, \alpha_1 = \sqrt{2}, \beta = \frac{1}{2}, \beta_1 = \frac{1}{4}, T = 1, n = k = 1, f(t,y,x) = \frac{(t+1)^2}{100}\Big(\sin y(t) + \frac{|x|}{1+|x|} + 3\Big)$, $g(t,y,x) = t^2 \sin x(t) + \cos(I^{\frac{1}{4}}x(t)) + 1$, $h(t,y,x) = \frac{2te^{-3t}}{15(3+t)}(\sin y(t) + \frac{x^2(t)+9|x(t)|}{|x(t)|+5})$, $m(t) = \frac{2t}{15(3+t)}$ and $p(t) = \frac{(t+1)^2}{100}$ for $t \in [0,1]$. Note that, $\|g(t,y,x)\| \leq t^2 + 2$, and

$$|f(t,x,y) - f(t,x',y')| \leq \frac{(t+1)^2}{100}(|x-x'| + |y-y'|)$$

and

$$|h(t,x,y) - h(t,x',y')| \leq \Big(\frac{2t}{15(3+t)}\Big)(|x-x'| + |y-y'|)$$

We have

$$\Big(1 + \frac{T^{\alpha_1}}{\Gamma(\alpha_1+1)}\Big)\Big(\frac{2\|p\|\|h\|_{L^1}T^{\alpha}}{\Gamma(\alpha+1)} + |x_0| + |x_1| + |d| + \|m\|\frac{T^{\beta}}{\Gamma(\beta+1)}\Big)$$
$$\simeq 0.18957628293 < 1.$$

By using the theorem 1, the problem 6 has a solution.

References

1. A.A. Kilbas, H.M. Srivastava, J.J. Trujillo, *Theory and Applications of Fractional Differential Equations, North-Holland Mathematics Studies, 204* (Elsevier Science B.V, Amsterdam, 2006)
2. I. Podlubny, *Fractional Differential Equations* (Academic Press, San Diego, 1999)
3. J. Sabatier, O.P. Agrawal, J.A.T. Machado (eds.), *Advances in Fractional Calculus: Theoretical Developments and Applications in Physics and Engineering* (Springer, Dordrecht, 2007)
4. K.S. Miller, B. Ross, *An Introduction to the Fractional Calculus and Fractional Differential Equations* (Wiley, New York, 1993)
5. B. Ahmad, S. Sivasundaram, On four-point nonlocal boundary value problems of nonlinear integro-differential equations of fractional order. Appl. Math. Comput. **217**, 480–487 (2010)
6. B. Ahmad, S.K. Ntouyas, A. Alsaedi, Existence theorems for nonlocal multi-valued Hadamard fractional integro-differential boundary value problems. J. Inequal. Appl. **2014**, 454 (2014)
7. W. Chen, Y. Zhao, Solvability of boundary value problems of nonlinear fractional differential equations. Adv. Differ. Equ. **2015**, 36 (2015)
8. Y. Zhao, S. Sun, Z. Han, Q. Li, The existence of multiple positive solutions for boundary value problems of nonlinear fractional differential equations. Commun. Nonlinear Sci. Numer. Simul. **16**, 2086–2097 (2011)
9. Y. Zhao, S. Sun, Z. Han, Q. Li, Theory of fractional hybrid differential equations. Comput. Math. Appl. **62**, 1312–1324 (2011)
10. S. Sun, Y. Zhao, Z. Han, Y. Li, The existence of solutions for boundary value problem of fractional hybrid differentialequations. Commun. Nonlinear Sci. Numer. Simul. **17**, 4961–4967 (2012)
11. B. Ahmad, S.K. Ntouyas, An existence theorem for fractional hybrid differential inclusions of Hadamard type with Dirichlet boundary conditions. Abstr. Appl. Anal. 2014, Article ID 705809 (2014)
12. B.C. Dhage, S.K. Ntouyas, Existence results for boundary value problems for fractional hybrid differential inclusions. Topol. Methods Nonlinear Anal. **44**, 229–230 (2014)
13. Y. Zhao, Y. Wang, Existence of solutions to boundary value problem of a class of nonlinear fractional differential equations. Adv. Differ. Equ. **2014**, 174 (2014)
14. B. Ahmad, S.K. Ntouyas, A. Alsaedi, Existence results for a system of coupled hybrid fractional differential equations. Sci. World J. 2014, Article ID 426438 (2014)
15. B.C. Dhage, V. Lakshmikantham, Basic results on hybrid differential equations. Nonlinear Anal. Hybrid **4**, 414–424 (2010)
16. B.C. Dhage, Basic results in the theory of hybrid differential equations with mixed perturbations of second type. Funct. Differ. Equ. **19**, 1–20 (2012)
17. S. Sitho, S.K. Ntouyas, J. Tariboon, Existence results for hybrid fractional integro-differential equations, Boundary Value Problems (2015)
18. K. Hilal, A. Kajouni, Boundary value problems for hybrid differential equations with fractional order. Advances in Difference Equations (2015)
19. Y. Zhao, S. Suna, Z. Han, Q. Li, Theory of fractional hybrid differential equations. Comput. Math. Appl. **62**, 1312–1324 (2011)
20. M. Benchohra, S. Hamani, S.K. Ntouyas, Boundary value problems for differential equations with fractional order. Surveys Math. Appl. **3**, 1–12 (2008)
21. B.C. Dhage, A fixed point theorem in Banach algebras with applications to functional integral equations. Kyungpook Math. J. **44**, 145–155 (2004)
22. V. Lakshmikantham, S. Leela, J. Vasundhara Devi, *Theory of Fractional Dynamic Systems* (Cambridge Academic Publishers, Cambridge, 2009)
23. V. Lakshmikantham, A.S. Vatsala, Basic theory of fractional differential equations. Nonlinear Anal. **69**(8), 2677–2682 (2008)
24. I. Podlubny, *Fractional Differential Equations* (Academic Press, San Diego, 1999)
25. J. Tariboon, S.K. Ntouyas, W. Sudsutad, Fractional integral problems for fractional differential equations via Caputo derivative. Adv. Differ. Equ. **181** (2014)

26. B. Ahmad, S.K. Ntouyas, A four-point nonlocal integral boundary value problem for fractional differential equations of arbitrary order. Electron. J. Qual. Theory Differ. Equ. **2011**, 22 (2011)
27. B. Ahmad, S. Sivasundaram, Existence and uniqueness results for nonlinear boundary value problems of fractional differential equations with separated boundary conditions. Commun. Appl. Anal. **13**, 121–228 (2009)
28. D. Baleanu, V. Hedayati, S. Rezapour, M. Mohamed Al Qurashi, On two fractional differential inclusions. Springer Plus **5**, 882 (2016)

Convergence on Intuitionistic Fuzzy Metric Space

M. El Hassnaoui, Said Melliani and L. S. Chadli

Abstract Using the idea of intuitionistic fuzzy metric space, due to George and Veeramani [fuzzy sets and systems 90(1997) 365–368], and the results of metric space by Jin Han Park [intuitionistic fuzzy metric spaces (2004) 1036–1046] we define a hausdroff topology on a fuzzy metric space. Also we prove an equivalence between the convergence in a fuzzy separable metric space and the adhesion of intuitionistic fuzzy set.

1 Introduction

One of the most important problems in fuzzy topology is to obtain an appropriate concept of fuzzy metric space, this problem has been investigated by many authors [1–3] from different points of views. In particular, George and Veerimani [4] have introduced and studied a notion of fuzzy metric space with help of continuous t-norms, Atanassov [1] introduced and studied the concept of intuitionistic fuzzy sets as a generalization of fuzzy sets, and later there has been much progress in the study of intuitionistic fuzzy sets by many author [1–3, 5]. In this paper, using the idea of intuitionistic fuzzy set. We define the notion of intuitionistic fuzzy metric spaces with help of continuous t-norms and continuous t-conorms due to George and Veermani, Also the Hausdorff topology on this intuitionistic fuzzy metric space and show that every metric induces an intuitionistic fuzzy metric, more we define an intuitionistic fuzzy topological space and the closure set in this space, finaly we prove the tow implications between the convergence on an intuitionistic fuzzy metric space and the closure set.

M. El Hassnaoui · S. Melliani (✉) · L. S. Chadli
Sultan Moulay Slimane University, BP 523 Beni Mellal, Morocco
e-mail: s.melliani@usms.ma

M. El Hassnaoui
e-mail: mariam.hassnaoui@gmail.com

L. S. Chadli
e-mail: sa.chadli@yahoo.fr

S. Melliani and O. Castillo (eds.), *Recent Advances in Intuitionistic Fuzzy Logic Systems*, Studies in Fuzziness and Soft Computing 372,
https://doi.org/10.1007/978-3-030-02155-9_18

2 Preliminaries

Definition 1 A binary operation *: $[0, 1] \times [0, 1] \longrightarrow [0, 1]$ is a continuous t-norm if * is satisfying the following conditions:

1. * is commutative and associative.
2. * is continuous.
3. $a * 1 = a$ for all $a \in [0, 1]$.
4. $a * b \le c * d$ whenever $a \le c$, $c \le d$ and $a, b, c \in [0, 1]$.

Definition 2 A binary operation $\Diamond : [0, 1] \times [0, 1] \longrightarrow [0, 1]$ is continuous t-conorm if $\Diamond$ is satisfying the following conditions:

1. $\Diamond$ is commutative and associative.
2. $\Diamond$ is continuous.
3. $a \Diamond 0 = a$ for all $a \in [0, 1]$.
4. $a \Diamond b \le c \Diamond d$ whenever $a \le c$, $c \le d$ and $a, b, c \in [0, 1]$.

Definition 3 A 5-tuple $(X, M, N, *, \Diamond)$ is said to be an intuitionistic fuzzy space if X is an arbitrary set, $*$ is a continuous t-norm, $\Diamond$ is a continuous t-conorm and M, N are fyzzy sets on $X^2 \times (0, \infty)$ satisfying the following conditions: for all $x, y, z \in X$, $s, t > 0$,

1. $M(x, y, t) + N(x, y, t) \le 1$.
2. $M(x, y, t) > 0$.
3. $M(x, y, t) = 1$ if and only if $x = y$.
4. $M(x, y, t) = M(y, x, t)$.
5. $M(x, y, t) * M(y, z, s) \le M(x, z, t + s)$.
6. $M(x, y, \cdot) : (0, \infty) \longrightarrow (0, 1]$ is continuous.
7. $N(x, y, t) > 0$.
8. $N(x, y, t) = 0$ if and only if $x = y$.
9. $N(x, y, t) = N(y, x, t)$.
10. $N(x, y, t) \Diamond N(y, z, s) \le N(x, z, t + s)$.
11. $N(x, y, \cdot) : (0, \infty) \longrightarrow (0, 1]$ is continuous.

Then (M, N) is called an intuitionistic fuzzy metric space on X. The function $M(x, y, t)$ and $N(x, y, t)$ denote the degree of nearness and the degree of non-nearness between x and y with respect to t, respectively.

Remark 1 Every fuzzy metric space $(X, M, *)$ is an intuitionistic fuzzy metric space of the form $(X, M, 1 - M, *, \Diamond)$ such that t-norm $*$ and t-conorm $\Diamond$ are associated, i.e $x \Diamond y = 1 - ((1 - x) * (1 - y))$ for any $x, y \in X$.

Remark 2 In intuitionistic fuzzy metric space X, $M(x, y, \cdot)$ is non-decreasing and $N(x, y, \cdot)$ is non-increasing for all $x, y \in X$.

Example 1 (Induced intuitionistic fuzzy metric). Let (X, d) be a metric space. denote $a * b = ab$ and $a \Diamond b = min\{1, a + b\}$ for all $a, b \in [0, 1]$ and let M_d and N_d be fuzzy sets on $X^2 \times (0, \infty)$ defined as follows:

$$M_d(x, y, t) = \frac{ht^m}{ht^n + md(x, y)}, \quad N_d(x, y, t) = \frac{d(x, y)}{kt^n + md(x, y)} \quad \text{for all } h, k, m, n \in \mathbb{R}^+.$$

Then $(X, M_d, N_d, *, \Diamond)$ is an intuitionistic fuzzy metric space.

Remark 3 Note the above example holds even with the t-norm $a * b = min\{a, b\}$ and the t-conorom
$a \Diamond b = max\{a, b\}$ and hence (M, N) is an intuitionistic fuzzy metric with respect to any continuous t-norm and continuous t-conorm. In the above example by taking $h = k = m = n = 1$, We get $M_d(x, y, t) = \frac{t}{t+d(x,y)}$, $N_d = \frac{d(x,y}{t+d(x,y)}$. We call this intuitionistic fuzzy metric induced by *a* metric *d* the standard intuitionistic fuzzy metric.

3 Topology Induced by an Intuitionistic Fuzzy Metric

Definition 4 Let $(X, M, N, *, \Diamond)$ be an intuitionistic fuzzy metric space, and let $r \in (0, 1)$, $t > 0$ and $x \in X$. The set $B(x, r, t) = \{y \in X : M(x, y, t) > 1 - r, N(x, y, t) < r\}$ is called the open ball with center x and radius r with respect to t.

Theorem 1 *Every open ball $B(x, r, t)$ is an open set.*

Proof Let $B(x, r, t)$ be an open ball with centre x and radius r with respect to t. Let $y \in B(x, r, t)$. Then $M(x, y, t) > 1 - r$ and $N(x, y, t) < r$. Since $M(x, y, t) > 1 - r$, there exists $t_0 \in (0, t)$ such that $M(x, y, t_0) > 1 - r$ and $N(x, y, t_0) < r$. Put $r_0 = M(x, y, t_0)$, since $r_0 > 1 - r$, there exists $s \in (0, 1)$ such that $r_0 > 1 - s > 1 - r$. Now for given r_0 and s such that $r_0 > 1 - s$, there exist $r_1, r_2 \in (0, 1)$ such that $r_0 * r_1 > 1 - s$ and $(1 - r_0) \Diamond (1 - r_2) \leq s$. Put $r_3 = max\{r_1, r_2\}$ and consider the open ball $B(y, 1 - r_3, t - t_0)$.
We claim $B(y, 1 - r_3, t - t_0) \subset B(x, r, t)$. Now, let $z \in B(y, 1 - r_3, t - t_0)$. Then $M(y, z, t - t_0) > r_3$ and $N(y, z, t - t_0) < r_3$. Therefore
$M(x, z, t) \geq M(x, y, t_0) * M(y, z, t - t_0) \geq r_0 * r_3 \geq r_0 * r_1 \geq 1 - s > 1 - r$ and
$N(x, z, t) \leq N(x, y, t_0) \Diamond N(y, z, t - t_0) \leq (1 - r_0) \Diamond (1 - r_3) \leq (1 - r_0) \Diamond (1 - r_2)$
$\leq s < r$.
Thus $z \in B(x, r, t)$ and hence $B(y, 1 - r_3, t - t_0) \subset B(x, r, t)$.

Remark 4 Let $(X, M, N, *, \Diamond)$ be an intuitionistic fuzzy metric space. Define $\tau_{(M,N)} = \{A \in X$: for each $x \in A$, there exist t > 0 and $r \in (0, 1)$ such that $B(x, r, t) \subset A\}$. Then $\tau_{(M,N)}$ is a topology on X.

Remark 5 1. From Theorem 1 and Remark 3, every intuitionistic fuzzy metric space (N, M) on X generates a topology $\tau_{(M,N)}$ on X which has as a base the family of open sets of open sets of the form $\{B(x, r, t) : x \in, r \in (0, 1), t > 0\}$.

2. Since $\{B(x, \frac{1}{n}, \frac{1}{n}) : n = 1, 2, \ldots\}$ is a local base at x,the topology $\tau_{(M,N)}$ is first countable.

Theorem 2 *Every intuitionistic fuzzy metric space is Hausdorff.*

Proof Let $(X, M, N, *, \diamondsuit)$ be an intuitionistic fuzzy metric space. Let x and y be two distinct points in X then $0 < M(x, y, t) < 1$ and $0 < N(x, y, t) < 1$. Put $r_1 = M(x, y, t), r_2 = M(x, y, t)$ and $r = max\{r_1, 1 - r_2\}$. For each $r_0 \in$(r,1), there exist r_3 and r_4 such that $r_3 * r_3 \geq r_0$ and $(1 - r_4)\diamondsuit(1 - r_4) \leq 1 - r_0$ put $r_5 = max\{r_3, 1 - r_4\}$ and consider the open balls $B(x, 1 - r_5, \frac{t}{2})$ and $B(y, 1 - r_5, \frac{t}{2})$. Then clearly $B(x, 1 - r_5, \frac{t}{2}) \cap B(y, 1 - r_5, \frac{t}{2}) = \emptyset$. If there exists $z \in B(x, 1 - r_5, \frac{t}{2}) \cap B(y, 1 - r_5, \frac{t}{2})$ then

$$r_1 = M(x, y, t) \geq M(x, z, \frac{t}{2}) * M(z, y, \frac{t}{2}) \geq r_5 * r_5 \geq r_3 * r_3 \geq r_0 > r_1$$

and

$$\begin{aligned} r_2 = N(x, y, t) &\leq N(x, z, \frac{t}{2})\diamondsuit N(z, y, \frac{t}{2}) \leq (1 - r_5)\diamondsuit(1 - r_5) \\ &\leq (1 - r_4)\diamondsuit(1 - r_4) \leq 1 - r_0 < r_2 \end{aligned}$$

which is a contradiction. Hence $(X, M, N, *, \diamondsuit)$ is Hausdorff.

Remark 6 Let (X, d) be a metric space.
Let $M(x, y, t) = \frac{t}{t+d(x,y)}$, $N(x, y, t) = \frac{d(x,y)}{kt+d(x,y)}$ $k \in \mathbb{R}^+$ be the intuitionistic fuzzy metric defined on X. Then the topology τ_d induced by the metric d and the topology $\tau_{(M,N)}$ induced by the intuitionistic fuzzy metric (M, N) are the same.

Definition 5 Let $(X, M, N, *, \diamondsuit)$ be an intuitionistic fuzzy metric space. A subset A of X is said to be **IF**-bounded if there exist $t > 0$ and $r \in (0, 1)$ such that $M(x, y, t) > 1 - r$ and $N(x, y, t) < r$ for all $x, y \in A$.

Theorem 3 *Let $(X, M, N, *, \diamondsuit)$ be a fuzzy metric space and $\tau_{(M,N)}$ be the topology in X induced by the fuzzy metric.*
Then for a sequence $\{x_n\}$ in X, $x_n \longrightarrow x$ if and only if $M(x_n, x, t) \longrightarrow 1$ and $N(x_n, x, t) \longrightarrow 0$ as $n \longrightarrow \infty$.

Proof Fix $t > 0$, suppose $x_n \longrightarrow x$ then for $r \in (0, 1)$, there exists $n_0 \in \mathbb{N}$ such that $x_n \in B(x, r, t)$ for all $n \geq n_0$. Then $1 - M(x_n, x, t) < r$ and $N(x_n, x, t) < r$ and hence $M(x_n, x, t) \longrightarrow 1$ $N(x_n, x, t) \longrightarrow 0$ as $n \longrightarrow \infty$.
Conversely, if for each $t > 0$, $M(x_n, x, t) \longrightarrow 1$ and $N(x_n, x, t) \longrightarrow 0$ as $n \longrightarrow \infty$, then for $r \in (0, 1)$, there exists $n_0 \in \mathbb{N}$ $n \geq n_0$. Thus $x_n \in B(x, r, t)$ for all $n \geq n_0$ and hence $x_n \longrightarrow x$.

4 Intuitionistic Fuzzy Topological Space

Definition 6 An intuitionistic fuzzy topology (IFT) on a nonempty set X is a family τ of IFS in X satisfying the folloying axioms.

T_1. $0_{\sim}, 1_{\sim} \in \tau$.
T_2. $G_1 \cap G_2 \in \tau$, for any $G_1, G_2 \in \tau$.
T_3. $\cup G_i \in \tau$, for any arbitrary family $\{G_i : G_i \in \tau. i \in I\}$.

In this case the pair (X, τ) is called an intuitionistic fuzzy topological space and any IFS in τ is known as intuitionistic fuzzy open set in X.

Definition 7 Let (X, τ) be an IFTS and $A =< x, \mu_A, \gamma_A >$ be an IFS in X. Then the fuzzy interior and fuzzy closure of A are defined by: $cl(A) = \cap\{K$: K is an IFCS in X and $A \subseteq K\}$.
Denoted by $\bar{A}$ the intuitionistic fuzzy closed set of A in X.

5 Main Results

Theorem 4 *Let $(X, M, N, *, \diamondsuit)$ be an intuitionistic fuzzy metric space, for all $A \subset X$ and for all $x \in X$ we have the follwing result:*

$$x \in \bar{A} \iff \text{there exist a sequence } (x_n) \in A \text{ such that } x = \lim x_n.$$

Proof Suppose that $x \in \bar{A}$, Since A is first countable, there exists a sequence $x_n \in A$ and we have.
• $x_n \longrightarrow x$.
• Conversely it is easy to show if $x_n \longrightarrow x$ and using the definition of limit on intuitionistic fuzzy metric space the result holds.
indeed:
Suppose that $x \in \bar{A}$, Since X is first countable, there exists a sequence $x_n \in A$ and we have

$$\begin{aligned}
\forall \varepsilon > 0,\ \exists N > 0,\ n > N,\ d(x_n, x) < \varepsilon &\iff t + d(x_n, x) < \varepsilon + t \\
&\iff \frac{1}{t + d(x_n, x)} > \frac{1}{\varepsilon + t} \\
&\iff \frac{t}{t + d(x_n, x)} > \frac{t}{\varepsilon + t} = \frac{t + \varepsilon - \varepsilon}{\varepsilon + t} = 1 - \frac{\varepsilon}{\varepsilon + t} \\
&\iff M(x_n, x, t) > 1 - \varepsilon
\end{aligned}$$

and

$$N(x_n, x, t) \longrightarrow 0\ (N(x_n, x, t) = \frac{d(x_n, x)}{kt + d(x_n, x)} \leq d(x_n, x) < \varepsilon)$$

hence

$$x_n \in B(x, \varepsilon, t) \qquad \text{i.e.} \qquad \lim_{n\to\infty} x_n = x$$

Now we claim that $x_n \in \bar{A}$, suppose that $x_n \longrightarrow x$ then for $r \in (0, 1)$ there exists $n_0 \in \mathbb{N}$ for all $n \geq n_0$ such that $x_n \in B(x, r, t)$ and $x_n \in A$ then $B(x, r, t) \cap A \neq \varnothing$ hence $x_n \in \bar{A}$.

Remark 7 The second implication is true for all intuitionistic fuzzy topological space (X, τ), (even if (X, τ) not a separate space).
On the other hand, for the first implication we need that all point have a countable base of neighbourhood which is true on intuitionistic fuzzy metric space.

References

1. K. Atanassov, Intuitionistic fuzzy sets. Fuzzy Sets Syst. **2**, 87–96 (1989)
2. D. Coker, An introduction to intuitionistic fuzzy topological spaces. Fuzzy Sets Syst. **88**, 81–89 (1997)
3. M.A. Erceg, Metric spaces in fuzzy set theory. J. Math. Anal. Appl. **69**, 316–328 (1979)
4. J.H. Park, Intuitionistic fuzzy metric spaces. Chaos Solitons Fractals **22**(5), 1039–1046 (2004)
5. A. George, P. Veeramani, On some results in fuzzy metric spaces. Fuzzy Sets Syst. **64**(3), 395–399 (1994)
6. E.P. Klement, Operations on fuzzy sets: an axiomatic approach. Inform. Sci. **27**, 221–232 (1984)
7. O. Kramosil, J. Michalek, Fuzzy metric and statistical metric spaces. Kybernetica **11**, 326–334 (1975)
8. R. Owen, *Fuzzy Set Theory* (Kluwer Academic Publishers, 1996)
9. K. Menger, Statistical metrics. Proc. Nat. Acad. Sci. **28**, 535–537 (1942)
10. J.R. Munkres, *Topologya First Course* (Prentice-Hall, 1975)
11. J. Nagata, *Modern General Topology* (North-Holland, 1974)
12. B. Schweizer, A. Sklar, Statistical metric spaces. Pac. J. Math. **10**, 314–334 (1960)
13. R.R. Yager, On a general class of fuzzy connectives. Fuzzy Sets Syst. **4**, 235–242 (1980)
14. L.A. Zadeh, Fuzzy sets. Inform. Control **8**, 338–353 (1965)

Graphical Representation of Intuitionistic Membership Functions for Its Efficient Use in Intuitionistic Fuzzy Systems

Amaury Hernandez-Aguila, Mario Garcia-Valdez and Oscar Castillo

Abstract This work proposes an approach for representing intuitionistic fuzzy sets for their efficient use in Mamdani type intuitionistic fuzzy inference systems. The proposed approach is used, and plots for several membership and non-membership functions are presented, including: triangular, Gaussian, trapezoidal, generalized bell, sigmoidal, and left-right functions. Plots of some operators used in fuzzy logic are also presented, i.e., union, intersection, implication and alpha-cut operators. The proposed approach should produce plots that are clear to understand in the design of an intuitionistic fuzzy inference system, as the membership and non-membership functions are clearly separated and can be plotted in the same figure and still be recognized with ease.

Keywords Fuzzy inference systems · Intuitionistic fuzzy logic
Membership functions

1 Introduction

Fuzzy sets have been used as the building blocks of many other areas and applications since they were conceived by Zadeh in 1965 [1]. One of the most prominent areas is fuzzy logic and its wide area of application of control. As a consequence of its usefulness in the industry, many works have been dedicated to improving the architectures and theory used in the construction of control systems. A remarkable example of this is the extension of fuzzy sets to the concept of type-2 fuzzy sets by Zadeh in 1975 [2] and the extension from traditional fuzzy sets to intuitionistic

A. Hernandez-Aguila · M. Garcia-Valdez · O. Castillo (✉)
Tijuana Institute of Technology, Tijuana, BC, Mexico
e-mail: ocastillo@tectijuana.mx

A. Hernandez-Aguila
e-mail: amherag@tectijuana.mx

M. Garcia-Valdez
e-mail: mario@tectijuana.mx

S. Melliani and O. Castillo (eds.), *Recent Advances in Intuitionistic Fuzzy Logic Systems*, Studies in Fuzziness and Soft Computing 372,
https://doi.org/10.1007/978-3-030-02155-9_19

fuzzy sets (IFS) by Atanassov in 1986 [3]. These extensions focus on increasing the uncertainty a fuzzy set can model, and as a consequence these fuzzy sets can help in the creation of better controls where the input data has high levels of noise. In particular, type-2 fuzzy sets enable the membership of an element in a fuzzy set to be described with another fuzzy set, and IFSs enable the description of an element in terms of both membership and non-membership.

A common application of fuzzy sets is as a part of inference systems, and these systems are called fuzzy inference systems (FIS). FISs use fuzzy sets to associate different degrees of membership to the inputs of the system and work as the antecedents of the inference system. Other fuzzy sets are used to model the consequents of the inference system (as in a Mamdani type FIS). Traditional fuzzy sets (also called type-1 fuzzy sets), are commonly defuzzified in order to obtain a scalar value as the output of the system, instead of a fuzzy set. When a FIS uses type-2 fuzzy sets as its antecedents and consequents, the output of the system needs to be reduced to a type-1 fuzzy set first, and then this type-1 fuzzy set is defuzzified to a scalar value.

One of the drawbacks of working with type-2 fuzzy sets in a FIS is that the type reduction procedure is very time consuming, due to the high quantity of steps involved in the process, and the use of a type-2 FIS (T2-FIS) is slower than the use of a type-1 FIS (T1-FIS). The type reduction and defuzzification processes are described in detail by Karnik and Mendel in [4]. This drawback is a possible explanation of why many controllers still rely on the use of T1-FIS instead of T2-FIS. Furthermore, most of the T2-FIS use a special case of type-2 fuzzy sets called interval type-2 fuzzy sets, as these fuzzy sets require less steps in their type reduction procedure, and thus, interval type-2 FIS (IT2-FIS) are faster than a general type-2 FIS (GT2-FIS). Type-2 fuzzy logic has found many applications, like in [5–8].

IFSs can also be used to construct a Mamdani type FIS, as in [9]. In contrast to a T2-FIS, an intuitionistic FIS (IFIS) doesn't require a type reduction procedure, and an implementation of an IFIS should work nearly as fast as a T1-FIS. The advantage of an IFIS over a T2-FIS is that the inference system can handle more uncertainty without a high penalty in time.

Actually, there are not many works involving Mamdani type intuitionistic inference systems yet, and the authors of this work are only aware of the work by Castillo et al. [10], and Hernandez-Aguila and Garcia-Valdez [9]. As a consequence of this lack of works involving Mamdani FISs, there is not a common way to graphically represent IFS for its use as membership and non-membership functions for an IFIS. Additionally, there is not a common way to graphically represent the architecture of an IFIS, as the one presented in the Fuzzy Logic Toolbox of Matlab. Having a standardized way of representing these components of an IFIS should ease the description of such systems in future research.

This work proposes an approach to construct graphical representations of IFSs for their use in IFISs, which is described in detail in Sect. 4. Some preliminaries can be found in Sect. 2, which are needed in order to understand the proposed approach. In Sect. 3, one can find a number of related works, which describe other ways of representing IFSs. Finally, in Sect. 5 one can find the conclusions and the future work.

2 Preliminaries

An IFS, as defined by Atanassov in [11], is represented by a capital letter with superscript star. An example of this notation is A^*. The definition of an IFS is also defined in [11], and is described in Eq. (1). In this equation, x is an element in set A^*, $\mu_{A^*}(x)$ is the membership of x, and $\upsilon_{A^*}(x)$ is the non-membership of x. For every triplet in A^*, (2) must be satisfied.

$$A^* = \{\langle x, \mu_A(x), \upsilon_A(x)\rangle | x \in E\} \quad (1)$$

$$0 \leq \mu_A(x) + \upsilon_A(x) \leq 1 \quad (2)$$

An IFS is a generalization of a traditional fuzzy set, meaning that a traditional fuzzy set can be expressed using the terminology of an IFS. An example of this is found in (3).

$$A^* = \{\langle x, \mu_A(x), 1 - \mu_A(x)\rangle | x \in E\} \quad (3)$$

If $0 \leq \mu_A(x)+\upsilon_A(x) < 1$ is true for an IFS, it is said that indeterminacy exists in the set. This concept of indeterminacy can also be found in the literature as hesitancy or non-determinacy, and it is described in Eq. (4).

$$\pi_A(x) = 1 - \mu_A(x) - \upsilon_A(x) \quad (4)$$

3 Related Work

The most common approach to graphically representing an IFS is by lattices. Examples of this type of representation can be found in the works by Despi et al. [12], and Deschrijver et al. [13]. This is a popular approach to graphically represent an IFS as it enables more compact and concise mathematical expressions. Another representation that is suitable for mathematical processes is that of a matrix, and is discussed in detail in the works by Parvathi et al. [14], Çuvalcioglu et al. [15], and Yilmaz et al. [16].

IFSs have been graphically represented like membership functions are usually represented in Mamdani FISs, and some example works are the ones by Angelov [17], Atanassov [11], and Davarzani and Khorheh [18]. This notation can be suitable for representing an architecture of an IFIS, but if the plot is in black and white, or in grayscale, the reader can get confused by the membership and non-membership plots. This problem can be alleviated by plotting the membership and non-membership functions in separate plots, is in the works by Castillo et al. [10], and Akram et al. [19].

There are several other graphical representations of IFSs, such as by radar charts, as in the work by Atanassova [20], and by geometrical representations, orthogonal projections and three-dimensional representations, as can be found in the work by Szmidt and Kacprzyk [21].

Some applications of IFSs in the area of medical sciences can be found in the works by Szmidt and Kacprzyk [22], Own [23], and Chakarska and Antonov [24]. In the area of group decision making, we have an example in the work by Xu [25]. IFSs have also been used in word recognition, in the area of artificial vision, as in the example work of Baccour et al. [26].

This work proposes that IFSs, in a Mamdani IFIS, should follow an approach similar to that found in the work by Atanassov [11], where the membership is plotted as is commonly done in a traditional FIS, but the non-membership function should be plotted as $1 - \upsilon_A$. The reason behind this decision is that the non-membership function should be easily differentiated from the membership function, while seeing both functions in the same plot. An implementation of an IFIS that uses this approach for representing IFSs for a Mamdani IFIS can be found in the work by Hernandez-Aguila and Garcia-Valdez [9].

4 Proposed Approach

What follows is a series of graphical representations of several commonly used membership functions in FISs, as well as graphical representations of common operators used in the construction of these systems, such as the union, intersection, and implication between two IFSs.

In Fig. 1, one can appreciate how a traditional fuzzy set can be constructed using the proposed approach. A Gaussian membership function with mean of 50 and a standard deviation of 15 is depicted.

Figure 2 is the first case of an IFS that cannot be considered a traditional fuzzy set. The red line represents the membership function, while the blue line represents the non-membership function. As can be seen, the Gaussian membership function does not have a kernel, meaning that its highest valued member does not equal to 1. In this case, its highest valued member equals to 0.7, and for the non-membership function, its highest valued member equals to 0.3. The Gaussian membership function is constructed with a mean of 50 and a standard deviation of 15. For the non-membership function, it is constructed with a mean of 30 and standard deviation of 30.

The triangular membership function is depicted in Fig. 3. The membership function is constructed with the following points: 30, 50, and 80, meaning that, from left to right, the last 0 valued member is at 30, the first and only 1 valued member is at 50, and the first 0 valued member after the previous series of non-0 valued members is at 80. In the same fashion, the non-membership function is constructed with the following points: 40, 60, and 80. The highest valued member for the membership function is equal to 0.8, and for the non-membership function it equals to 0.2.

The trapezoidal membership function is illustrated in Fig. 4. The membership function is constructed using the following points: 20, 40, 60, and 80. These points mean that, from left to right, the first non-0 valued member will be at 20, and a line

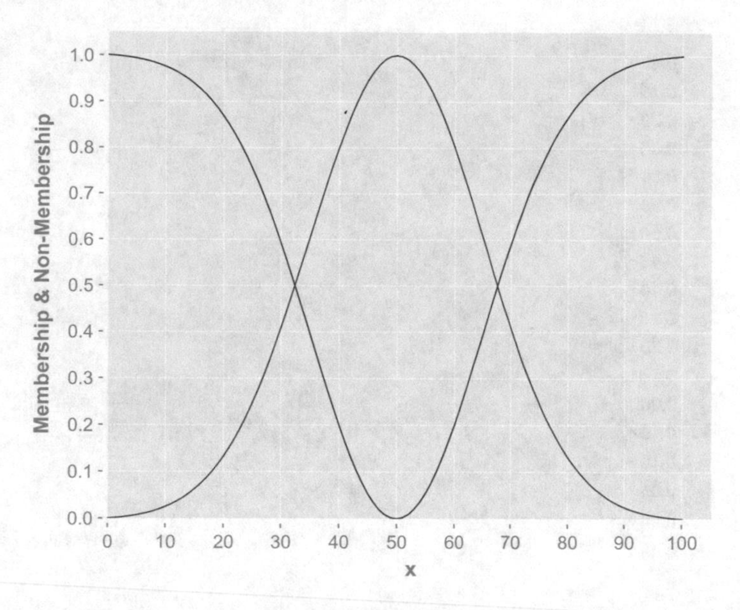

Fig. 1 A traditional fuzzy set represented as an intuitionistic fuzzy set

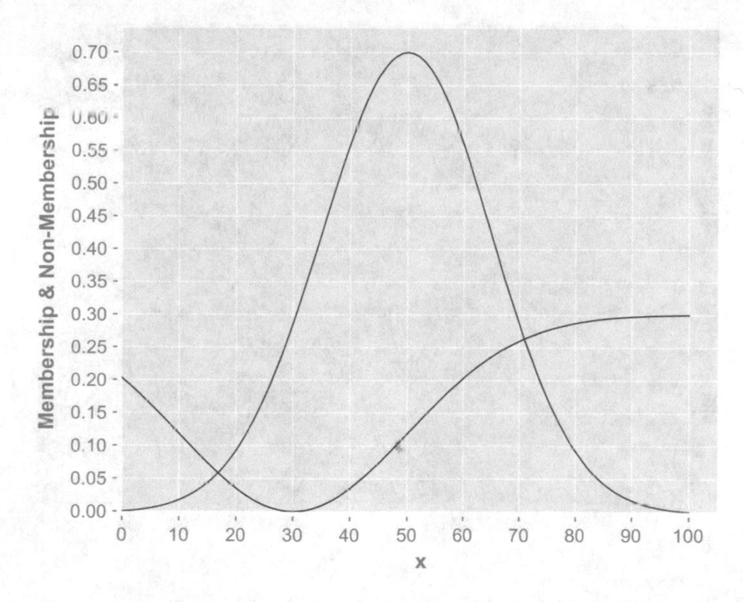

Fig. 2 Example of an intuitionistic fuzzy set

will be drawn from 20 to 40. All members from 40 to 60 will have a value of 1, and then a line will be drawn to 80. In the same fashion, the non-membership function is drawn using the following points: 40, 60, 80, and 100. The highest valued members

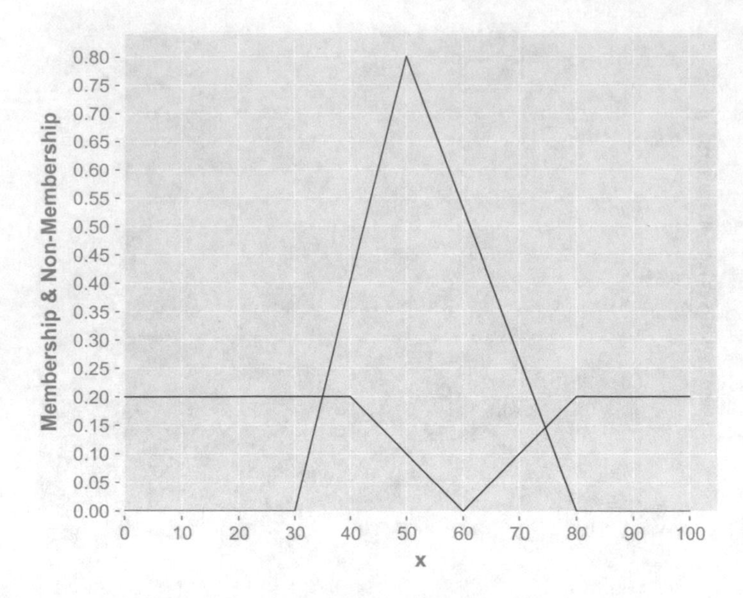

Fig. 3 Example of triangular membership and non-membership functions

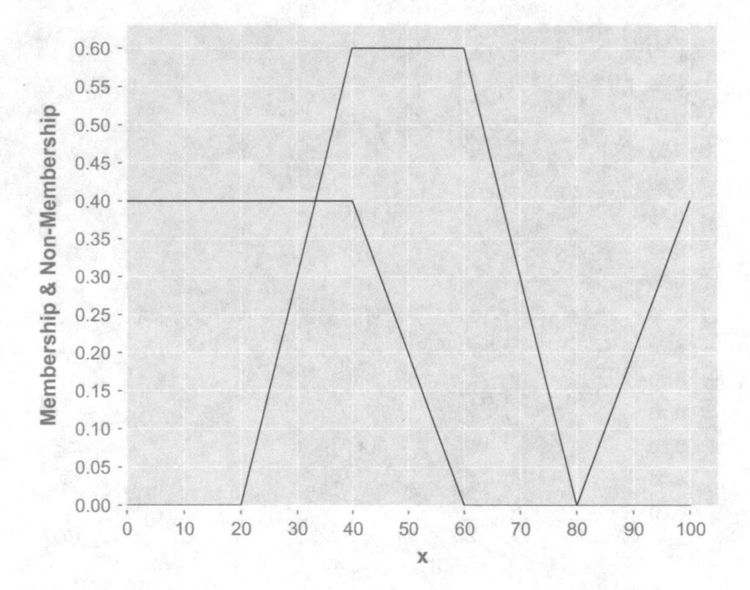

Fig. 4 Example of trapezoidal membership and non-membership functions

in the membership function equal to 0.6, while in the non-membership function these members equal to 0.4.

In Fig. 5 a generalized bell membership function is plotted. The membership function is constructed with a center of 50, a width of 20, and the parameter that

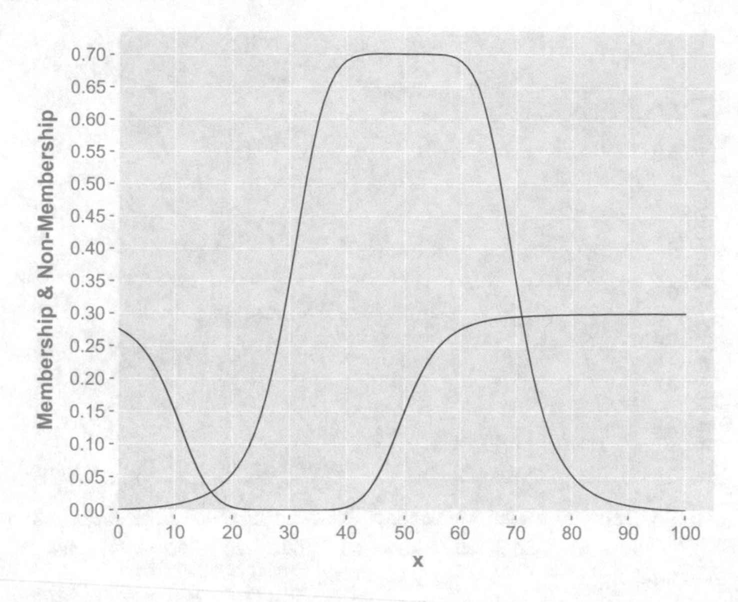

Fig. 5 Example of generalized bell membership and non-membership functions

determines the roundness of the corners of the bell is set at 3. Considering (5), which is the equation to generate a generalized bell, the membership function would be constructed with $a = 20, b = 3, c = 50$. In the case of the non-membership function, it would be constructed with $a = 20, b = 3, c = 30$. The highest valued member in the membership function is equal to 0.7, while the highest valued member in the non-membership function is equal to 0.3.

$$f(x; a, b, c) = \frac{1}{1 + \left|\frac{x-c}{a}\right|^{2b}} \tag{5}$$

Figure 6 shows an example of a left-right membership function and a non-membership function. Considering (6), the membership function is constructed with the following parameters: $c = 60, \alpha = 10, \beta = 65$, and the non-membership function is constructed with the same parameters. Both the membership and the non-membership functions have highest valued members that equal to 1, and in this case we are depicting a traditional fuzzy set represented as an IFS.

$$LF(x; c, \alpha, \beta) = \begin{cases} F_L\left(\frac{c-x}{\alpha}\right), & x \leq c \\ F_R\left(\frac{x-c}{\beta}\right), & x \geq c \end{cases} \tag{6}$$

The last membership function presented in this Section is the sigmoidal membership function. Figure 7 presents a sigmoidal membership function that is constructed with the equation presented in (7). The membership and non-membership functions

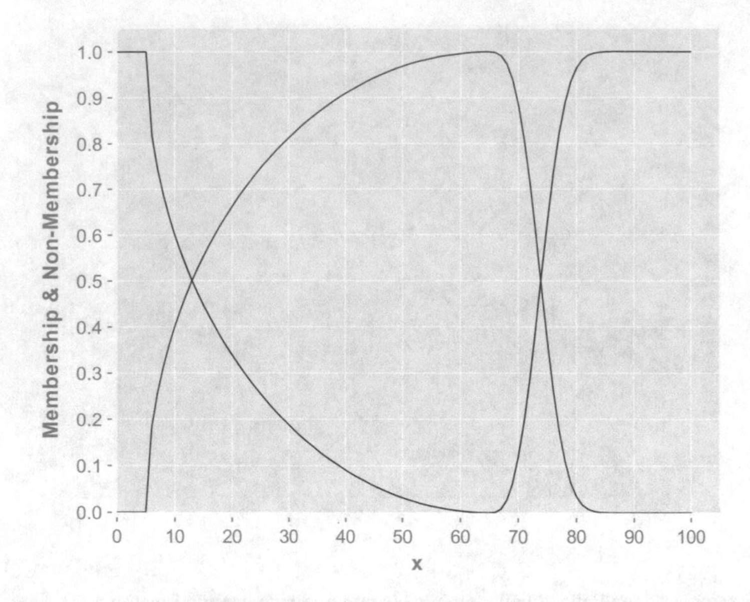

Fig. 6 Example of left-right membership and non-membership functions

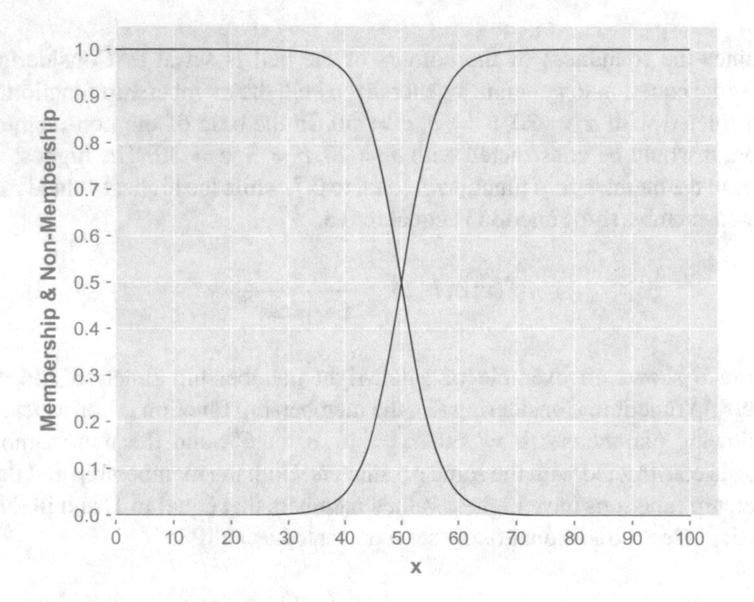

Fig. 7 Example of sigmoidal membership and non-membership functions

are constructed by using the same parameters, which are $a = 0.3, b = 50$. As in the example of the left-right membership and non-membership functions, we are depicting a traditional fuzzy set represented as an intuitionistic fuzzy set.

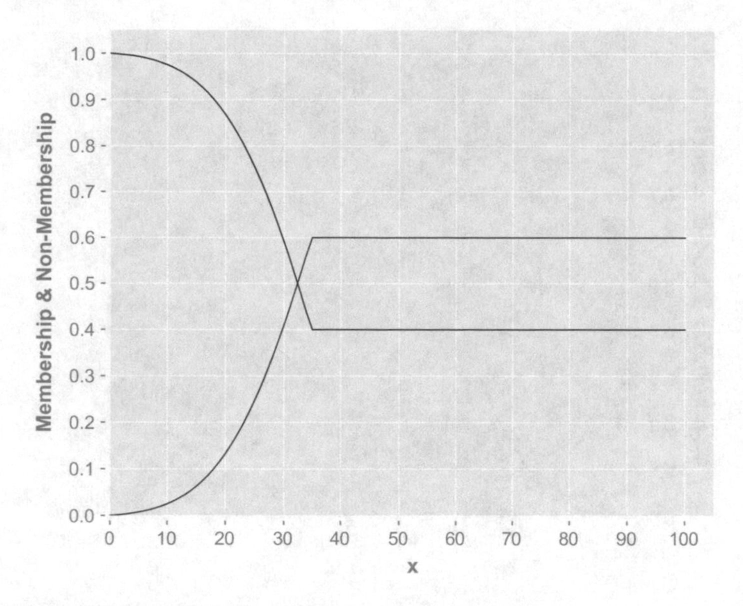

Fig. 8 Example of the union of two intuitionistic fuzzy sets

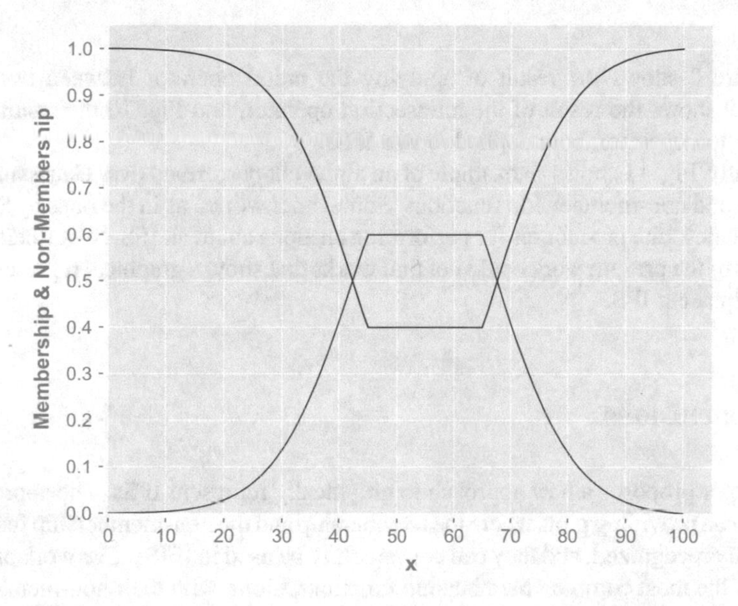

Fig. 9 Example of the intersection of two intuitionistic fuzzy sets

$$sig(x; a, c) = \frac{1}{1 + \exp[-a(x - c)]} \tag{7}$$

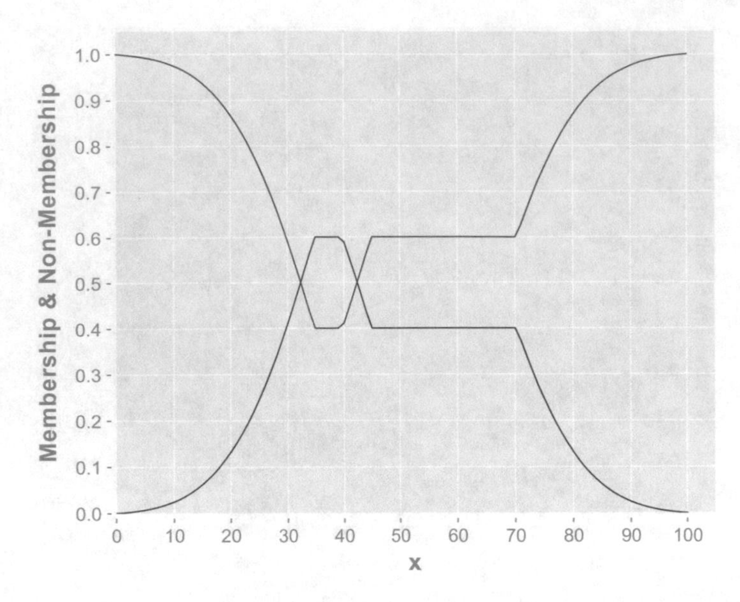

Fig. 10 Example of implication of two intuitionistic fuzzy sets

Figure 8 shows the result of applying the union operator between two IFSs. Figure 9 shows the result of the intersection operator, and Fig. 10 the result of the implication operator, both applied to two IFSs.

Finally, Fig. 11 shows an example of an alpha-cut performed over Gaussian membership and non-membership functions. Some other works, as in the ones by Sharma [27, 28], describe procedures for performing an alpha cut to an IFS. Nevertheless, the authors of the present work could not find works that show a graphical representation of an alpha cut IFS.

5 Conclusions

This paper proposes a new approach to graphically represent IFSs. The approach is focused on providing plots where the membership and the non-membership functions are easily recognized, and they can conveniently be used in IFISs. The work presents plots of the most common membership functions, along with their non-membership functions, and this way the reader can then decide if the use of this approach is convenient to represent the antecedents and consequents in the construction of an IFIS. The authors of this paper are currently working on an implementation of a graphical user interface (GUI) to create IFISs.

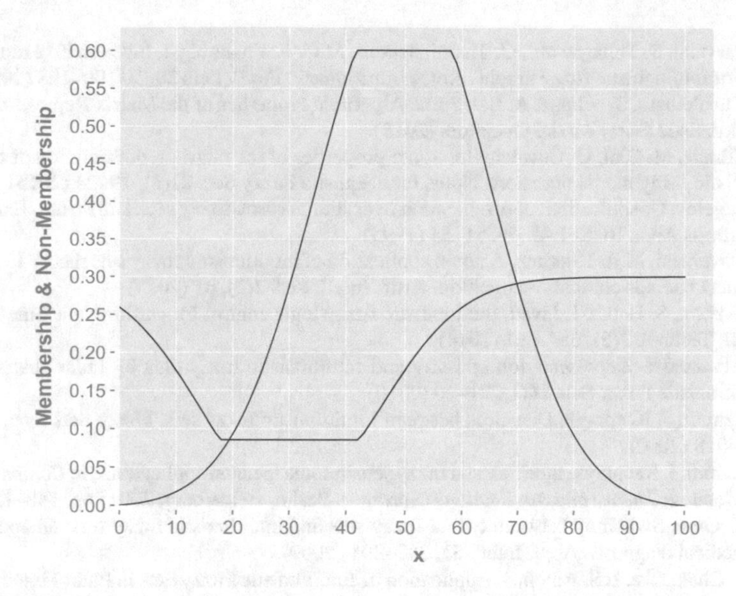

Fig. 11 Example of an alpha cut of an intuitionistic fuzzy set

References

1. L.A. Zadeh, Fuzzy sets. Inf. Control **8**(3), 338–353 (1965)
2. L.A. Zadeh, The concept of a linguistic variable and its application to approximate reasoning—I. Inf. Sci. **8**(3), 199–249 (1975)
3. K.T. Atanassov, Intuitionistic fuzzy sets. Fuzzy Sets Syst. **20**(1), 87–96 (1986)
4. N.N. Karnik, J.M. Mendel, Centroid of a type-2 fuzzy set. Inf. Sci. **132**(1), 195–220 (2001)
5. O. Castillo, P. Melin, Intelligent systems with interval type-2 fuzzy logic. Int. J. Innov. Comput. Inf. Control **4**(4), 771–783 (2008)
6. M.A. Sanchez, O. Castillo, J.R. Castro, Generalized Type-2 fuzzy systems for controlling a mobile robot and a performance comparison with interval Type-2 and Type-1 fuzzy systems. Expert Syst. Appl. **42**(14), 5904–5914 (2015)
7. N.R. Cázarez-Castro, L.T. Aguilar, O. Castillo, Designing Type-1 and Type-2 fuzzy logic controllers via fuzzy lyapunov synthesis for nonsmooth mechanical systems. Eng. Appl. Artif. Intell. **25**(5), 971–979 (2012)
8. C. Leal-Ramírez, O. Castillo, P. Melin, A. Rodríguez-Díaz, Simulation of the bird age-structured population growth based on an interval type-2 fuzzy cellular structure. Inf. Sci. **181**(3), 519–535 (2011)
9. A. Hernandez-Aguila, M. Garcia-Valdez, A proposal for an intuitionistic fuzzy inference system, in *IEEE World Congress on Computational Intelligence*, pp. 1294–1300 (2016)
10. O. Castillo, A. Alanis, M. Garcia, H. Arias, An intuitionistic fuzzy system for time series analysis in plant monitoring and diagnosis. Appl. Soft Comput. **7**(4), 1227–1233 (2007)
11. K.T. Atanassov, Intuitionistic fuzzy sets: past, present and future, in *EUSFLAT Conference*, pp. 12–19 (2003)
12. I. Despi, D. Opris, E. Yalcin, Generalised atanassov intuitionistic fuzzy sets, in *Proceeding of the Fifth International Conference on Information, Process, and Knowledge Management* (2013)
13. G. Deschrijver, C. Cornelis, E.E. Kerre, On the representation of intuitionistic fuzzy t-norms and t-conorms. Fuzzy Syst. IEEE Trans. **12**(1), 45–61 (2004)

14. R. Parvathi, S. Thilagavathi, G. Thamizhendhi, M.G. Karunambigai, Index matrix representation of intuitionistic fuzzy graphs. Notes Intuitionistic Fuzzy Sets **20**(2), 100–108 (2014)
15. G. Cuvalcioglu, S. Yilmaz, A. Bal, Some Algebraic Properties of the Matrix Representation of Intuitionistic Fuzzy Modal Operators (2015)
16. S. Yilmaz, M. Citil, G. Cuvalcioglu, Some properties of the matrix representation of the intuitionistic fuzzy modal operators. Notes Intuitionistic Fuzzy Sets **21**(2), 19–24 (2015)
17. P. Angelov, Crispification: defuzzification over intuitionistic fuzzy sets. Bull. Stud. Exchanges Fuzziness Appl. BUSEFAL **64**, 51–55 (1995)
18. H. Davarzani, M.A. Khorheh, A novel application of intuitionistic fuzzy sets theory in medical science: bacillus colonies recognition. Artif. Intell. Res. **2**(2), p1 (2013)
19. M. Akram, S. Habib, I. Javed, Intuitionistic fuzzy logic control for washing machines. Indian J. Sci. Technol. **7**(5), 654–661 (2014)
20. V. Atanassova, Representation of fuzzy and intuitionistic fuzzy data by radar charts. Notes Intuitionistic Fuzzy Sets **16**(1), 21–26 (2010)
21. E. Szmidt, J. Kacprzyk, Distances between intuitionistic fuzzy sets. Fuzzy Sets Syst. **114**(3), 505–518 (2000)
22. E. Szmidt, J. Kacprzyk, Intuitionistic fuzzy sets in some medical applications, in *Computational Intelligence. Theory and Applications* (Springer, Berlin, Heidelberg, 2001) pp. 148–151
23. C.M. Own, Switching between type-2 fuzzy sets and intuitionistic fuzzy sets: an application in medical diagnosis. Appl. Intell. **31**, 283–291 (2009)
24. D.D. Chakarska, L.S. Antonov, Application of Intuitionistic Fuzzy Sets in Plant Tissue Culture and in Invitro Selection (1995)
25. Z. Xu, Intuitionistic preference relations and their application in group decision making. Inf. Sci. **177**(11), 2363–2379 (2007)
26. L. Baccour, S. Kanoun, V. Maergner, A.M. Alimi, An application of intuitionistic fuzzy information for handwritten Arabic word recognition, in *12th International Conference on IFSs (NIFS08)* (vol. 14, No. 2, pp. 67–72) (2008)
27. P.K. Sharma, Cut of intuitionistic fuzzy groups. Int. Math. Forum **6**(53), 2605–2614 (2011)
28. P.K. Sharma, Cut of intuitionistic fuzzy modules. Int. J. Math. Sci. Appl. **1**(3), 1489–1492 (2011)

Existence of Mild Solutions for an Impulsive Fractional Integro-differential Equations with Non-local Condition

K. Hilal, L. Ibnelazyz, K. Guida and Said Melliani

Abstract In this paper we are interested in studying the existence of solutions for a controlled impulsive fractional evolution equations. We use several tools such as fractional calculus, fixed point theorems and the theory of semigroup. We first give some preliminaries and notations, the second part of the work we provide an existence result for our problem and in the final section, we give some examples to show the importance of our results.

1 Introduction

The theory of non-integer calculus has been introduced in 1695 by Leibniz and L'Hopital. Fractional calculus characterize the memory in the evolution process and it is an important tool in describing real-world phenomena and it is used in many fields, such as physics, biology and economics.

Fractional differential equations with instantaneous impulses have been introduced as a new exciting and interesting branch. But, this type of equations cannot describe the dynamics of evolution process in many areas of research, such as pharmacotherapy. That is why, Hernandez and O'Regan [1] introduced a new model which is the impulsive differential equations with non-instantaneous impulses.

K. Hilal · L. Ibnelazyz · K. Guida · S. Melliani (✉)
Sultan Moulay Slimane University, BP 523 Beni Mellal, Morocco
e-mail: s.melliani@usms.ma

K. Hilal
e-mail: hilalkhalid@yahoo.fr

L. Ibnelazyz
e-mail: ibelazyzlahcen@gmail.com

K. Guida
e-mail: guida.karim@gmail.com

S. Melliani and O. Castillo (eds.), *Recent Advances in Intuitionistic Fuzzy Logic Systems*, Studies in Fuzziness and Soft Computing 372,
https://doi.org/10.1007/978-3-030-02155-9_20

In [2] the authors studied the following non-instantaneous model

$$\begin{cases} {}^{c}D^{\alpha}x(t) = Ax(t) + f(t, x(t)) + B(t)u(t), \quad t \in \cup_{i=0}^{N}(s_i, t_{i+1}], \quad u \in U_{ad} \\ x(t) = g_i(t, x(t)), \quad t \in (t_i, s_i], i = 1, 2, \ldots, N, \\ x(s_i^{+}) = x(s_i^{-}), i = 1, 2, \ldots, N, \\ x(0) = x_0 \in X \end{cases} \tag{1}$$

where ${}^{c}D^{\alpha}$ denotes the Caputo fractional derivatives of order $\alpha \in (0, 1)$ with the lower limit zero,
$A : D(A) \subset X \longrightarrow X$ is the generator of a C_0-semigroup of bounded operators $T(t)_{t\geq 0}$ on a Banach space $(X, \|\,.\,\|)$, $x_0 \in X, 0 = t_0 = s_0 < t_1 < s_1 < t_2 < s_2 < \cdots < t_m \leq s_m < t_{m+1} = T$ are fixed numbers, $g_i \in C(J \times X; X)$.

The symbols $x(s_i^{+})$ and $x(s_i^{-})$ represents the limits of $x(t)$ at $t = s_i$.

Motivated by the work of Liu and Wang in [2], we study the impulsive differential equation

$$\begin{cases} {}^{c}D^{\alpha}x(t) = Ax(t) + f(t, x(t), Fx(t), Bx(t)) + \int_0^t q(t-s)k(s, x(s))ds + \\ C(t)u(t), \quad t \in (s_i, t_{i+1}], \quad i = 0, 1, 2, \ldots, m, \quad u \in U_{ad}, \\ {}^{c}D^{\beta}x(t) = g_i(t, x(t)), \quad t \in (t_i, s_i], \quad i = 1, 2, \ldots, m, \\ x(0) = x_0 + h(x). \end{cases} \tag{2}$$

where ${}^{c}D^{\alpha}, {}^{c}D^{\beta}$ are the Caputo fractional derivatives of order $\alpha \in (0, 1)$ and $\beta \in (0, 1)$ respectively with the lower limit zero, $A : D(A) \subset X \longrightarrow X$ is the generator of a C_0-semigroup of bounded operators $T(t)_{t\geq 0}$ on a Banach space $(X, \|\,.\,\|)$, $x_0 \in X$, $0 = t_0 = s_0 < t_1 < s_1 < t_2 < s_2 < \cdots < t_m \leq s_m < t_{m+1} = T$ are fixed numbers, $g_i \in C(J \times X; X)$, h, f, k are given functions, $F, B, q : C(J; X) \to C(J; X)$ are given by $Bx(t) = \int_0^t B(t, s)x(s)ds$, $Fx(t) = \int_0^t F(t, s)x(s)ds$ and $\{F(t, s); t, s \in J\}$ $\{B(t, s); t, s \in J\}$ are a set of bounded linear operators on X such that:
$F(t, .)x \in C([0, t]; X)$, $F(., s) \in C([s, T]; X)$ for all $t, s \in J$, $B(t, .)x \in C([0, t]; X)$, $B(., s) \in C([s, T]; X)$ for all $t, s \in J$ and $F^* = \sup_{t\in J} \int_0^t \| F(t, s) \|_{L(X)}\, ds$,

$B^* = \sup_{t\in J} \int_0^t \| B(t, s) \|_{L(X)}\, ds, \quad q^* = \sup_{s\in J} \int_0^s \|q(s-t)\|dt.$

In our work we will need the continuity of $x(t)$ at both points $t = s_i$ and $t = t_i$.
The rest of the paper is organized as follows. In Sect. 2 we present the notations, definitions and preliminaries results needed in the following sections. In Sect. 3, a suitable concept of mild solutions for our problems is introduced. Section 4 is concerned with the existence results of our problems.

2 Preliminaries

Let us set $J = [0, T]$, $J_0 = [0, t_1]$, $J_1 = (t_1, t_2], \ldots, J_{m-1} = (t_{m-1}, t_m]$, $J_m = (t_m, t_{m+1}]$ and introduce the space $PC(J, X) := \{u : J \to X | u \in C(J_k, X),\ k = 0, 1, 2, \ldots, m$, and there exists $u(t_k^+)$ and $u(t_k^-)$, $k = 1, 2, \ldots, m$, with $u(t_k^-) = u(t_k^+)\}$. It is clear that $PC(J, X)$ is a Banach space with the norm $\|u\|_{PC} = sup\,\{\|u(t)\| : t \in J\}$. Let Y be a separable reflexive Banach space where controls u takes values, and $P_f(Y)$ is a class of nonempty closed and convex subsets of Y. We suppose that the multi-valued map $w : [0, a] \longrightarrow P_f(Y)$ is measurable, $w(.) \subset E$, where E is bounded set of Y, and the admissible control set
$U_{ad} = \{c \in L^p(E) : c(t) \in w(t), a.e\}$, $p > \frac{1}{\tau}$, $(\tau \in (0, \alpha))$, for more detail about admissible control set, we refer the readers to [3].

Lemma 1 (Theorem 2.1 in [4]) *Suppose* $W \subseteq PC(J, X)$. *If the following conditions are satisfied:*

(1) W is uniformly bounded subset of PC(J,X);
(2) W is equicontinuous in (t_i, t_{i+1}), $i = 0, 1, 2, \ldots, m$, *where* $t_0 = 0$, $t_{m+1} = T$;
(3) $W(t) = \{u(t) : u \in W, t \in J \setminus \{t_1, t_2, \ldots, t_m\}\}$, $W(t_i^+) = \{u(t_i^+) : u \in W\}$ *and* $W(t_i^-) = \{u(t_i^-) : u \in W\}$, $i = 1, 2, \ldots, m$, *are relatively compact subsets of X. Then W is a relatively compact subset of* $PC(J, X)$.

Let us recall the following well-known definitions.

Definition 1 ([5]) The Riemann Liouville fractional integral of order q with lower limit zero for a function f is defined as
$I^q f(t) = \frac{1}{\Gamma(q)} \int_0^t (t-s)^{q-1} f(s)ds, \quad q > 0,$
provided the integral exists, where $\Gamma(.)$ is the gamma function.

Definition 2 ([5]) The Riemann-Liouville derivative of order q with the lower limit zero for a function $f : [0, \infty) \to \mathbb{R}$ can be written as
${}^LD^q f(t) = \frac{1}{\Gamma(n-q)} \frac{d^n}{dt^n} \int_0^t (t-s)^{n-q-1} f(s)ds, \quad n-1 < q < n, t > 0.$

Definition 3 ([5]) The Caputo derivative of order q for a function $f : [0, \infty) \longrightarrow \mathbb{R}$ can be written as
${}^cD^q f(t) = {}^LD^q \left(f(t) - \sum_{k=0}^{k} \frac{t^k}{k!} f^k(0) \right), n-1 < q < n, t > 0.$

Definition 4 ([6, 7]) A function $x \in C(J, X)$ is said to be a mild solution of the following problem:

$$\begin{cases} {}^cD^\alpha x(t) = Ax(t) + y(t), & t \in (0, T], \\ x(0) = x_0. \end{cases}$$

If it satisfies the integral equation
$x(t) = P_\alpha(t)x_0 + \int_0^t (t-s)^{\alpha-1} Q_\alpha(t-s)y(s)ds.$

Here

$$P_\alpha(t) = \int_0^\infty \xi_\alpha(\theta)T(t^\alpha\theta)d\theta, \quad Q_\alpha(t) = \alpha\int_0^\infty \theta\xi_\alpha(\theta)T(t^\alpha\theta)d\theta,$$

$$\xi_\alpha(\theta) = \frac{1}{\alpha}\theta^{-1-\frac{1}{\alpha}}\overline{\omega}_\alpha(\theta^{-\frac{1}{\alpha}}) \geq 0,$$

$$\overline{\omega_\alpha}(\theta) = \frac{1}{\pi}\sum_{n=1}^\infty(-1)^{n-1}\theta^{-n\alpha-1}\frac{\Gamma(n\alpha+1)}{n!}sin(n\pi\alpha), \theta \in (0,\infty),$$

and $\xi_\alpha(\theta) \geq 0, \theta \in (0,\infty), \int_0^\infty \xi_\alpha(\theta)d\theta = 1.$

It is easy to verify that

$$\int_0^\infty \theta\xi_\alpha(\theta)d\theta = \frac{1}{\Gamma(1+\alpha)}.$$

We make the following assumption on A in the whole paper.

H(1): The operator A generates a strongly continuous semigroup $\{T(t) : t \geq 0\}$ in X, and there is a constant $M_A \geq 1$ such that $\sup_{t\in[0,\infty)} \| T(t) \|_{L(X)} \leq M_A$. For any $t > 0$, $T(t)$ is compact.

Lemma 2 ([6, 7]) *Let $H(A)$ hold, then the operator P_α and Q_α have the following properties:*

(1) For any fixed $t \geq 0$, $P_\alpha(t)$ and $Q_\alpha(t)$ are linear and bounded operators, and for any $x \in X$,
$\| P_\alpha(t)x \| \leq M_A \| x \|, \quad \| Q_\alpha(t)x \| \leq \frac{\alpha M_A}{\Gamma(1+\alpha)} \| x \|,$
(2) $\{P_\alpha(t), t \geq 0\}$ and $\{Q_\alpha(t), t \geq 0\}$ are strongly continuous,
(3) for every $t > 0$, $P_\alpha(t)$ and $Q_\alpha(t)$ are compact operators.

We recall a fixed point theorem which will be needed in the sequel.

Theorem 1 *(Krasnoselskii fixed point theorem) Let M be a closed, convex, and non-empty subset of a Banach space X. Let A,B be the operators such that:*

(a) $Ax + By \in M$ for all $x, y \in M$,
(b) A is compact and continuous,
(c) B is a contraction.

Then there exists a $x \in M$ such that $x = Ax + Bx$.

3 The Construction of Mild Solutions

Let $x \in PC(J, X)$. We first consider the following fractional impulsive problem:

$$\begin{cases} {}^{c}D^{\alpha}x(t) = Ax(t) + f(t, x(t), Fx(t), Bx(t)) + \int_0^t q(t-s)k(s, x(s))ds \\ \qquad + C(t)u(t),\ t \in (s_i, t_{i+1}], \\ i = 0, 1, 2, \ldots, m,\ u \in U_{ad}, \\ {}^{c}D^{\beta}x(t) = g_i(t, x(t)),\ t \in (t_i, s_i],\ i = 1, 2, \ldots, m, \\ x(0) = x_0 + h(x). \end{cases}$$

From the property of the Caputo derivative, a general solution can be written as

$$x(t) = \begin{cases} x_0 + h(x) + \frac{1}{\Gamma(\alpha}\int_0^t (t-s)^{\alpha-1}[Ax(s) + f(s, x(s), Fx(s), Bx(s)) \\ + \int_0^s q(s-\tau)k(\tau, x(\tau))d\tau + C(s)u(s)]ds,\ t \in (0, t_1], \\ d_{1x} + \frac{1}{\Gamma(\beta)}\int_0^t (t-s)^{\beta-1} g_1(s, x(s))ds,\ t \in (t_1, s_1], \\ K_{1x} + \frac{1}{\Gamma(\alpha)}\int_0^t (t-s)^{\alpha-1}[Ax(s) + f(s, x(s), Fx(s), Bx(s)) \\ + \int_0^s q(s-\tau)k(\tau, x(\tau))d\tau + C(s)u(s)]ds,\ t \in (s_1, t_2], \\ \vdots \\ d_{ix} + \frac{1}{\Gamma(\beta)}\int_0^t (t-s)^{\beta-1} g_i(s, x(s))ds,\ t \in (t_i, s_i], \\ K_{ix} + \frac{1}{\Gamma(\alpha)}\int_0^t (t-s)^{\alpha-1}[Ax(s) + f(s, x(s), Fx(s), Bx(s)) \\ + \int_0^s q(s-\tau)k(\tau, x(\tau))d\tau + C(s)u(s)]ds,\ t \in (s_i, t_{i+1}], \end{cases}$$

where d_{ix} and K_{ix}, $i = 1, 2, \ldots, m$, are elements of X.
Using [8] pages 5 and 6 we obtain:

$$x(t) = \begin{cases} d_{ix} + \frac{1}{\Gamma(\beta)}\int_0^t (t-s)^{\beta-1} g_i(s, x(s))ds,\ t \in (t_i, s_i], 1 \le i \le m, \\ P_\alpha(t-s_i)K_{ix} + \int_0^t (t-s)^{\alpha-1}Q_\alpha(t-s)[f(s, x(s), Fx(s), Bx(s)) \\ + \int_0^s q(s-\tau)k(\tau, x(\tau))d\tau + C(s)u(s)]ds,\ t \in (s_i, t_{i+1}], 0 \le i \le m, \\ K_{0x} = x(0). \end{cases}$$

And using the fact that x is continuous at the points t_i, we get:

$$\begin{aligned} x(t_i) &= P_\alpha(t_i - s_{i-1})K_{(i-1)x} + \int_0^{t_i}(t_i-s)^{\alpha-1}Q_\alpha(t_i-s)[f(s, x(s), Fx(s), Bx(s)) \\ &\quad + \int_0^s q(s-\tau)k(\tau, x(\tau))d\tau + C(s)u(s)]ds \\ &= d_{ix} + \frac{1}{\Gamma(\beta)}\int_0^{t_i}(t_i-s)^{\beta-1}g_i(s, x(s))ds. \end{aligned}$$

Which implies that:

$$d_{ix} = P_\alpha(t_i - s_{i-1})K_{(i-1)x} + \int_0^{t_i} (t_i - s)^{\alpha-1} Q_\alpha(t_i - s)[f(s, x(s), Fx(s), Bx(s))+ \\ + \int_0^s q(s-\tau)k(\tau, x(\tau))d\tau + C(s)u(s)]ds - \frac{1}{\Gamma(\beta)} \int_0^{t_i} (t_i - s)^{\beta-1} g_i(s, x(s))ds.$$

Using the fact that x is continuous at the points s_i, we get:

$$\begin{aligned} x(s_i) &= d_{ix} + \tfrac{1}{\Gamma(\beta)} \textstyle\int_0^{s_i} (s_i - s)^{\beta-1} g_i(s, x(s))ds \\ &= K_{ix} + \textstyle\int_0^{s_i} (s_i - s)^{\alpha-1} Q_\alpha(s_i - s) \left[f(s, x(s), Fx(s), Bx(s)) \right. \\ &\quad \left. + \textstyle\int_0^s q(s-\tau)k(\tau, x(\tau))d\tau + C(s)u(s) \right] ds. \end{aligned}$$

Which implies that:

$$\begin{aligned} K_{ix} &= d_{ix} + \tfrac{1}{\Gamma(\beta)} \textstyle\int_0^{s_i} (s_i - s)^{\beta-1} g_i(s, x(s))ds- \\ &\quad - \textstyle\int_0^{s_i} (s_i - s)^{\alpha-1} Q_\alpha(s_i - s) \left[f(s, x(s), Fx(s), Bx(s)) \right. \\ &\quad \left. + \textstyle\int_0^s q(s-\tau)k(\tau, x(\tau))d\tau + C(s)u(s) \right] ds. \end{aligned}$$

Therefore, a mild solution of problem (1.2) is given by

$$x(t) = \begin{cases} P_\alpha(t)(x_0 + h(x)) + \displaystyle\int_0^t (t-s)^{\alpha-1} Q_\alpha(t-s)[f(s, x(s), Fx(s), Bx(s)) \\ \quad + \displaystyle\int_0^s q(s-\tau)k(\tau, x(\tau))d\tau + C(s)u(s)]ds, \quad t \in (0, t_1], \\ d_{1x} + \frac{1}{\Gamma(\beta)} \int_0^t (t-s)^{\beta-1} g_1(s, x(s))ds, \quad t \in (t_1, s_1], \\ P_\alpha(t - s_1)K_{1x} + \displaystyle\int_0^t (t-s)^{\alpha-1} Q_\alpha(t-s)[f(s, x(s), Fx(s), Bx(s)) \\ \quad + \displaystyle\int_0^s q(s-\tau)k(\tau, x(\tau))d\tau + C(s)u(s)]ds, \quad t \in (s_1, t_2], \\ \vdots \\ d_{ix} + \frac{1}{\Gamma(\beta)} \int_0^t (t-s)^{\beta-1} g_i(s, x(s))ds, \quad t \in (t_i, s_i], \ 1 \le i \le m, \\ P_\alpha(t - s_i)K_{ix} + \displaystyle\int_0^t (t-s)^{\alpha-1} Q_\alpha(t-s)[f(s, x(s), Fx(s), Bx(s)) \\ \quad + \displaystyle\int_0^s q(s-\tau)k(\tau, x(\tau))d\tau + C(s)u(s)]ds, \\ \quad t \in (s_i, t_{i+1}], \ 1 \le i \le m, \end{cases}$$

where

$$\begin{aligned} K_{0x} &= x_0 + h(x), \\ d_{ix} &= P_\alpha(t_i - s_{i-1})K_{(i-1)x} + \textstyle\int_0^{t_i} (t_i - s)^{\alpha-1} Q_\alpha(t_i - s)[f(s, x(s), Fx(s), Bx(s)) \\ &\quad + \textstyle\int_0^s q(s-\tau)k(\tau, x(\tau))d\tau + C(s)u(s)]ds - \tfrac{1}{\Gamma(\beta)} \textstyle\int_0^{t_i} (t_i - s)^{\beta-1} g_i(s, x(s))ds, \\ K_{ix} &= d_{ix} - \textstyle\int_0^{s_i} (s_i - s)^{\alpha-1} Q_\alpha(s_i - s) \left[f(s, x(s), Fx(s), Bx(s)) \right. \\ &\quad \left. + \textstyle\int_0^s q(s-\tau)k(\tau, x(\tau))d\tau + C(s)u(s) \right] ds + \tfrac{1}{\Gamma(\beta)} \textstyle\int_0^{s_i} (s_i - s)^{\beta-1} g_i(s, x(s))ds. \end{aligned}$$

Definition 5 A function $x \in PC(J;X)$ is said to be a mild solution of problem (1.2) if it satisfies the following relation:

$$x(t) = \begin{cases} P_\alpha(t)K_{0x} + \int_0^t (t-s)^{\alpha-1} Q_\alpha(t-s)[f(s,x(s),Fx(s),Bx(s)) \\ \quad + \int_0^s q(s-\tau)k(\tau,x(\tau))d\tau + C(s)u(s)]ds, \ \ t \in (0,t_1], \\ d_{ix} + \frac{1}{\Gamma(\beta)} \int_0^t (t-s)^{\beta-1} g_i(s,x(s))ds, \ \ t \in (t_i,s_i], 1 \le i \le m, \\ P_\alpha(t-s_i)K_{ix} + \int_0^t (t-s)^{\alpha-1} Q_\alpha(t-s)[f(s,x(s),Fx(s),Bx(s)) \\ \quad + \int_0^s q(s-\tau)k(\tau,x(\tau))d\tau + C(s)u(s)]ds, \ \ t \in (s_i,t_{i+1}], 1 \le i \le m. \end{cases}$$

where

$$\begin{aligned} K_{0x} &= x_0 + h(x), \\ d_{ix} &= P_\alpha(t_i - s_{i-1})K_{(i-1)x} + \int_0^{t_i} (t_i-s)^{\alpha-1} Q_\alpha(t_i-s)[f(s,x(s),Fx(s),Bx(s)) \\ &\quad + \int_0^s q(s-\tau)k(\tau,x(\tau))d\tau + C(s)u(s)]ds - \frac{1}{\Gamma(\beta)} \int_0^{t_i} (t_i-s)^{\beta-1} g_i(s,x(s))ds, \\ K_{ix} &= d_{ix} - \int_0^{s_i} (s_i-s)^{\alpha-1} Q_\alpha(s_i-s)\,[f(s,x(s),Fx(s),Bx(s)) \\ &\quad + \int_0^s q(s-\tau)k(\tau,x(\tau))d\tau + C(s)u(s)]\,ds + \frac{1}{\Gamma(\beta)} \int_0^{s_i} (s_i-s)^{\beta-1} g_i(s,x(s))ds. \end{aligned}$$

Here

$P_\alpha(t) = \int_0^\infty \xi_\alpha(\theta)T(t^\alpha\theta)d\theta, \quad Q_\alpha(t) = \alpha\int_0^\infty \theta\xi_\alpha(\theta)T(t^\alpha\theta)d\theta,$

$\xi_\alpha(\theta) = \frac{1}{\alpha}\theta^{-1-\frac{1}{\alpha}}\varpi_\alpha(\theta^{-\frac{1}{\alpha}}) \ge 0,$

$\overline{\omega_\alpha}(\theta) = \frac{1}{\pi}\sum_{n=1}^\infty (-1)^{n-1}\theta^{-n\alpha-1}\frac{\Gamma(n\alpha+1)}{n!} sin(n\pi\alpha), \theta \in (0,\infty),$

and $\xi_\alpha(\theta) \ge 0, \theta \in (0,\infty), \int_0^\infty \xi_\alpha(\theta)d\theta = 1.$

It is easy to verify that

$\int_0^\infty \theta\xi_\alpha(\theta)d\theta = \frac{1}{\Gamma(1+\alpha)}.$

We make the following assumption on A in the whole paper.

H(1): The operator A generates a strongly continuous semigroup $\{T(t) : t \ge 0\}$ in X, and there is a constant $M_A \ge 1$ such that $\sup_{t\in[0,\infty)} \| T(t) \|_{L(X)} \le M_A$. For any $t > 0$, $T(t)$ is compact.

Lemma 3 [6, 7] *Let $H(A)$ hold, then the operator P_α and Q_α have the following properties:*

(1) For any fixed $t \ge 0$, $P_\alpha(t)$ and $Q_\alpha(t)$ are linear and bounded operators, and for any $x \in X$,
$\| P_\alpha(t)x \| \le M_A \| x \|, \quad \| Q_\alpha(t)x \| \le \frac{\alpha M_A}{\Gamma(1+\alpha)} \| x \|,$
(2) $\{P_\alpha(t), t \ge 0\}$ and $\{Q_\alpha(t), t \ge 0\}$ are strongly continuous,

(3) *for every $t > 0$, $P_\alpha(t)$ and $Q_\alpha(t)$ are compact operators.*

We recall a fixed point theorem which will be needed in the sequel.

Theorem 2 (Krasnoselskii fixed point theorem) *Let M be a closed, convex, and non-empty subset of a Banach space X. Let A,B be the operators such that:*

(a) $Ax + By \in M$ for all $x, y \in M$,
(b) A is compact and continuous,
(c) B is a contraction.

Then there exists a $x \in M$ such that $x = Ax + Bx$.

4 Existence and Uniqueness of Mild Solution

This section deals with the existence results for the problem (1.2).

From Definition (3.1), we define an operator $P : PC(J, X) \to PC(J, X)$ as

$$Px(t) = \begin{cases} P_\alpha(t)(x_0 + h(x)) + \displaystyle\int_0^t (t-s)^{\alpha-1} Q_\alpha(t-s)[f(s, x(s), Fx(s), Bx(s)) \\ + \displaystyle\int_0^s q(s-\tau)k(\tau, x(\tau))d\tau + C(s)u(s)]ds, \ t \in [0, t_1], \\ d_{ix} + \frac{1}{\Gamma(\beta)} \int_0^t (t-s)^{\beta-1} g_i(s, x(s))ds, \ \ t \in (t_i, s_i], \\ P_\alpha(t-s_i)K_{ix} + \displaystyle\int_0^t (t-s)^{\alpha-1} Q_\alpha(t-s)[f(s, x(s), Fx(s), Bx(s)) \\ + \displaystyle\int_0^s q(s-\tau)k(\tau, x(\tau))d\tau + C(s)u(s)]ds, \ t \in (s_i, t_{i+1}]. \end{cases}$$

To prove our first existence result we introduce the following assumptions.
(H_2) $C : [0, T] \longrightarrow L(Y, X)$ is essentially bounded, ie $C \in L^\infty([0, T], L(Y, X))$,
(H_3) The functions $f \in C(J \times X \times X \times X; X)$,and $k \in C(J \times X; X)$,
(H_4) There exists $C_f, L_f, M_f, L_k > 0$ such that

- $\| f(t, x_1, y_1, z_1) - f(t, x_2, y_2, z_2) \| \le L_f \| x_1 - x_2 \| + C_f \| y_1 - y_2 \| + M_f \| z_1 - z_2 \|$, for all $x_1, x_2, y_1, y_2, z_1, z_2 \in X$, and $t \in J$,
- $\| k(t, x_1) - k(t, x_2) \| \le L_k \| x_1 - x_2 \|$, for all $x_1, x_2 \in X$, and $t \in J$,

(H_5) There exists a constant $D, D' > 0$ such that

- $\| f(t, x, y, z) \| \le D\left(1 + \| x \|^\mu + \| y \|^\nu + \| z \|^\phi\right)$ for every $t \in J$ and $x, y, z \in X$, $\mu, \nu, \phi \in [0, 1]$,
- $\| k(t, x) \| \le D' (1 + \| x \|^\mu)$ for every $t \in J$ and $x \in X$, $\mu \in [0, 1]$,

(H_6)

- For $i = 1, 2, \ldots, m$, $g_i \in C(J \times X; X)$, and there exists $L_g > 0$ such that $\| g_i(t, x) - g_i(t, y) \| \le L_g \| x - y \|$, for all $x, y \in X$,

- There exists a function $t \longrightarrow \varphi_i(t)$, $i = 1, 2, \ldots, m$ such that $\| g_i(t, x(t)) \| \leq \varphi_i(t)$ for all $t \in J, x \in X$ and $L_{g_i} = \sup_{t \in J} \varphi_i(t)$ and $L'_g = \max_{1 \leq i \leq m} L_{g_i}$,
- $h : PC(J; X) \to X$ and there exists a constant $L_h > 0$ and $\varphi_h \in C([0, \infty); \mathbb{R}^+)$ such that for $x, y \in PC(J; X)$, $\| h(x) - h(y) \| \leq L_h \| x - y \|_{PC}$; $\| h(x) \| \leq \varphi_h(t)$ and $L'_h = \sup_{t \in J} \varphi_h(t)$.

Theorem 3 *Assume that* $(H_1) - (H_4), (H_6)$ *are satisfied and*

$$[M^{m+1} L_h + \frac{MT^\alpha + M^2(t_m^\alpha + s_m^\alpha) + \cdots + M^{m+1}(t_1^\alpha + s_1^\alpha)}{\Gamma(\alpha + 1)}(L_f + q^* L_k + M_f B^* + C_f F^*)$$
$$+ \frac{ML_g(t_m^\beta + s_m^\beta) + \cdots + M^m L_g(t_1^\beta + s_1^\beta)}{\Gamma(\beta + 1)}] < 1.$$

Then there exists a unique mild solution of problem (1.2).

Proof From the assumptions it is easy to show that the operator P is well defined on $PC(J, X)$.
Let $x, y \in PC(J, X)$.
Case 1. For $t \in [0, t_1]$, we have

$$\|(Px)(t) - (Py)(t)\| \leq \|P_\alpha(t)\,(x_0 + h(x) - x_0 - h(y))\,\|$$
$$+ \| \int_0^t (t - s)^{\alpha - 1} Q_\alpha(t - s)[f(s, x(s), Fx(s), Bx(s)) + \int_0^s q(s - \tau)k(\tau, x(\tau))d\tau$$
$$+ C(s)u(s) - f(s, y(s), Fy(s), By(s)) - \int_0^s q(s - \tau)k(\tau, y(\tau))d\tau - C(s)u(s)]ds\|$$
$$\leq M_A \|h(x) - h(y)\| + \frac{\alpha M_A}{\Gamma(\alpha + 1)} \int_0^t (t - s)^{\alpha - 1} \parallel f(s, x(s), Fx(s), Bx(s))$$
$$- f(s, y(s), Fy(s), By(s)) + \int_0^s q(s - \tau)k(\tau, x(\tau))d\tau - \int_0^s q(s - \tau)k(\tau, y(\tau))d\tau \parallel ds$$
$$\leq M_A L_h \parallel x - y \parallel_{PC} + \frac{\alpha M_A}{\Gamma(\alpha + 1)} \int_0^t (t - s)^{\alpha - 1} (L_f \parallel x(s) - y(s) \parallel$$
$$+ C_f \parallel Fx(s) - Fy(s) \parallel + M_f \parallel Bx(s) - By(s) \parallel + q^* L_k \parallel x(s) - y(s) \parallel) ds$$
$$\leq M_A L_h \parallel x - y \parallel_{PC} + \frac{M_A t_1^\alpha}{\Gamma(\alpha + 1)} \left(L_f + q^* L_k + M_f B^* + C_f F^*\right) \parallel x - y \parallel_{PC}$$
$$\leq [M_A L_h + \frac{M_A t_1^\alpha}{\Gamma(\alpha + 1)} \left(L_f + q^* L_k + M_f B^* + C_f F^*\right)] \parallel x - y \parallel_{PC}.$$

Case 2. For $t \in (t_i, s_i] \cup (s_i, t_{i+1}]$.
We prove that for $t \in (t_i, s_i]$, $i = 1, 2, \ldots, m$:

$$\begin{aligned}
&\| (Px)(t) - (Py)(t) \| \\
&\leq [M_A^i L_h + \frac{M_A t_i^\alpha + M_A^2(t_{i-1}^\alpha + s_{i-1}^\alpha) + M_A^3(t_{i-2}^\alpha + s_{i-2}^\alpha) + \cdots + M_A^i(t_1^\alpha + s_1^\alpha)}{\Gamma(\alpha+1)} \times \\
&(L_f + q^* L_k + M_f B^* + C_f F^*) + \\
&\frac{L_g(t_i^\beta + s_i^\beta) + M_A L_g(t_{i-1}^\beta + s_{i-1}^\beta) + \cdots + M_A^{i-2} L_g(t_2^\beta + s_2^\beta) + M_A^{i-1} L_g(t_1^\beta + s_1^\beta)}{\Gamma(\beta+1)}] \times \\
&\| x - y \|_{PC},
\end{aligned}$$

and for $t \in [s_i, t_{i+1}],\ i = 1, 2, \ldots, m,$

$$\begin{aligned}
&\| (Px)(t) - (Py)(t) \| \\
&\leq [M_A^{i+1} L_h + \frac{M_A t_{i+1}^\alpha + M_A^2(t_i^\alpha + s_i^\alpha) + \cdots + M_A^i(t_2^\alpha + s_2^\alpha) + M_A^{i+1}(t_1^\alpha + s_1^\alpha)}{\Gamma(\alpha+1)} \times \\
&(L_f + q^* L_k + M_f B^* + C_f F^*) + \\
&\frac{M_A L_g(t_i^\beta + s_i^\beta) + \cdots + M_A^{i-1} L_g(t_2^\beta + s_2^\beta) + M_A^i L_g(t_1^\beta + s_1^\beta)}{\Gamma(\beta+1)}] \| x - y \|_{PC}.
\end{aligned}$$

For $t \in (t_1, s_1]$

$$\begin{aligned}
&\| (Px)(t) - (Py)(t) \| \\
&\leq \| d_1 x + \frac{1}{\Gamma(\beta)} \int_0^t (t-s)^{\beta-1} g_1(s, x(s))ds - d_1 y - \frac{1}{\Gamma(\beta)} \int_0^t (t-s)^{\beta-1} g_1(s, y(s))ds \| \\
&\leq \| d_1 x - d_1 y \| + \frac{1}{\Gamma(\beta)} \int_0^t (t-s)^{\beta-1} \| g_1(s, x(s)) - g_1(s, y(s)) \| ds \\
&\leq \| P_\alpha(t_1)(h(x) - h(y)) \| + \| \int_0^{t_1} (t_1 - s)^{\alpha-1} Q_\alpha(t_1 - s)[f(s, x(s), Fx(s), Bx(s)) - \\
&- f(s, y(s), Fy(s), By(s)) + \int_0^s q(s-\tau) k(\tau, x(\tau)) d\tau - \int_0^s q(s-\tau) \\
&k(\tau, y(\tau)) d\tau] ds \| + \| \frac{1}{\Gamma(\beta)} \int_0^{t_1} (t-s)^{\beta-1} (g_1(s, x(s)) - g_1(s, y(s)))ds \| + \\
&+ \| \frac{1}{\Gamma(\beta)} \int_0^t (t-s)^{\beta-1} (g_1(s, x(s)) - g_1(s, y(s)))ds \| \\
&\leq M_A L_h \| x - y \|_{PC} + \frac{M_A t_1^\alpha}{\Gamma(\alpha+1)} (L_f + q^* L_k + M_f B^* + C_f F^*) \| x - y \|_{PC} + \\
&\frac{(s_1^\beta + t_1^\beta)}{\Gamma(\beta+1)} L_g \| x - y \|_{PC} \\
&\leq \left[M_A L_h + \frac{M_A t_1^\alpha}{\Gamma(\alpha+1)} (L_f + q^* L_k + M_f B^* + C_f F^*) + \frac{(s_1^\beta + t_1^\beta)}{\Gamma(\beta+1)} L_g \right] \| x - y \|_{PC}.
\end{aligned}$$

For $t \in (s_1, t_2]$

$$
\begin{aligned}
&\| Px(t) - Py(t) \| \leq \| P_\alpha(t-s_1)K_{1x} - P_\alpha(t-s_1)K_{1y} \| \\
&\quad + \| \int_0^t (t-s)^{\alpha-1} Q_\alpha(t-s)[\| f(s,x(s),Fx(s),Bx(s)) - f(s,y(s),Fy(s),By(s)) \\
&\quad + \int_0^s q(s-\tau)k(\tau,x(\tau))d\tau - \int_0^s q(s-\tau)k(\tau,y(\tau))d\tau]ds \| \\
&\leq M_A \| d_{1x} - d_{1y} \| + M_A \| \int_0^{s_1} (s_1-s)^{\alpha-1} Q_\alpha(s_1-s)[f(s,x(s),Fx(s),Bx(s)) - \\
&\quad f(s,y(s),Fy(s),By(s)) + \int_0^s q(s-\tau)k(\tau,x(\tau))d\tau - \int_0^s q(s-\tau) \\
&\quad k(\tau,y(\tau))d\tau]ds \| + M_A \| \frac{1}{\Gamma(\beta)} \int_0^{s_1} (s_1-s)^{\beta-1} \left[g_1(s,x(s)) - g_1(s,y(s))\right] ds \| \\
&\quad + \| \int_0^t (t-s)^{\alpha-1} Q_\alpha(t-s)[f(s,x(s),Fx(s),Bx(s)) - f(s,y(s),Fy(s),By(s)) + \\
&\quad + \int_0^s q(s-\tau)k(\tau,x(\tau))d\tau - \int_0^s q(s-\tau)k(\tau,y(\tau))d\tau]ds \| \\
&\leq \left[M_A^2 L_h + \frac{M_A^2 t_1^\alpha + M_A^2 s_1^\alpha + M_A t_2^\alpha}{\Gamma(\alpha+1)} (L_f + q^* L_k + M_f B^* + C_f F^*) \right. \\
&\quad \left. + \frac{M_A L_g (s_1^\beta + t_1^\beta)}{\Gamma(\beta+1)} \right] \| x - y \|_{PC} .
\end{aligned}
$$

We suppose that for $1 \leq j \leq i$ we have:
for $t \in (t_j, s_j]$

$$
\begin{aligned}
\| \quad & Px(t) - Py(t) \| \\
& \leq [M_A^j L_h + \frac{M_A t_j^\alpha + M_A^2 (t_{j-1}^\alpha + s_{j-1}^\alpha) + M_A^3 (t_{j-2}^\alpha + s_{j-2}^\alpha) + \cdots + M_A^j (t_1^\alpha + s_1^\alpha)}{\Gamma(\alpha+1)} \times \\
& (L_f + q^* L_k + M_f B^* + C_f F^*) \\
+ \quad & \frac{L_g (t_j^\beta + s_j^\beta) + M_A L_g (t_{j-1}^\beta + s_{j-1}^\beta) + \cdots + M_A^{j-1} L_g (t_1^\beta + s_1^\beta)}{\Gamma(\beta+1)}] \| x - y \|_{PC},
\end{aligned}
$$

and for $t \in (s_j, t_{j+1}]$,

$$
\begin{aligned}
& \| (Px)(t) - (Py)(t) \| \\
& \leq M_A [M_A^j L_h + \frac{t_{j+1}^\alpha + M_A (t_j^\alpha + s_j^\alpha) + \cdots + M_A^j (t_1^\alpha + s_1^\alpha)}{\Gamma(\alpha+1)} \times \\
& (L_f + q^* L_k + M_f B^* + C_f F^*) + \\
& \frac{L_g (t_j^\beta + s_j^\beta) + \cdots + M_A^{j-2} L_g (t_2^\beta + s_2^\beta) + M_A^{j-1} L_g (t_1^\beta + s_1^\beta)}{\Gamma(\beta+1)}] \| x - y \|_{PC} .
\end{aligned}
$$

We prove the relations for $j = i + 1$.
For $t \in (t_{i+1}, s_{i+1}]$

$$
\begin{aligned}
& \| Px(t) - Py(t) \| \\
& \leq \| d_{i+1}x + \frac{1}{\Gamma(\beta)} \int_0^t (t-s)^{\beta-1} g_{i+1}(s,x(s))ds - d_{i+1}y - \\
& \frac{1}{\Gamma(\beta)} \int_0^t (t-s)^{\beta-1} g_{i+1}(s,y(s))ds \| \\
\leq & \| d_{i+1}x - d_{i+1}y \| + \frac{1}{\Gamma(\beta)} \int_0^t (t-s)^{\beta-1} \| g_{i+1}(s,x(s)) - g_{i+1}(s,y(s)) \| ds. \\
\leq & \| P_\alpha(t_{i+1} - s_i)(K_{ix} - K_{iy}) \| + \| \int_0^{t_{i+1}} (t_{i+1} - s)^{\alpha-1} Q_\alpha(t_{i+1} - s) \times \\
& [f(s,x(s),Fx(s),Bx(s)) - - f(s,y(s),Fy(s),By(s)) + \\
& \int_0^s q(s-\tau)k(\tau,x(\tau))d\tau - \int_0^s q(s-\tau)k(\tau,y(\tau))d\tau]ds \| \\
& + \| \frac{1}{\Gamma(\beta)} \int_0^{t_{i+1}} (t_{i+1} - s)^{\beta-1} (g_{i+1}(s,x(s)) - g_{i+1}(s,y(s))ds \| + \\
& + \frac{1}{\Gamma(\beta)} \int_0^t (t-s)^{\beta-1} \| (g_{i+1}(s,x(s)) - g_{i+1}(s,y(s))) \| ds \quad \text{For } t \in (s_{i+1}, t_{i+2}] \\
\leq & [M_A^{i+1} L_h + \frac{M_A t_{i+1}^\alpha + M_A^2 (t_i^\alpha + s_i^\alpha) + \cdots + M_A^{i+1} (t_1^\alpha + s_1^\alpha)}{\Gamma(\alpha+1)} \times \\
& (L_f + q^* L_k + M_f B^* + C_f F^*) + \\
& \frac{L_g (t_{i+1}^\beta + s_{i+1}^\beta) + M_A^i L_g (t_1^\beta + s_1^\beta) + \cdots + M_A L_g (t_i^\beta + s_i^\beta)}{\Gamma(\beta+1)}] \| x - y \|_{PC} .
\end{aligned}
$$

$$
\begin{aligned}
&\| Px(t)-Py(t)\| \leq \| P_\alpha(t-s_{i+1})K_{(i+1)x}-P_\alpha(t-s_{i+1})K_{(i+1)y}\| \\
&\quad +\| \int_0^t (t-s)^{\alpha-1}Q_\alpha(t-s)[f(s,x(s),Fx(s),Bx(s))-f(s,y(s),Fy(s),By(s)) \\
&\quad +\int_0^s q(s-\tau)k(\tau,x(\tau))d\tau-\int_0^s q(s-\tau)k(\tau,y(\tau))d\tau]ds\| \\
&\leq\leq M_A[\| d_{(i+1)x}-d_{(i+1)y}\| +\tfrac{s_{i+1}^{\alpha}M_A}{\Gamma(\alpha+1)}(L_f+q^*L_k+M_fB^*+C_fF^*)\| x-y\|_{PC}+ \\
&\quad +\tfrac{s_{i+1}^{\beta}}{\Gamma(\beta+1)}L_g\| x-y\|_{PC}]+\tfrac{t_{i+2}^{\alpha}M_A}{\Gamma(\alpha+1)}(L_f+q^*L_k+M_fB^*+C_fF^*)\| x-y\|_{PC} \\
&\leq \Bigg[M_A\Bigg[M_A^{i+1}L_h+\tfrac{t_{i+2}^{\alpha}+M_A^{i+1}(t_1^{\alpha}+s_1^{\alpha})+\cdots+M_A^2(t_i^{\alpha}+s_i^{\alpha})+M_A(t_{i+1}^{\alpha}+s_{i+1}^{\alpha})}{\Gamma(\alpha+1)}\times \\
&\quad \times(L_f+q^*L_k+M_fB^*+C_fF^*)\Bigg]+ \\
&\quad +\tfrac{M_AL_g(s_{i+1}^{\beta}+t_{i+1}^{\beta})+M_A^2L_g(s_i^{\beta}+t_i^{\beta})+\cdots+M_A^{i+1}L_g(t_1^{\beta}+s_1^{\beta})}{\Gamma(\beta+1)}\Bigg]\| x-y\|_{PC}\,.
\end{aligned}
$$

Then it follows that P is a contraction on the space $PC(J,X)$. Hence by the Banach contraction mapping principale, P has a unique fixed point $x\in PC(J,X)$ which is the unique mild solution of problem (1.2).

The next result is based on Krasnoselskii fixed point theorem.

Theorem 4 *Assume that* $(H_1)-(H_3)$, *and* $(H_5)-(H_6)$ *hold. In addition, let's suppose that the following condition is verified:*

$$
\max\Bigg\{M_A^{m+1}L_h+\tfrac{M_AL_g(s_m^{\beta}+t_m^{\beta})+M_A^2L_g(s_{m-1}^{\beta}+t_{m-1}^{\beta})+\cdots+M_A^mL_g(s_1^{\beta}+t_1^{\beta})}{\Gamma(\beta+1)};
\tfrac{[M_AT^{\alpha}+M_A^2(t_m^{\alpha}+s_m^{\alpha})+\cdots+M_A^{m+1}(t_1^{\alpha}+s_1^{\alpha})]}{\Gamma(\alpha+1)}(D(1+B^*+F^*)+D'q^*)\Bigg\}<1.
$$

Then the problem (1.2) *has at least one mild solution.*

Proof We introduce the composition $Q=Q_1+Q_2$ where:

$$
Q_1x(t)=\begin{cases}
P_\alpha(t)(x_0+h(x))+\int_0^t(t-s)^{\alpha-1}Q_\alpha(t-s)C(s)u(s)ds, & t\in[0,t_1],\\
d_{i1x}+\frac{1}{\Gamma(\beta)}\int_0^t(t-s)^{\beta-1}g_i(s,x(s))ds, & t\in(t_i,s_i],\ i=1,2,\ldots,m,\\
P_\alpha(t-s_i)K_{i1x}+\int_0^t(t-s)^{\alpha-1}Q_\alpha(t-s)C(s)u(s)ds, & t\in(s_i,t_{i+1}],\\
i=1,2,\ldots,m, &
\end{cases}
$$

$$
Q_2x(t)=\begin{cases}
\int_0^t(t-s)^{\alpha-1}Q_\alpha(t-s)\left[f(s,x(s),Fx(s),Bx(s))\right.\\
\left.+\int_0^s q(s-\tau)k(\tau,x(\tau))d\tau\right]ds, \quad t\in[0,t_1],\\
d_{i2x},\quad t\in(t_i,s_i],\quad i=1,2,\ldots,m,\\
P_\alpha(t-s_i)K_{i2x}+\int_0^t(t-s)^{\alpha-1}Q_\alpha(t-s)\left[f(s,x(s),Fx(s),Bx(s))\right.\\
\left.+\int_0^s q(s-\tau)k(\tau,x(\tau))d\tau\right]ds,\\
t\in(s_i,t_{i+1}],\ i=1,2,\ldots,m,
\end{cases}
$$

with

$$\begin{cases} d_{i1x} = P_\alpha(t_i - s_{i-1})K_{(i-1)1x} + \int_0^{t_i}(t_i - s)^{\alpha-1}Q_\alpha(t_i - s)C(s)u(s)ds \\ -\frac{1}{\Gamma(\beta)}\int_0^{t_i}(t_i - s)^{\beta-1}g_i(s, x(s))ds,\ i = 1, 2, \dots, m, \\ K_{i1x} = d_{i1x} + \frac{1}{\Gamma(\beta)}\int_0^{s_i}(s_i - s)^{\beta-1}g_i(s, x(s))ds \\ -\int_0^{s_i}(s_i - s)^{\alpha-1}Q_\alpha(s_i - s)C(s)u(s)ds,\ i = 1, 2, \dots, m, \\ K_{01x} = x_0 + h(x), \end{cases}$$

and

$$\begin{cases} d_{i2x} = P_\alpha(t_i - s_{i-1})K_{(i-1)2x} + \int_0^{t_i}(t_i - s)^{\alpha-1}Q_\alpha(t_i - s)[f(s, x(s), Fx(s), Bx(s)) + \\ \int_0^s q(s-\tau)k(\tau, x(\tau))d\tau]ds,\ i = 1, 2, \dots, m, \\ K_{i2x} = d_{i2x} - \int_0^{s_i}(s_i - s)^{\alpha-1}Q_\alpha(s_i - s)\left[f(s, x(s), Fx(s), Bx(s))\right. \\ \left. + \int_0^s q(s-\tau)k(\tau, x(\tau))d\tau\right]ds, i = 1, 2, \dots, m, \\ K_{02x} = 0. \end{cases}$$

Our proof will be divided into several steps.

- We show that $QB_r(J) \subset B_r(J)$
  where $B_r = \{x \in PC(J, X);\ \| x \| \leq r\}$ the ball with radius $r > 0$,
  and $K_{\alpha,\tau} = \alpha\left(\frac{1-\tau}{\alpha-\tau}\right)^{1-\tau} \| Cu \|_{L^{1/\tau}}, D^{''} = D + D'q^*$

  $\gamma_1 = M_A^{m+1}\left[\| x_0 \| + L_h'\right] + \frac{M_A[D^{''}T^\alpha + K_{\alpha,\tau}T^{\alpha-\tau}]}{\Gamma(\alpha+1)}$,

  $$\gamma_2 = \frac{\begin{array}{l} M_A^2[D^{''}(t_m^\alpha + s_m^\alpha) + K_{\alpha,\tau}(t_m^{\alpha-\tau} + s_m^{\alpha-\tau})] + \cdots + M_A^{m+1}[D^{''}(t_1^\alpha + s_1^\alpha) \\ + K_{\alpha,\tau}(t_1^{\alpha-\tau} + s_1^{\alpha-\tau})] \end{array}}{\Gamma(\alpha+1)},$$

  $\gamma_3 = \frac{M_A L_g'(t_i^\beta + s_m^\beta) + M_A^2 L_g'(t_{m-1}^\beta + s_{m-1}^\beta) + \cdots + M_A^m L_g'(t_1^\beta + s_1^\beta)}{\Gamma(\beta+1)}$,

  $\gamma_4 = \frac{[M_A T^\alpha + M_A^2(t_m^\alpha + s_m^\alpha) + \cdots + M_A^{m+1}(t_1^\alpha + s_1^\alpha)]}{\Gamma(\alpha+1)}(D(1 + B^* + F^*) + D'q^*)$.

  Here $\frac{\gamma_1+\gamma_2+\gamma_3}{1-\gamma_4} < r$.
  For any $x \in B_r$, we have:
  Case1. For $t \in [0, t_1]$,

  $$\begin{aligned} &\| Qx(t) \| \\ &\leq M_A\left[\| x_0 \| + L_h'\right] + \frac{M_A t_1^\alpha}{\Gamma(\alpha+1)}\left[D\left(1 + r(1 + B^* + F^*)\right) + D'q^*(1+r)\right] + \\ &\frac{K_{\alpha,\tau}t_1^{\alpha-\tau}}{\Gamma(\alpha+1)}M_A \\ &\leq M_A\left[\| x_0 \| + L_h' + \frac{D^{''}t_1^\alpha + K_{\alpha,\tau}t_1^{\alpha-\tau}}{\Gamma(\alpha+1)}\right] + M_A\frac{(D(1 + B^* + F^*) + D'q^*)t_1^\alpha}{\Gamma(\alpha+1)}r \\ &\leq r. \end{aligned}$$

  Similar to the proof of Theorem (4.1), we prove that:

  Case 2. For $t \in (t_i, s_i]$, $i = 1, 2, \dots, m$,

$$\begin{aligned}
&\| Qx(t) \| \\
&\leq M_A^i \left[\| x_0 \| + L'_h\right] + \frac{M_A[D'' t_i^\alpha + K_{\alpha,\tau} t_i^{\alpha-\tau}]}{\Gamma(\alpha+1)} \\
&+ \frac{M_A^2[D''(t_{i-1}^\alpha + s_{i-1}^\alpha) + K_{\alpha,\tau}(t_{i-1}^{\alpha-\tau} + s_{i-1}^{\alpha-\tau})] + \cdots + M_A^i[D''(t_1^\alpha + s_1^\alpha) + K_{\alpha,\tau}(t_1^{\alpha-\tau} + s_1^{\alpha-\tau})]}{\Gamma(\alpha+1)} \\
&+ \frac{L'_g(t_i^\beta + s_i^\beta) + M_A L'_g(t_{i-1}^\beta + s_{i-1}^\beta) + \cdots + M_A^{i-1} L'_g(t_1^\beta + s_1^\beta)}{\Gamma(\beta+1)} \\
&+ \frac{[M_A t_i^\alpha + M_A^2(t_{i-1}^\alpha + s_{i-1}^\alpha) + M_A^3(t_{i-1}^\alpha + s_{i-1}^\alpha) + \cdots + M_A^i(t_1^\alpha + s_1^\alpha)]}{\Gamma(\alpha+1)} \times \\
&(D(1 + B^* + F^*) + D' q^*))r \leq r.
\end{aligned}$$

Case 3. For $t \in (s_i, t_{i+1}]$,

$$\begin{aligned}
&\| Qx(t) \| \\
&\leq M_A^{i+1} \left[\| x_0 \| + L'_h\right] + \frac{M_A[D'' t_{i+1}^\alpha + K_{\alpha,\tau} t_{i+1}^{\alpha-\tau}]}{\Gamma(\alpha+1)} \\
&+ \frac{M_A^2[D''(t_i^\alpha + s_i^\alpha) + K_{\alpha,\tau}(t_i^{\alpha-\tau} + s_i^{\alpha-\tau})] + \cdots + M_A^{i+1}[D''(t_1^\alpha + s_1^\alpha) + K_{\alpha,\tau}(t_1^{\alpha-\tau} + s_1^{\alpha-\tau})]}{\Gamma(\alpha+1)} + \\
&+ \frac{M_A L'_g(t_i^\beta + s_i^\beta) + M_A^2 L'_g(t_{i-1}^\beta + s_{i-1}^\beta) + \cdots + M_A^i L'_g(t_1^\beta + s_1^\beta)}{\Gamma(\beta+1)} + \\
&+ \frac{[M_A t_{i+1}^\alpha + M_A^2(t_i^\alpha + s_i^\alpha) + \cdots + M_A^{i+1}(t_1^\alpha + s_1^\alpha)]}{\Gamma(\alpha+1)} (D(1 + B^* + F^*) + D' q^*)r \\
&\leq r.
\end{aligned}$$

- Q_1 is a contraction on B_r. Let $x, y \in B_r$
  Case 1. For $t \in [0, t_1]$, we have:
  $\| Q_1 x(t) - Q_1 y(t) \| \leq M_A L_h \| x - y \|_{PC} < \| x - y \|_{PC}$.
  Similar to the proof of Theorem (4.1), we prove that:
  Case 2. For $t \in [t_i, s_i]$, $1 \leq i \leq m$,

$$\begin{aligned}
&\| Q_1 x(t) - Q_1 y(t) \| \\
&\leq \left[M_A^i L_h + \frac{L_g(s_i^\beta + t_i^\beta) + M_A L_g(s_{i-1}^\beta + t_{i-1}^\beta) + \cdots + M_A^{i-1} L_g(s_1^\beta + t_1^\beta)}{\Gamma(\beta+1)} \right] \| x - y \|_{PC} \\
&quad < \| x - y \|_{PC} .
\end{aligned}$$

For $t \in (s_i, t_{i+1}]$, $1 \leq i \leq m$,

$$\| Q_1x(t) - Q_1y(t) \|$$
$$\leq \left[M_A^{i+1}L_h + \frac{M_AL_g(s_i^\beta + t_i^\beta) + M_A^2L_g(s_{i-1}^\beta + t_{i-1}^\beta)\cdots M_A^iL_g(s_1^\beta + t_1^\beta)}{\Gamma(\beta+1)} \right] \| x - y \|_{PC}$$
$$< \| x - y \|_{PC} .$$

This implies that Q_1 is a contraction.

- Q_2 is continuous.
  Let $(x_n)_{n\geq 0}$ be a sequence such that $\lim_{x\to\infty} \| x_n - x \|_{PC} = 0$, we have :
  Case 1. For $t \in [0, t_1]$

$$\| Q_2x_n(t) - Q_2x(t) \| \leq \| \int_0^t (t-s)^{\alpha-1}Q_\alpha(t-s)([f(s, x_n(s), Fx_n(s), Bx_n(s)) +$$
$$\int_0^s q(s-\tau)k(\tau, x_n(\tau))d\tau - f(s, x(s), Fx(s), Bx(s)) -$$
$$\int_0^s q(s-\tau)k(\tau, x(\tau))d\tau)]ds \|$$
$$\leq \frac{M_At_1^\alpha}{\Gamma(\alpha+1)}\Big(\| f(., x_n(.), Fx_n(.), , Bx_n(.)) - f(., x(.), Fx(.), Bx(.)) \|_{PC} +$$
$$+ \| q^*[k(., x_n(.)) - k(., x(.))] \|_{PC} \Big).$$

Similar to the proof of Theorem (4,1), we prove that:

Case 2. For $t \in (t_i, s_i],\ i = 1, 2, \ldots, m,$

$$\| Q_2x_n(t) - Q_2x(t) \| \leq \left[\frac{M_At_i^\alpha + M_A^2(t_{i-1}^\alpha + s_{i-1}^\alpha) + \cdots + M_A^i(t_1^\alpha + s_1^\alpha)}{\Gamma(\alpha+1)} \right] \times$$
$$\Big(\| f(., x_n(.), Fx_n(.), , Bx_n(.)) - f(., x(.), Fx(.), Bx(.)) \|_{PC} +$$
$$\| q^*[k(., x_n(.)) - k(., x(.))] \|_{PC} \Big).$$

Case 3. For $t \in (s_i, t_{i+1}]$,

$$\| Q_2x_n(t) - Q_2x(t) \| \leq \left[\frac{M_At_{i+1}^\alpha + M_A^2(t_i^\alpha + s_i^\alpha) + \cdots + M_A^{i+1}(t_1^\alpha + s_1^\alpha)}{\Gamma(\alpha+1)} \right] \times$$
$$\times\Big(\| f(., x_n(.), Fx_n(.), , Bx_n(.)) - f(., x(.), Fx(.), Bx(.)) \|_{PC} +$$
$$\| q^*[k(., x_n(.)) - k(., x(.))] \|_{PC} \Big).$$

- Q_2 is compact.
  1. We have $Q_2B_r \subseteq B_r$, then Q_2 is uniformly bounded on B_r,
  2. For $x \in B_r$, we have the following:
  for $0 \le t' < t'' \le t_1$, we have:

$$\begin{aligned}
&\| Q_2x(t'') - Q_2x(t') \| \\
&\le \| \int_0^{t''} (t''-s)^{\alpha-1} Q_\alpha(t''-s)\Big(f(s,x(s),Fx(s),Bx(s)) + \\
&\int_0^s q(s-\tau)k(\tau,x(\tau))d\tau \Big) ds - \int_0^{t'} (t'-s)^{\alpha-1} Q_\alpha(t'-s) \times \\
&\Big(f(s,x(s),Fx(s),Bx(s)) + \int_0^s q(s-\tau)k(\tau,x(\tau))d\tau \Big) ds \| \\
&\le I_1 + I_2 + I_3,
\end{aligned}$$

where

$$\begin{aligned}
I_1 &= \| \int_{t'}^{t''} (t''-s)^{\alpha-1} Q_\alpha(t''-s)\Big(f(s,x(s),Fx(s),Bx(s)) \\
&\quad + \int_0^s q(s-\tau)k(\tau,x(\tau))d\tau \Big) ds \| \\
I_2 &= \| \int_0^{t'} (t'-s)^{\alpha-1} \left[Q_\alpha(t''-s) - Q_\alpha(t'-s) \right] \Big(f(s,x(s),Fx(s),Bx(s)) + \\
&\int_0^s q(s-\tau)k(\tau,x(\tau))d\tau \Big) ds \| \\
I_3 &= \| \int_0^{t'} \left[(t''-s)^{\alpha-1} - (t'-s)^{\alpha-1} \right] Q_\alpha(t''-s)\Big(f(s,x(s),Fx(s),Bx(s)) + \\
&\int_0^s q(s-\tau)k(\tau,x(\tau))d\tau \Big) ds \| .
\end{aligned}$$

$I_1 \le \frac{\alpha M_A[D(1+r(1+B^*+F^*))+D'q^*(1+r)]}{\Gamma(\alpha+1)} \int_{t'}^{t''} (t''-s)^{\alpha-1} ds$
$\le \frac{M_A[D(1+r(1+B^*+F^*))+D'q^*(1+r)]}{\Gamma(\alpha+1)} (t''-t')^\alpha \longrightarrow 0,\ t''-t' \longrightarrow 0.$
$I_2 \le [D(1+r(1+B^*+F^*)) + D'q^*(1+r)] \int_0^{t'} (t'-s)^{\alpha-1} \| Q_\alpha(t''-s)^{\alpha-1}$
$- Q_\alpha(t'-s)^{\alpha-1} \| ds \longrightarrow 0,\ t''-t' \longrightarrow 0.$
$I_3 \le \frac{M_A[D(1+r(1+B^*+F^*))+D'q^*(1+r)]}{\Gamma(\alpha)} \int_0^{t'} \left((t''-s)^{\alpha-1} - (t'-s)^{\alpha-1}\right) ds$
$\le \frac{M_A[D(1+r(1+B^*+F^*))+D'q^*(1+r)]}{\Gamma(\alpha+1)} (t''-t')^\alpha \longrightarrow 0;\ t''-t' \longrightarrow 0.$
Case 1. For $t_i \le t' < t'' \le s_i$,
$\| Q_2x(t'') - Q_2x(t') \| = 0.$
Case 2. For $s_i \le t' < t'' \le t_{i+1}$,

$$\| Q_2x(t'') - Q_2x(t') \| \le I_1 + I_2 + I_3 + \| \left(P_\alpha(t''-s_i) - P_\alpha(t'-s_i)\right) K_{i2x} \| . \quad (3)$$

From (H_1) and the proof of lemma 3.4 in [7] we have that the continuity of $P_\alpha(t)$ and $Q_\alpha(t)$ $(t > 0)$ in t is in the uniform operator topology, we deduce that the right-hand side of (4.1) tends to 0 independently of $x \in B_r$ as $t'' \to t'$.

Case 3. For $t_i \le t' < s_i < t'' \le t_{i+1}$,

$$\begin{aligned}\| Q_2x(t'') - Q_2x(t') \| &\le \| P_\alpha(t'' - s_i)K_{i2x} \\ &+ \int_0^{t''} (t'' - s)^{\alpha-1} Q_\alpha(t'' - s)\Big(f(s, x(s), Fx(s), Bx(s)) + \\ &+ \int_0^s q(s - \tau)k(\tau, x(\tau))d\tau \Big)ds - d_{i2x} \| \longrightarrow 0\end{aligned}$$

independently of $x \in B_r$, as $t'' \longrightarrow t'$ we have $(t'' \longrightarrow s_i)$.

In conclusion, $\| Q_2x(t'') - Q_2x(t') \| \longrightarrow 0$, as $t'' - t' \longrightarrow 0$, which implies that $Q_2(B_r(J))$ is equicontinuous.
We have $Q_2B_r \subseteq B_r$, let $Q_2B_r(t) = \{Q_2x(t); x \in B_r\}$ for $t \in J$.

3. $Q_2B_r(t)$ is relatively compact.
$Q_2B_r(0) = \{0\}$ is compact.
For $t \in [0, t_1]$.
For each $\varepsilon \in (0, t)$ and $\delta > 0$, we define a set:
$Q_2^{\varepsilon,\delta}(B_r)(t) = \left\{Q_2^{\varepsilon,\delta}x(t); x \in B_r\right\}$
with

$$\begin{aligned}Q_2^{\varepsilon,\delta}x(t) &= \alpha \int_0^{t-\varepsilon} \int_\delta^\infty \theta(t - s)^{\alpha-1} \xi_\alpha(\theta) T((t - s)^\alpha \theta)\Big(f(s, x(s), Fx(s), Bx(s)) + \\ &\int_0^s q(s - \tau)k(\tau, x(\tau))d\tau \Big)d\theta ds \\ &= \alpha \int_0^{t-\varepsilon} \int_\delta^\infty \theta(t - s)^{\alpha-1} \xi_\alpha(\theta) \big[T(\varepsilon^\alpha \delta) T((t - s)^\alpha \theta \\ &-\varepsilon^\alpha \delta)\big] \Big(f(s, x(s), Fx(s), Bx(s)) + \int_0^s q(s - \tau)k(\tau, x(\tau))d\tau \Big)d\theta ds \\ &= \alpha T(\varepsilon^\alpha \delta) \int_0^{t-\varepsilon} \int_\delta^\infty \theta(t - s)^{\alpha-1} \xi_\alpha(\theta) T((t - s)^\alpha \theta \\ &-\varepsilon^\alpha \delta)\Big(f(s, x(s), Fx(s), Bx(s)) + \int_0^s q(s - \tau)k(\tau, x(\tau))d\tau \Big)d\theta ds.\end{aligned}$$

(Observe that $\theta \ge \delta$ and $t - \varepsilon \ge s$, hence $(t - s)^\alpha \theta - \varepsilon^\alpha \delta \ge 0$) Since the operator $T(\varepsilon^\alpha \delta)$ $(\varepsilon^\alpha \delta > 0)$ is compact, the set $Q_2^{\varepsilon,\delta} B_r(t)$ is relatively compact in X. Moreover, for every $x \in B_r$ we have

$$\| Q_2x(t) - Q_2^{\varepsilon,\delta}x(t) \| \leq \alpha \| \int_0^t \int_0^\delta \theta(t-s)^{\alpha-1}\xi_\alpha(\theta)T((t-s)^\alpha\theta)$$
$$\left(f(s,x(s),Fx(s),Bx(s)) + \int_0^s q(s-\tau)k(\tau,x(\tau))d\tau \right) d\theta ds$$
$$+ \int_0^t \int_\delta^\infty \theta(t-s)^{\alpha-1}\xi_\alpha(\theta)T((t-s)^\alpha\theta) \times$$
$$\times \left(f(s,x(s),Fx(s),Bx(s)) + \int_0^s q(s-\tau)k(\tau,x(\tau))d\tau \right) d\theta ds$$
$$- \int_0^{t-\varepsilon} \int_\delta^\infty \theta(t-s)^{\alpha-1}\xi_\alpha(\theta)T((t-s)^\alpha\theta)\Big(f(s,x(s),Fx(s),Bx(s)) +$$
$$\int_0^s q(s-\tau)k(\tau,x(\tau))d\tau \Big) d\theta ds \|$$
$$\leq G_1 + G_2.$$

With
$G_1 = \alpha \| \int_0^t \int_0^\delta \theta(t-s)^{\alpha-1}\xi_\alpha(\theta)T((t-s)^\alpha\theta)\Big(f(s,x(s),Fx(s),Bx(s)) + \int_0^s q$
$(s-\tau)k(\tau,x(\tau))d\tau \Big) d\theta ds \|$
and
$G_2 = \alpha \| \int_{t-\varepsilon}^t \int_\delta^\infty \theta(t-s)^{\alpha-1}\xi_\alpha(\theta)T((t-s)^\alpha\theta)\Big(f(s,x(s),Fx(s),Bx(s)) + \int_0^s q$
$(s-\tau)k(\tau,x(\tau))d\tau \Big) d\theta ds \|$.
We have

$$G_1 \leq \alpha M_A \int_0^t (t-s)^{\alpha-1} \| \Big(f(s,x(s),Fx(s),Bx(s))$$
$$+ \int_0^s q(s-\tau)k(\tau,x(\tau))d\tau \Big) \| ds \int_0^\delta \theta\xi_\alpha(\theta)d\theta$$
$$\leq M_A t_1^\alpha [D\left(1 + r(1+B^*+F^*)\right) + D'q^*(1+r)] \int_0^\delta \theta\xi_\alpha(\theta)d\theta,$$

and

$$G_2 \leq \alpha M_A \int_{t-\varepsilon}^t (t-s)^{\alpha-1} \| \Big(f(s,x(s),Fx(s),Bx(s))$$
$$+ \int_0^s q(s-\tau)k(\tau,x(\tau))d\tau \Big) \| ds \int_\delta^\infty \theta\xi_\alpha(\theta)d\theta$$
$$\leq M_A \varepsilon^\alpha [D\left(1 + r(1+B^*+F^*)\right) + D'q^*(1+r)] \int_0^\infty \theta\xi_\alpha(\theta)d\theta$$
$$\leq \frac{M_A[D(1+r(1+B^*+F^*)) + D'q^*(1+r)]}{\Gamma(\alpha+1)}\varepsilon^\alpha.$$

Then $\| (Q_2x(t) - Q_2^{\varepsilon,\delta}x(t) \| \longrightarrow 0$, as $\varepsilon \longrightarrow 0; \delta \longrightarrow 0$
This means that there are relatively compact sets arbitrarily close to the set $Q_2B_r(t)$.
Hence the set $Q_2B_r(t)$ is also relatively compact in X.
For $t_i < t \leq s_i, i = 1, 2, \ldots, m$, in such case
$Q_2B_r(t) = \{d_{i2x}, x \in B_r\}$ is compact.
For $s_i < t \leq t_{i+1}, i = 1, 2, \ldots, m$,

$$Q_2B_r(t) = \Big\{P_\alpha(t-s_i)K_{i2x} + \int_0^t (t-s)^{\alpha-1}Q_\alpha(t-s)\Big(f(s, x(s), Fx(s), Bx(s)) + \int_0^s q(s-\tau)k(\tau, x(\tau))d\tau\Big)ds, x \in B_r\Big\}.$$

By the same argument in case 1 ($t \in [0, t_1]$) and $P_\alpha(t - s_i)$ is a compact operator, we know that $Q_2B_r(t)$ is relatively compact.

Example In this section, we give two examples which illustrate the applicability of our results.
Throughout this section, we let $X = L^2(0, 1)$, $J = [0, 1]$, $t_0 = s_0 = 0$, $t_1 = \frac{1}{3}$, $s_1 = \frac{2}{3}$, $T = 1$.
Define $Ax = \frac{\partial^2}{\partial^2 v}x$ for $x \in D(A) = \Big\{x \in X : \frac{\partial x}{\partial v}, \frac{\partial^2 x}{\partial^2 v} \in X, x(0) = x(1) = 0\Big\}$.
Then A is the infinitesimal generator of strongly continuous semigroup $\{T(t), t \geq 0\}$ on X. In addition $T(t)$ is compact and $\| T(t) \| \leq 1 = M_A$, for all $t \geq 0$. See [9]

Example Consider

$$\begin{cases} \frac{\partial^\alpha}{\partial t^\alpha}y(t,v) = \frac{\partial^2}{\partial v^2}y(t,v) + \frac{1}{24}sin\Big[y(t,v) + \int_0^t \frac{e^{-(s-t)}}{80}y(s,v) \\ \qquad + \int_0^t \frac{e^{-2(s-t)}}{160}y(s,v)ds\Big] + \\ \int_0^t e^{-(s-t)}\frac{1}{24}cos(y(s,v))ds + C(t,v), \quad v \in (0,1), \quad t \in [0, \frac{1}{3}] \cup (\frac{2}{3}, 1], \\ \frac{\partial^\beta}{\partial t^\beta}y(t,v) = \frac{1}{8}cos(y(t,v)), \quad t \in (\frac{1}{3}, \frac{2}{3}], \\ y(t,v) = y_0 + \frac{1}{8e^t}\left(y(s_1,v) + y(t_1,v)\right). \end{cases}$$

Denote $x(t)(v) = y(t, v)$ and $C(t, v) = C(t)u(t)(v)$
This problem can be abstracted into;

$$(P)\begin{cases} {}^cD^\alpha x(t)(v) = Ax(t)(v) + f(t, x(t), Fx(t), Bx(t))(v) \\ \qquad + \int_0^t q(t-s)k(s, x(s)(v))ds + C(t)u(t)(v) \\ t \in [0, \frac{1}{3}] \cup (\frac{2}{3}, 1], \ u \in U_{ad}, \\ {}^cD^\beta x(t) = g_1(t, x(t)), \ t \in (t_i, s_i], \ t \in (\frac{1}{3}, \frac{2}{3}], \\ x(0) = x_0 + h(x). \end{cases}$$

where: $Bx(t)(v) = \int_0^t \frac{e^{-2(s-t)}}{160}y(s,v)ds$

$$Fx(t)(v) = \int_0^t \frac{e^{-(s-t)}}{80} y(s,v)ds,\ f(t,x(t),Fx(t),Bx(t))(v) = \frac{1}{24} sin[y(t,v)$$
$$+ \int_0^t \frac{e^{-(s-t)}}{80} y(s,v) + \int_0^t \frac{e^{-2(s-t)}}{160} y(s,v)ds]$$
$$k(t,x(t))(v) = \frac{1}{24} cos(x(t)(v)),\ g_1(t,x(t))(v) = \frac{1}{8} cos(x(t)(v)),\ h(t,x(t))(v)$$
$$= \frac{1}{8}\left(x(s_1)(v) + x(t_1)(v)\right),\ q(t) = \frac{e^t}{4} and \alpha = 0.85,\ \beta = 0.95,$$

In this case we have: $L_k = L_f = C_f = M_f = \frac{1}{24}, L_g = L_h = \frac{1}{8}$, and $F^* = B^* = \frac{1}{40}$, $q^* = \frac{e-1}{4}$,
$\left[L_h + \frac{1+(t_1^\alpha + s_1^\alpha)}{\Gamma(\alpha+1)}(L_f + q^* L_k + M_f B^* + C_f F^*) + \frac{L_g(t_1^\beta + s_1^\beta)}{\Gamma(\beta+1)}\right] \approx 0.393717$
< 1.

Which implies that all the assumptions of theorem 4.1 are satisfied. Therefore, there exists a unique mild solution to our problem.

Example Consider

$$\begin{cases} \frac{\partial^\alpha}{\partial t^\alpha} y(t,v) = \frac{\partial^2}{\partial v^2} y(t,v) + \frac{1}{24e^t} \frac{|y(t,v)+By(t,v)+Fy(t,v)|}{1+|y(t,v)+By(t,v)+Fy(t,v)|} \\ \qquad + \int_0^t \frac{1}{24e^{-(t-s)}} \frac{e^s|y(s,v)|}{1+|y(s,v)|} ds + C(t,v), \\ v \in (0,1),\ t \in [0, \frac{1}{3}] \cup (\frac{2}{3}, 1], \\ \frac{\partial^\beta}{\partial t^\beta} y(t,v) = \frac{1}{8e^t} \frac{|y(t,v)|}{1+|y(t,v)|},\ t \in (\frac{1}{3}, \frac{2}{3}], \\ y(t,v) = y_0 + \frac{1}{8e^t}\left(y(s_1,v) + y(t_1,v)\right). \end{cases}$$

Denote $x(t)(v) = y(t,v)$ and $C(t,v) = C(t)u(t)(v)$.

This problem can be abstracted into (P). Where: $Bx(t)(v) = \int_0^t \frac{e^{-2(s-t)}}{160} y(s,v)ds$
$Fx(t)(v) = \int_0^t \frac{e^{-(s-t)}}{80} y(s,v)ds, \quad f(t,x(t),Fx(t),Bx(t))(v)$
$= \frac{1}{24e^t} \frac{|x(t)(v)+Bx(t)(v)+Fx(t)(v)|}{1+|x(t)(v)+Bx(t)(v)+Fx(t)(v)|}$

$k(t,x(t))(v) = \frac{1}{24} \frac{e^t|y(t,v)|}{1+|y(t,v)|}, \qquad g_1(t,x(t))(v) = \frac{1}{8e^t} \frac{|y(t,v)|}{1+|y(t,v)|}, \qquad h(t,x(t))(v)$
$= \frac{1}{8e^t}\left(x(s_1)(v) + x(t_1)(v)\right),$
$q(t) = \frac{e^t}{4}$ and $\alpha = 0.75,\ \beta = 0.65$.

In this case we have: $L_h = L_g = \frac{1}{8}, D = D' = \frac{1}{24}$, and $B^* = F^* = \frac{1}{40}, q^* = \frac{e-1}{4}$
$L_h + L_g \frac{(t_1^\beta + s_1^\beta)}{\Gamma(\beta+1)} \approx 0.3$
and $\frac{1+(t_1^\alpha + s_1^\alpha)}{\Gamma(\alpha+1)}(D(1 + B^* + F^*) + D'q^*) \approx 0.145993$.
$\frac{1+(t_1^\alpha + s_1^\alpha)}{\Gamma(\alpha+1)}(D(1 + B^* + F^*) + D'q^*) \approx 0.145993$.
We have $max\{0.3; 0.145993\} < 1$.

This implies that all assumptions of Theorem 4.2 are satisfied. Then, our problem has at least one mild solution.

References

1. E. Hernandez, D. O'Regan, On a new class of abstract impulsive differential equations. Proc. Am. Math. Soc. **141**, 1641–1649 (2013)
2. S. Liu, J.R. Wang, Optimal controls of systems governed by semilinear fractional differential equations with not instantaneous impulses. J. Optim. Theory Appl. https://doi.org/10.1007/s10957-017-1122-3
3. P.L. Falb, Infinite dimensional control problems: on the closure of the set of attainable states for linear systems. Math. Anal. Appl. **9**,12–22 (1964)
4. W. Wei, X. Xiang, Y. Peng, Nonlinear impulsive integro-differential equation of mixed type and optimal controls. Optimization **55**, 141–156 (2006)
5. A.A. Kilbas, H.M. Srivastava, J.J. Trujillo, in Theory and Applications of Fractional Differential Equations. North-Holland Mathematics Studies, vol. 204 (Elsevier, Amesterdam, 2006)
6. Y. Zhou, F. Jiao, Nonlocal Cauchy problem for fractional evolution equations. Nonlinear Anal. Real World Appl. **11**, 4465–4475 (2010)
7. Y. Zhou, F. Jiao, Existence of mild solutions for fractional neutral evolution equations. Comput. Math. Appl. **59**, 1063–1077 (2010)
8. X. Fu, X. Liu, B. Lu, On a new class of impulsive fractional evolution equations. Adv. Differ. Equ. **2015**, 227 (2015)
9. A. Pazy, *Semigroups of Linear Operators and Applications to Partial Differential Equations* (Springer, Berlin, 1983)
10. A. Angaraj, K. Karthikeyan, Existence of solutions for impulsive neutral functional differential equations with non-local conditions. Nonlinear Anal. **70**, 2717–2721 (2009)
11. A. Anguraj, M. Lathamaheshwari, Existence results for fractional differential equations with infinite delay and interval impulsive conditions. Malaya J. Mat. **2**(1), 16–23 (2014)
12. J.H. Liu, Nonlinear impulsive evolution equations. Dynam. Contin. Discrete Impuls. Syst. **6**, 77–85 (1999)
13. V. Lakshmikantham, D.D. Bainov, P.S. Simeonov, *Theory of Impulsive Differential Equations* (World Scientific, Singapore, 1989)
14. J.J. Nieto, D. O'Regan, Variational approach to impulsive differential equations. Nonlinear Anal. Real World Appl. **10**, 680–690 (2009)

Fixed Point Theory, Contractive Mapping, Fuzzy Metric Space

Said Melliani, A. Moussaoui and L. S. Chadli

Abstract In this paper we introduce a new class of fuzzy contractive mapping and we show that such a class unify and generalize several existing concepts in the literature. We establish fixed point theorem for such mappings in complete strong fuzzy metric spaces and we give an illustrative example.

1 Introduction

One of the important theoretical development in the fuzzy sets theory is the way of defining the concept of fuzzy metric space, In 1975, Kramosil and Michalek [12] introduced the concept of fuzzy metric spaces, which constitutes a reformulation of the notion of probabilistic metric space, in 1988 Grabiec [1] introduced the Banach contraction in a fuzzy metric space in the sense of Kramosil and Michalek and extended the well-known fixed point theorems of Banach and Edelstein [2] to fuzzy metric spaces. In order to define a Hausdorff topology George and Veeramani [3] modified the concept of fuzzy metric space introduced by Kramosil and Michalek. Gregori and Sapena [4] reconsidered the Banach contraction principle by initiating a new concept of fuzzy contractive mapping. In this direction, Mihet [10] introduced the notion of fuzzy ψ-contractive mappings and generalized the definitions given in [13] and [4]. Wardowski [11] proposed a New family of contractive mappings in a new sense called $\mathcal{H}$-contractive and proved that the class of fuzzy contractive mappings is included in the class of fuzzy $\mathcal{H}$-contractive mappings and obtained a fixed point result in complete fuzzy metric spaces in the sense of George and Veeramani, Gregori and Miñana [6] pointed out some drawbacks on the conditions

S. Melliani (✉) · A. Moussaoui · L. S. Chadli
Sultan Moulay Slimane University, BP 523 Beni Mellal, Morocco
e-mail: s.melliani@usms.ma

A. Moussaoui
e-mail: a.moussaoui@usms.ma

L. S. Chadli
e-mail: sa.chadli@yahoo.fr

S. Melliani and O. Castillo (eds.), *Recent Advances in Intuitionistic Fuzzy Logic Systems*, Studies in Fuzziness and Soft Computing 372,
https://doi.org/10.1007/978-3-030-02155-9_21

in Wardowski Theorem, The main Wardowskis theorem is correct and it is different from the ones known in the literature. For other related concepts and results on the development of fixed point theory in fuzzy metric spaces and its applications the reader is referred to [7, 9, 13–15].

Building on this background and aiming to unifying different classes of fuzzy contractive mappings we introduce a new class of mappings called fuzzy $\mathcal{F}\mathcal{Z}$-contractive mappings. Moreover, we show that many existing concepts in the literature can be easily deduced from our definition and we provide a fixed point theorem for this class of contractive mapping in strong fuzzy metric spaces.

2 Preliminaries

For the sake of completeness, we briefly recall some basic concepts used in the following.

Definition 1 (Schweizer and Sklar [8]) A binary operation $\star : [0, 1] \times [0, 1] \longrightarrow [0, 1]$ is called a continuous triangular norm (in short, continuous t-norm) if it satisfies the following conditions

(T1) $a \star 1 = a \quad \forall a \in [0, 1]$,
(T2) $a \star b \leq c \star d \quad \forall a \leq c\ ,\ b \leq d\ \ and\ \ a, b, c, d \in [0, 1]$,
(T3) $\star$ is commutative and associative,
(T4) $\star$ is continuous,

Example 1 The following instances are classical examples of continuous t-norm:

(a) $a \star b = \min(a, b)$ Zadeh's t-norm
(b) $a \star b = \max[0, a + b - 1]$ Lukasiewicz's t-norm
(c) $a \star b = a.b$ Probabilistic t-norm.

Definition 2 (George and Veeramani [3]) The 3-tuple $(X, M, \star)$ is said to be a fuzzy metric space if X is an arbitrary set, $\star$ is a continuous t-norm and M is fuzzy set on $X \times X \times]0, \infty[$ satisfying the following conditions:
$\forall\ \ x, y, z \in X\ \ and\ \ s, t > 0$:

(GV1) $M(x, y, t) > 0$,
(GV2) $M(x, y, t) = 1 \Longleftrightarrow x = y$,
(GV3) $M(x, y, t) = M(y, x, t)$,
(GV4) $M(x, y, t) \star M(y, z, s) \leq M(x, z, t + s)$,
(GV5) $M(x, y, .) :]0, \infty[\rightarrow [0, 1]$ is continuous.

$M(x, y, t)$ can be thought of as the degree of nearness between x and y with respect to t. In the above definition, if we replace $(GV5)$ by $(GV5') : M(x, y, t) \star M(y, z, s) \leq M(x, z, \max\{t, s\}) \forall\ x, y, z \in X\ \ and\ \ s, t > 0$ Then the triple $(X, M, \star)$ is said to be a non-Archimedean fuzzy metric space. As $(GV5') \Rightarrow (GV5)$, each non-Archimedean fuzzy metric space is a fuzzy metric space, Further if $M(x, y, t) \star M(y, z, t) \leq M(x, z, t)$ $\forall t > 0$ M is said to be a strong fuzzy metric.

Let $(X, M, \star)$ be a fuzzy metric space. The open ball $B_M(x, r, t)$ for $t > 0$ with centre $x \in X$ and radius r, $0 < r < 1$ is defined by:

$$B_M(x, r, t) = \{y \in X : M(x, y, t) > 1 - r\}$$

A subset A of a fuzzy metric space $(X, M, \star)$ is said to be open if given any point $x \in A$, there exists $0 < r < 1$, and $t > 0$ such that $B(x, r, t) \subseteq A$.

The familly:

$$\tau_M = \{A \subset X : x \in A \textit{ if only if there exist } t > 0 \textit{ and } 0 < r < 1 \textit{ such that } B(x, r, t) \subset A\}$$

is a topology on X, τ_M is called the topology on X induced by the fuzzy metric M. (X, τ_M) is a Housdorff first countable topological space [3].

Example 2 [3] Let (X, d) be a metric space, Define $a \star b = \min(a, b)\, \forall\, a, b \in [0, 1]$. Define $M_d(x, y, t) = t/t + d(x, y)$ $\forall$ $t > 0$. Then $(X, M_d, \star)$ is a fuzzy metric space. M_d is called the standard fuzzy metric induced by d.
Moreover, The topology τ_{M_d} generated by the induced fuzzy metric M_d coincides with the topology generated by d.

Theorem 1 [1] *Let $(X, M, \star)$ be a fuzzy metric space. Then $M(x, y, .)$ is non-decreasing for all $x, y \in X$.*

Definition 3 [3] Let $(X, M, \star)$ be a fuzzy metric space. Then:

1. A sequence $\{x_n\}$ in X is said to be a Cauchy sequence if and only if for each ε, $0 < \varepsilon < 1$ and $t > 0$, there exists $n_0 \in \mathbb{N}$ such that $M(x_n, x_m, t) > 1 - \varepsilon$ for all $n, m \geq n_0$.
2. A sequence $\{x_n\}$ in X is said to convergent to x in X, denoted $x_n \to x$, if an only if $\lim_{n\to\infty} M(x_n, x, t) = 1$ for all $t > 0$, i.e. for each $r \in]0, 1[$ and $t > 0$, there exists $n_0 \in \mathbb{N}$ such that $M(x_n, x, t) > 1 - r$ for all $n \geq n_0$.
3. The fuzzy metric space $(X, M, \star)$ is called complete if every Cauchy sequence is convergent.

Definition 4 (Gregori and Sapena [4]) Let $(X, M, \star)$ be a fuzzy metric space. A mapping $T : X \to X$ is said to be fuzzy contractive if there exists $k \in]0, 1[$ such that

$$\frac{1}{M(T(x), T(y), t)} - 1 \leq k\left(\frac{1}{M(x, y, t)} - 1\right)$$

for each $x, y \in X$ and $t > 0$. A sequence $\{x_n\}$ in X is said to be fuzzy contractive if there exists $k \in]0, 1[$ such that

$$\frac{1}{M(x_{n+1}, x_{n+2}, t)} - 1 \leq k(\frac{1}{M(x_n, x_{n+1}, t)} - 1)$$

for all $t > 0 \quad n \in \mathbb{N}$.

Theorem 2 [4] *Let $(X, M, \star)$ be a complete fuzzy metric space in which fuzzy contractive sequences are Cauchy. Let $T : X \rightarrow X$ be a fuzzy contractive mapping being k the contractive constant. Then T has a unique fixed point.*

Definition 5 (Mihet [10]) Let Ψ be the class of all mapping $\psi :]0, 1] \longrightarrow]0, 1]$ such that ψ is continuous, nondecreasing and, $\psi(t) > t$ for all $t \in]0, 1[$.
Let $\psi \in \Psi$, A mapping $T : X \longrightarrow X$ is said to be fuzzy ψ-contractive mapping if:

$$M(T(x), T(y), t) \geq \psi(M(x, y, t)) \;\; for\ all \;\; x, y \in X, t > 0$$

In [11] Wardowski proposed a new type of contraction in a fuzzy metric space. We can read it as follows:

Definition 6 (Wardowski [11]) Let $\mathscr{H}$ be the family of the mappings $\eta :]0, 1] \longrightarrow [0, \infty[$ satisfying the following conditions:

($\mathscr{H}_1$) η transforms $]0, 1]$ onto $[0, \infty[$,
($\mathscr{H}_2$) η is strictly decreasing.

A mapping $T : X \longrightarrow X$. is said to be fuzzy $\mathscr{H}$-contractive with respect to $\eta \in \mathscr{H}$ if $\exists k \in]0, 1[$ satisfying the following condition:

$$\eta(M(T(x), T(y), t)) \leq k\eta((M(x, y, t)) \;\; for\ all \;\; x, y \in X, t > 0$$

Remark 1 (Gregori and Miñana [6]) If $\eta \in \mathscr{H}$ then the mapping $\eta.k :]0, 1] \longrightarrow [0, \infty[$ and $\eta^{-1} : [0, \infty[\longrightarrow]0, 1]$, defined in its obvious sense, are two bijective continuous mappings which are strictly decreasing.

3 Main Results

Definition 7 Let $\phi :]0, 1] \times]0, 1] \longrightarrow \mathbb{R}$ be a mapping satisfying the following conditions:

(i) $\phi(1, 1) = 0$,
(ii) $\phi(t, s) < \frac{1}{s} - \frac{1}{t}$ for all $t, s < 1$,
(iii) if $\{t_n\}$, $\{s_n\}$ are sequence in $[0, 1]$ such that $\lim_{n\to\infty} t_n = \lim_{n\to\infty} s_n < 1$ then $\lim_{n\to\infty} \sup \phi(t_n, s_n) < 0$.

We denote by $\mathscr{F}\mathscr{Z}$ the class of all functions which satisfies the above conditions.

Definition 8 Let $(X, M, \star)$ be a fuzzy metric space. We say that a mapping $T : X \to X$ is a $\mathcal{FZ}$-contractive mapping with respect to ϕ if the following condition is satisfied

$$\phi(M(T(x), T(y), t), M(x, y, t)) \geq 0 \ \ for \ \ all \ \ x, y \in X.$$

A simple example of $\mathcal{FZ}$-contraction is the fuzzy contraction given by Gregori and Sapena which can obtained by taking $k \in]0, 1[$ and $\phi(t, s) = k(\frac{1}{s} - 1) - \frac{1}{t} + 1$ for all $t, s \in]0, 1]$. Consequently, The class of fuzzy contractive mappings in the sense of Gregori and Sapena is included in the class of $\mathcal{FZ}$-contractive mappings.

Remark 2 It is clear from the definition that $\phi(t, s) < 0$ for all $s \geq t$. Therefore, if T is a $\mathcal{FZ}$-contraction mapping with respect to ϕ then $M(x, y, t) < M(T(x), T(y), t)$.

Proposition 1 *The class of fuzzy ψ-contractive mappings are included in the class of $\mathcal{FZ}$-contractive mappings.*

Proof Suppose that $T : X \to X$ is ψ-contractive with respect to $\psi \in \Psi$,
Define $\phi_\psi :]0, 1] \times]0, 1] \longrightarrow \mathbb{R}$ by

$$\phi_\psi = \frac{1}{\psi(s)} - \frac{1}{t} \ \ for \ \ all \ \ s, t \in]0, 1]$$

T is $\mathcal{FZ}$-contractive mapping with respect to $\phi_\psi \in \mathcal{FZ}$.

Remark 3 Taking in account the remark 2.11, every $\mathcal{H}$-contractive mapping with respect to $\eta \in \mathcal{H}$ is a fuzzy $\mathcal{FZ}$-contraction with respect to the function $\phi \in \mathcal{FZ}$ defined by $\phi_\eta(t, s) = \frac{1}{\eta^{-1}(k.\eta(s))} - \frac{1}{t} \ \ for \ \ all \ \ s, t \in]0, 1]$.

Lemma 1 *Let $(X, M, \star)$ be a fuzzy metric space and T be a $\mathcal{FZ}$-contraction with respect to $\phi \in \mathcal{FZ}$. Then the fixed point of T in X is unique, provided it exists.*

Proof (**proof**) Suppose $u \in X$ be a fixed point of T, If possible, let $v \in X$ be another fixed point of T and it is distinct from u, that is, $Tv = v$ and $u \neq v$. Now it follows from the definition that

$$0 \leq \phi(M(T(u), T(v), t), M(u, v, t)) = \phi(M(u, v, t), M(u, v, t)).$$

The remark 3.3 yields a contradiction, so $u = v$.

Theorem 3 *Let $(X, M, \star)$ be a complete strong fuzzy metric space and T be a $\mathcal{FZ}$-contraction with respect to $\phi \in \mathcal{FZ}$. Then the fixed point of T in X is unique.*

Proof (**proof**) Let $x_0 \in X$ be any arbitrary point in X. Now we Construct a sequence $\{x_n\} \in X$ Such that $x_n = Tx_{n-1}$ for all $n \in \mathbb{N}$.
Without loss of generality we can assume that $x_n \neq x_{n+1}$ for all $n \in \mathbb{N}$ Trivially, if there exists n_0 such that $x_{n_0} = x_{n_0+1}$,then the equalities $x_{n_0} = x_{n_0+1} = Tx_{n_0}$ implies that x_{n_0} is a fixed point of T.
we prove that $\lim_{n\to\infty} M(x_n, x_{n+1}, t) = 1$ for all $t > 0$, Suppose, to the contrary, that there exists some t_0 such that

$$\lim_{n\to\infty} M(x_n, x_{n+1}, t_0) < 1$$

Now, by (GV2) we have that $M(x_n, x_{n+1}, t_0) < 1$ for all $n \in N$.
Then, by using (i) and (ii), taking $x = x_{n-1}$ and $y = x_n$, we have

$$0 \leq \phi(M(x_n, x_{n+1}, t_0), M(x_{n-1}, x_n, t_0)) < \frac{1}{M(x_{n-1}, x_n, t_0)} - \frac{1}{M(x_n, x_{n+1}, t_0)}$$

for all $n \in \mathbb{N}$, this implies that $\{M(x_{n-1}, x_n, t_0), n \in N\}$ is nondecreasing sequence of positives reals numbers. therefore, there exists $l \leq 1$ such that $\lim_{n\to\infty} M(x_{n-1}, x_n, t_0) = l$

We shall show that $l = 1$, we suppose that $l < 1$ and using (iii)

$$t_n = M(x_n, x_{n+1}, t_0) \ \ and \ \ s_n = M(x_{n-1}, x_n, t_0)$$

we conclude that:

$$0 \leq \phi(M(x_n, x_{n+1}, t_0), M(x_{n-1}, x_n, t_0)) < 0$$

which is a contradiction, Hence $l = 1$, That is:

$$\lim_{n\to\infty} M(x_{n-1}, x_n, t_0) = 1$$

The crucial point of the proof is in establishing that the sequence $\{x_n\}$ is Cauchy in X. Assuming it is not true, Then there exist $0 < \varepsilon < 1$ and two subsequences $\{x_{m_k}\}$ and $\{x_{n_k}\}$ of $\{x_n\}$ such that n_k is the smallest index for which $n_k > m_k \geq k$

$$M(x_{m_k}, x_{n_k}, t_0) \leq 1 - \varepsilon \tag{1}$$

and

$$M(x_{m_k}, x_{n_k-1}, t_0) > 1 - \varepsilon \tag{2}$$

Using (1) and (2) and triangular inequality we obtain

$$1 - \varepsilon \geq M(x_{m_k}, x_{n_k}, t_0) \geq M(x_{m_k}, x_{n_k-1}, t_0) \star M(x_{n_k-1}, x_{n_k}, t_0)$$
$$\geq (1 - \varepsilon) \star M(x_{n_k-1}, x_{n_k}, t_0)$$

and by taking limit as $k \to \infty$

$$1 - \varepsilon \geq \lim_{n\to\infty} M(x_{m_k}, x_{n_k}, t_0) \geq (1 - \varepsilon)$$

We deduce that $\lim_{n\to\infty} M(x_{m_k}, x_{n_k}, t_0) = 1 - \varepsilon$

Applying the same reasoning as above, we obtain

$$1 - \varepsilon \geq M(x_{m_k}, x_{n_k}, t_0) \geq M(x_{m_k}, x_{m_k-1}, t_0) \star M(x_{m_k-1}, x_{n_k-1}, t_0) \star M(x_{n_k-1}, x_{n_k}, t_0)$$

and

$$M(x_{m_k-1}, x_{n_k-1}, t_0) \geq M(x_{m_k-1}, x_{m_k}, t_0) \star M(x_{m_k}, x_{n_k}, t_0) \star M(x_{n_k}, x_{n_k-1}, t_0)$$

Taking limit as $k \to \infty$, we get $\lim_{n\to\infty} M(x_{m_k-1}, x_{n_k-1}, t_0) = 1 - \varepsilon$ hence using (iii) with $\tau_k = M(x_{m_k}, x_{n_k}, t_0)$ *and* $\delta_k = M(x_{m_k-1}, x_{n_k-1}, t_0)$ we obtain

$$0 \leq \lim_{n\to\infty} \sup \phi(M(x_{m_k}, x_{n_k}, t_0, M(x_{m_k-1}, x_{n_k-1}, t_0)) < 0$$

i.e.

$$0 \leq \phi(1 - \varepsilon, 1 - \varepsilon) < 0$$

Obviously, this inequality is not true and $\{x_n\}$ is a cauchy sequence in X.

The completeness of $(X, M, \star)$ ensures that the sequence $\{x_n\}$ converges to some $u \in X$, that is $\lim_{n\to\infty} M(x_n, u, t) = 1 \quad \forall t > 0$ we shall show that the point u is a fixed point of T, suppose that $Tu \neq u$ then $M(u, Tu, t) < 1$, we have

$$0 \leq \lim_{n\to\infty} \sup \phi(M(Tx_n, Tu, t), M(x_n, u, t))$$

$$\leq \lim_{n\to\infty} \sup[\frac{1}{M(x_n, u, t)} - \frac{1}{M(Tx_n, Tu, t)}]$$

$$= \lim_{n\to\infty} \sup[\frac{1}{M(x_n, u, t)} - \frac{1}{M(x_{n+1}, Tu, t)}]$$

$$= 1 - \frac{1}{M(u, Tu, t)}$$

Finally, from the above we have $1 \leq M(u, Tu, t)$ hence $M(u, Tu, t) = 1$ Which is contradiction, thus u is a fixed point of T.

Example 3 Let $X =]0, \infty[$, $a \star b = ab \ \forall a, b \in [0, 1]$ and

$$M(x, y, t) = \frac{min(x, y)}{max(x, y)} \quad \forall t \in]0, \infty[\ \forall x, y > 0$$

$(X, M, \star)$ is an complete strong fuzzy metric space [7].
The mapping $T : X \longrightarrow X$, $T(x) = \sqrt{x}$ is $\mathscr{F}\mathscr{Z}$-contractive mapping with respect to the function defined by

$$\Phi(t, s) = \frac{1}{\sqrt{s}} - \frac{1}{t} \quad \forall t, s \in]0, 1]$$

Indeed, since $\frac{1}{\sqrt{s}} - \frac{1}{t} < \frac{1}{s} - \frac{1}{t} \ \forall t, s \in]0, 1[$ with $\Phi(1, 1) = 0$;

Note that, all the condition of the previous Theorem are satisfied and T has a unique fixed point $x = 1 \in X$.

References

1. M. Grabiec, Fixed points in fuzzy metric spaces. Fuzzy Sets Syst. **27**, 385–389 (1988)
2. M. Edelstein, On fixed and periodic points under contractive mappings. J. Lond. Math. Soc. **37**, 74–79 (1962)
3. A. George, P. Veeramani, On some results in fuzzy metric spaces. Fuzzy Sets Syst. **64**, 395–399 (1994)
4. V. Gregori, A. Sapena, On fixed-point theorems in fuzzy metric spaces. Fuzzy Sets Syst. **125**, 245–252 (2002)
5. V. Gregori, S. Morillas, A. Sapena, Examples of fuzzy metrics and applications. Fuzzy Sets Syst. **170**, 95–111 (2011)
6. V. Gregori, J.-J. Miñana, Some remarks on fuzzy contractive mappings. Fuzzy Sets Syst. **251**, 101–103 (2014). https://doi.org/10.1016/j.fss.2014.01.002
7. V. Radu, Some remarks on the probabilistic contractions on fuzzy Menger spaces. Automat. Comput. Appl. Math. **11**, 125–131 (2002)
8. B. Schweizer, A. Sklar, Statistical metric spaces. Pacific J. Math. **10**, 313–334 (1960)
9. D. Mihet, On fuzzy contractive mappings in fuzzy metric spaces. Fuzzy Sets Syst. **158**, 915–921 (2007)
10. D. Mihet, Fuzzy ψ-contractive mappings in non-Archimedean fuzzy metric spaces. Fuzzy Sets Syst. **159**, 739–744 (2008)
11. D. Wardowski, Fuzzy contractive mappings and fixed points in fuzzy metric spaces. Fuzzy Sets Syst. **222**, 108–114 (2013)
12. I. Kramosil, J. Michalek, Fuzzy metric and statistical metric spaces. Kybernetica **15**, 326–334 (1975)
13. L. Ćirić, Some new results for Banach contractions and Edelstein contractive mappings on fuzzy metric spaces. Chaos Solitons Fractals **42**, 146–154 (2009)

14. Y. Shen, D. Qiu, W. Chen, Fixed point theorems in fuzzy metric spaces. Appl. Math. Lett. **25**, 138–141 (2012)
15. A. Roldan, J. Martinez, C. Roldan, Y.J. Cho, Multidimensional coincidence point results for compatible mappings in partially ordered fuzzy metric spaces. Fuzzy Sets Syst. **251**, 71–82 (2014)

Common Fixed Point Theorems in b-Menger Spaces

Abderrahim Mbarki and Rachid Oubrahim

Abstract In this work, we prove a common fixed point theorem in b-Menger spaces for nonlinear contractions. An example is provided to illustrate this result.

1 Introduction

Probabilistic metric spaces are introduced in 1942 by Menger [7]. In such spaces, the notion of distance between two points p and q is replaced by a distribution function $F_{pq}(x)$. Thus one thinks of the distance between points as being probabilistic with $F_{pq}(x)$ representing the probability that the distance between p and q is less than x.

Fixed point theory plays one of the important roles in nonlinear analysis. It has been applied in physical sciences, computing sciences and engineering. The first result from the fixed point theory in probabilistic metric spaces was obtained by Sehgal and Bharucha-Reid [9] in 1972 and their fixed point theorem is further generalized by many authors, for example see [4, 6, 8].

In 1989, Bakhtin [1] defined b-metric spaces as a generalization of metric spaces, and an extension of Banach's contraction [2] in these spaces was showed by Czerwik [3].

Recently, Mbarki et al. [6] introduced a probabilistic b-metric spaces (b-Menger spaces) as a generalization of probabilistic metric spaces (Menger spaces) and they studied topological structures and properties and showed the fixed point property for nonlinear contractions in these spaces.

In this paper, we prove the existence and uniqueness of the common fixed point for nonlinear contraction in b-Menger spaces.

A. Mbarki
Mohammed First University, P.O. BOX 669, Oujda, Morocco
e-mail: dr.mbarki@gmail.com

R. Oubrahim (✉)
ANO Laboratory, Faculty of Sciences, Mohammed First University, 60000 Oujda, Morocco
e-mail: rchd.oubrahim@gmail.com

S. Melliani and O. Castillo (eds.), *Recent Advances in Intuitionistic Fuzzy Logic Systems*, Studies in Fuzziness and Soft Computing 372,
https://doi.org/10.1007/978-3-030-02155-9_22

2 Preliminaries

Throughout this paper, the standard notations and terminologies of probabilistic b-metric theory are used. For more details, we refer the reader to [6].

A nonnegative real function f defined on $\mathbb{R}^+ \cup \{\infty\}$ is called a distance distribution function (briefly, a d.d.f) if it is nondecreasing, left continuous on $(0, \infty)$, with $f(0) = 0$ and $f(\infty) = 1$. The set of all d.d.f's will be noted by Δ^+; and the set of all F in Δ^+ for which $\lim_{t\to\infty} f(t) = 1$ by D^+.

A simple example of distribution function is Heavyside function in D^+

$$H(t) = \begin{cases} 0 \text{ if } t \leq 0, \\ 1 \text{ if } t > 0. \end{cases}$$

A commutative, associative and nondecreasing mapping $T : [0, 1] \times [0, 1] \to [0, 1]$ is called a t-norm if and only if

1. $T(a, 1) = a$, for all $a \in [0, 1]$,
2. $T(0, 0) = 0$.

As examples we mention the three typical examples of continuous t-norms as follows:
$T_P(a, b) = ab$, $T_M(a, b) = Min(a, b)$ and $T_L(a, b) = Max\{a + b - 1, 0\}$.

Definition 1 [6] A b-Menger space is a quadruple (X, F, T, s) where X is a nonempty set, F is a function from $X \times X$ into Δ^+, T is a t-norm, $s \geq 1$ is a real number, and the following conditions are satisfied: for all $p, q, r \in X$ and $x, y > 0$,

1. $F_{pp} = H$,
2. $F_{pq} = H \Rightarrow p = q$,
3. $F_{pq} = F_{qp}$,
4. $F_{pq}(s(x + y)) \geq T(F_{pr}(x), F_{rq}(y))$.

Definition 2 Let (X, F) be a probabilistic semimetric space (i.e., (1), (2) and (3) of Definition 2.1 are satisfied). For p in X and $t > 0$, the strong t-neighborhood of p is the set

$$N_p(t) = \{q \in M : F_{pq}(t) > 1 - t\}.$$

The strong neighborhood system at p is the collection $\wp_p = \{N_p(t) : t > 0\}$, and the strong neighborhood system for X is the union $\wp = \bigcup_{p\in X} \wp_p$.

Let (X, F, T, s) be a b-Menger space. The family $\Im$ consisting of $\emptyset$ and all unions of elements of the strong neighborhood system for X determines a topology for X. Moreover, Mbarki et al. [6] proved that if T is continuous, (X, F, T, s) endowed with this topology is a Hausdorff space and the function F is in general not continuous.

Example 1 Let $X = \mathbb{R}$. Define $F : X \times X \to \Delta^+$ by

$$F_{pq}(t) = H(t - |p - q|^2).$$

By Lemma 3.1 in [6], $(X, F, T_M, 2)$ is a b-Menger space with continuous probabilistic b-metric F.

Example 2 [6] Let $X = \mathbb{N} \cup \{\infty\}$. Define $F : X \times X \to \Delta^+$ as follow:

$$F_{pq}(t) = \begin{cases} H(t) & \text{if } p = q, \\ H(t-7) & \text{if } p \text{ and } q \text{ are odd and } p \neq q, \\ H(t - |\frac{1}{p} - \frac{1}{q}|) & \text{if } p \text{ and } q \text{ are even or } pq = \infty, \\ H(t-3) & \text{otherwise}. \end{cases}$$

It easy to show that $(X, F, T_M, 4)$ is a b-Menger space with T_M is continuous.

Consider the sequence $x_n = 2n$, $n \in \mathbb{N}$. Then $F_{2n\infty}(t) = H(t - \frac{1}{2n})$. Therefore $x_n \to \infty$, but $F_{2n1}(t) = H(t-3) \neq H(t-1) = F_{1\infty}(t)$. Hence F is not continuous at ∞.

In b-Menger space, the convergence of sequence is defined as follows

Definition 3 Let $\{p_n\}$ be a sequence in a b-Menger space (X, F, T, s).

1. A sequence $\{p_n\}$ in X is said to be convergent to p in M if, for every $\varepsilon > 0$ and $\delta \in (0, 1)$, there exists a positive integer $N(\varepsilon, \delta)$ such that $F_{p_n p}(\varepsilon) > 1 - \delta$, whenever $n \geq N(\varepsilon)$.
2. A sequence $\{p_n\}$ in X is called a Cauchy sequence if, for every $\varepsilon > 0$ and $\delta \in (0, 1)$, there exists a positive integer $N(\varepsilon, \delta)$ such that $F_{p_n p_m}(\varepsilon) > 1 - \delta$, whenever $n, m \geq N(\varepsilon, \delta)$.
3. (X, F, T, s) is said to be complete if every Cauchy sequence has a limit.

The letter Ψ denotes the set of all function $\varphi : [0, \infty) \to [0, \infty)$ such that

$$0 < \varphi(t) < t \text{ and } \lim_{n\to\infty}\varphi^n(t) = 0 \text{ for each } t > 0.$$

Definition 4 [5] We say that a t-norm T is of H-type if the family $\{T^n(t)\}$ is equicontinuous at $t = 1$, that is,
$\forall \varepsilon \in (0, 1), \exists \lambda \in (0, 1) : t > 1 - \lambda \Rightarrow T^n(t) > 1 - \varepsilon, \ \forall n \geq 1,$
where $T^1(x) = T(x, x)$, $T^n(x) = T(x, T^{n-1}(x))$, for every $n \geq 2$.

The t-norm T_M is a trivial example of t-norm of H-type.

Lemma 1 [6] *Let (X, F, T, s) be a complete b-Menger space under a t-norm T of H-type and $RanF \subset D^+$. Let $\{x_n\}$ be a sequence in X. If there exists a function $\varphi \in \Psi$ such that*

$$F_{x_{m+1}x_{n+1}}(\varphi(t)) \geq F_{x_m x_n}(st) \quad (n, m \geq 0, t > 0),$$

then $\{x_n\}$ is a Cauchy sequence.

3 Main Results

Before stating the main common fixed point theorem, we need the following concept.

Definition 5 Let f and g be two selmaps on a complete b-Menger space (X, F, T, s). f and g are said to be compatible if, whenever $\{x_n\}$ is a sequence of point in X such that $\lim_{n\to\infty} fx_n = \lim_{n\to\infty} gx_n = x$ then $fx = gx$.

Theorem 1 *Let (X, F, T, s) be a complete b-Menger space under a continuous t-norm T of H-type and $RanF \subset D^+$, and let $f, g : X \to X$ be maps that satisfy the following conditions:*

1. $f(X) \subseteq g(X)$,
2. *f and g are compatible, g is continuous,*
3. *There exists $\varphi \in \Psi$ such that $F_{fxfy}(\varphi(t)) \geq F_{gxgy}(st)$, for all $x, y \in X$ and $t > 0$.*

Then f and g have a unique common fixed point in X.

Proof Let $x_0 \in X$. By condition (1), we can find x_1 such that $g(x_1) = f(x_0)$. By induction, we can define a sequence $\{x_n\}$ such that $g(x_n) = f(x_{n-1})$. We put $y_n = g(x_n) = f(x_{n-1})$. By condition (3), we have

$$F_{f(x_m)f(x_n)}(\varphi(t)) \geq F_{g(x_m)g(x_n)}(st).$$

So,

$$F_{y_{m+1}y_{n+1}}(\varphi(t)) \geq F_{y_m y_n}(st).$$

Then the sequence $\{y_n\}$ satisfies the Lemma 1, hence $\{y_n\}$ is a Cauchy sequence. Since X is complete, there is some $y \in X$ such that

$$y_n \to y \quad as\ n \to \infty. \tag{1}$$

Now we will show that y is a common fixed point of f and g. Let $\varepsilon > 0$ and $\delta \in (0, 1)$, since $\varphi(\varepsilon) < \varepsilon$, by the monotonicity of F and the condition (3), we get

$$\begin{aligned} F_{y_{n+1}f(y)}(\varepsilon) &\geq F_{y_{n+1}f(y)}(\varphi(\varepsilon)) \\ &= F_{f(y_n)f(y)}(\varphi(\varepsilon)) \\ &\geq F_{g(y_n)g(y)}(s\varepsilon). \end{aligned}$$

By condition (3) and since g is continuous, then f is also continuous. Then $\{y_n\}$ converges to y implies that $\{g(y_n)\}$ converges to $g(y)$, then there exists $n_0 \in \mathbb{N}$ such that

$$F_{g(y_n)g(y)}(s\varepsilon) > 1 - \delta \quad for\ each\ n \geq n_0.$$

So,

$$F_{y_{n+1}f(y)}(\varepsilon) > 1 - \delta \quad for\ each\ n \geq n_0.$$

Then

$$\lim_{n\to\infty} y_{n+1} = f(y). \tag{2}$$

Since X is a Hausdorff space, then

$$f(y) = y.$$

So by (1) and (2) we have

$$y_{n+1} = f(x_n) \to y \ and \ g(x_n) \to y.$$

And since f and g are compatible, then

$$f(y) = g(y).$$

Finally,

$$y = f(y) = g(y).$$

Then y is a common fixed point for f and g.
Suppose, to the contrary, that there exists another common fixed point z in X of f and g. Then let $t > 0$, from condition (3), since $\varphi(t) < t$ and the fact that the distance distribution functions are nondecreasing, we have

$$\begin{aligned} F_{yz}(\varphi(t)) &= F_{f(y)f(z)}(\varphi(t)) \\ &\geq F_{g(y)g(z)}(st) \\ &\geq F_{g(y)g(z)}(t) \\ &= F_{yz}(t) \\ &\geq F_{yz}(\varphi(t)). \end{aligned}$$

Thus

$$F_{yz}(\varphi(t)) = F_{yz}(t).$$

By the same arguments, we get

$$F_{yz}(\varphi^n(t)) = F_{yz}(t), \quad for\ all\ n \geq 1.$$

Suppose that there exists $\alpha > 0$ such that $F_{yz}(\alpha) < 1$. Since $F_{yz} \in D^+$, then $\lim_{t\to\infty} F_{yz}(t) = 1$, so there exists $\beta > \alpha$ such that

$$F_{yz}(\beta) > F_{yz}(\alpha).$$

There exists a positive integer $n > 1$ such that $\varphi^n(\beta) < \alpha$ because $\lim_{n\to\infty} \varphi^n(\beta) = 0$. Then, the fact that F_{yz} is nondecreasing function, we get

$$F_{yz}(\varphi^n(\beta)) \leq F_{yz}(\alpha).$$

Hence

$$F_{yz}(\beta) = F_{yz}(\varphi^n(\beta)) \leq F_{yz}(\alpha),$$

a contradiction. Therefore $F_{yz}(t) = 1$ for all $t > 0$, since $F_{yz} \in D^+$. Hence $F_{yz} = H$. Then $y = z$.

Example 3 Let $X = [0, \infty)$. Define $F : X \times X \to \Delta^+$ as follows

$$F_{xy}(t) = H(t - |x - y|^2).$$

By Examples 3.2, 3.3 and Lemma 3.1 in [6], $(X, F, T_M, 2)$ is a complete b-Menger and is not a standard Menger space.
Define $f(x) = \frac{x}{12}$ and $g(x) = \frac{x}{4}$ in X. It is evident that $f(X) \subseteq g(X)$. Also, for $\varphi(t) = \frac{16t}{77}$,

$$\begin{aligned}
F_{f(x)f(y)}(\frac{16t}{77}) &= H(\frac{16t}{77} - |\frac{x}{12} - \frac{y}{12}|^2) \\
&= H(\frac{\frac{16t}{77}}{\frac{1}{144}} - |x - y|^2) \\
&= H(\frac{2t}{\frac{1}{16}} - |x - y|^2) \\
&= H(2t - |\frac{x}{4} - \frac{y}{4}|^2) \\
&= F_{g(x)g(y)}(2t).
\end{aligned}$$

Thus all the conditions of Theorem 1 are satisfied and f and g have the common fixed point 0.

References

1. I.A. Bakhtin, Contracting mapping principle in an almost metric space, (Russian) Funkts. Anal. **30**, 26–37 (1989)
2. S. Banach, Sur les opérations dans les ensembles abstraits et leur applications aux équations intégrales. Fundam. Maths. **3**, 133–181 (1922)
3. S. Czerwik, Nonlinear set-valued contraction mappings in b-metric spaces. Atti Sem. Math Univ. Modena **46**(2), 263–276 (1998)

4. M. Elamrani, A. Mbarki, B. Mehdaoui, Nonlinear contarctions and semigroups in general complete probabilistic metric spaces. Panam. Math. J. **11**(4), 79–87 (2001)
5. O. Hadzić, A fixed point theorem in Menger spaces. Publ. Inst. Math. (Beograd) T **20**, 107–112 (1979)
6. A. Mbarki, R. Oubrahim, Probabilistic b-metric spaces and nonlinear contractions. Fixed Point Theor. Appl. **2017**, 29 (2017)
7. K. Menger, Statistical metrics. Proc. Natl. Acad. Sci. **28**, 535–537 (1942)
8. B. Schweizer, A.Sklar, in *Probabilistic Metric Spaces*, North-Holland Series in Probability and Applied Mathematics, vol 5 (1983)
9. V.M. Sehgal, A.T. Bharucha-Reid, Fixed point theorems of contractions mappings in probabilistic metric spaces. Math. Systemes Theor. **6**, 97–102 (1972)

On Intuitionistic Fuzzy Vector Spaces

H. Sadiki, Said Melliani, I. Bakhadach and L. S. Chadli

Abstract In this paper we investigate the notion of intuitionistic fuzzy vector spaces over an intuitionistic fuzzy subfield and some related results has been established using the intuitionistic fuzzy points.

1 Introduction

After the introduction of fuzzy sets by Zadeh [16], several generalizations have been made of this fundamental concept for various objectives. In 1986 the notion of intuitionistic fuzzy sets (*IFSs*) introduced by Atanassov [1, 2]. Jun and S.Z. Song introduced the notion of intuitionistic fuzzy points [7]. In [11] Moumita C. and S. K. Samanta introduce a notion of intuitionistic fuzzy vector space (*IFVS*) and intuitionistic fuzzy basis ($IF - basis$) of a *IFVS*. From this end idea we introduce in this paper the concept of intuitionistic fuzzy spanning for an intuitionistic fuzzy subspaces which is analogous to the fuzzy spanning.
Concerning notation, we let $lm(A)$ denote the image of A and $|Im(A)|$ denote the cardinality of $Im(A)$. We say that A is finite-valued if $|Im(A)| < \infty$. If X is a subset of V, we let $sp(X)$ denote the subspace of V generated by X.

H. Sadiki · S. Melliani (✉) · I. Bakhadach · L. S. Chadli
Sultan Moulay Slimane University, BP 523 Beni Mellal, Morocco
e-mail: saidmelliani@gmail.com

H. Sadiki
e-mail: razika.imi@gmail.com

I. Bakhadach
e-mail: bakhadach@gmail.com

L. S. Chadli
e-mail: sa.chadli@yahoo.fr

S. Melliani and O. Castillo (eds.), *Recent Advances in Intuitionistic Fuzzy Logic Systems*, Studies in Fuzziness and Soft Computing 372,
https://doi.org/10.1007/978-3-030-02155-9_23

2 Preliminaries

In this section we recall some definitions and results which will be used in the sequel.

Definition 1 [1] Let X be a non−empty set. An intuitionistic fuzzy set (IFS for short) of X defined as an object having the form $A = \{\langle x,\ \mu_A(x),\ \nu_A(x)\rangle | x \in X\}$, where $\mu_A : X \to [0, 1]$ and $\nu_A : X \to [0, 1]$ denote the degree of membership (namely $\mu_A(x)$) and the degree of non-membership (namely $\nu_A(x)$) of each element $x \in X$ it to the set A, respectively and $0 \le \mu_A(x) + \nu_A(x) \le 1$ for each $x \in X$. For the sake of simplicity we shall use the symbol $A = (\mu_A,\ \nu_A)$ for the intuitionistic fuzzy set $A = \{\langle x,\ \mu_A(x),\ \nu_A(x)\rangle | x \in X\}$.

In this paper we use the symbols $a \wedge b = \min\{a, b\}$ *and* $a \vee b = \max\{a, b\}$.

Definition 2 [1] Let $A = (\mu_A,\ \nu_A)$ and $B = (\mu_B,\ \nu_B)$ be intuitionistic fuzzy sets of a set X. Then

(1) $A \subseteq B$ iff $\mu_A(x) \le \mu_B(x)$ and $\nu_A(x) \ge \nu_B(x)$ for all $x \in X$.
(2) $A = B$ iff $A \subseteq B$ and $B \subseteq A$.
(3) $A^c = \{\langle x,\ \nu_A(x), \mu_A(x)\rangle | x \in X\}$
(4) $A \cap B = \{\langle x,\ \mu_A(x) \wedge \mu_B(x),\ \nu_A(x) \vee \nu_B(x)\rangle | x \in X\}$.
(5) $A \cup B = \{\langle x,\ \mu_A(x) \vee \mu_B(x), \nu_A(x) \wedge \nu_B(x)\rangle | x \in X\}$.
(6) $\Box A = \{\langle x,\ \mu_A(x), 1 - \mu_A(x)\rangle | x \in X\}$.
(7) $\diamond A = \{\langle x,\ 1 - \nu_A(x), \nu_A(x)\rangle | x \in X\}$.

Definition 3 [6] Let A be Intuitionistic fuzzy set of a universe set X. Then $(\alpha,\ \beta)$-cut ofA is acrisp subset $\mathrm{C}_{\alpha,\beta}(\mathrm{A})$ of the IFS A is given by

$$\mathrm{C}_{\alpha,\beta}(\mathrm{A}) = \{\mathrm{x} \in \mathrm{X} \ :\ \mu_{\mathrm{A}}(\mathrm{x}) \ge \alpha,\ \nu_{\mathrm{A}}(\mathrm{x}) \le \beta\}, \quad \text{where}$$
$$\alpha, \beta \in [0, 1] \quad \text{with} \quad \alpha + \beta \le 1$$

Proposition 1 [6] *Let A be an IFS in a set X and* $(\alpha_1,\ \beta_1)$, $(\alpha_2,\ \beta_2) \in Im(A)$. *If* $\alpha_1 \le \alpha_2$ *and* $\beta_1 \ge \beta_2$, *then* $C_{\alpha_1,\beta_1}(A) \supseteq C_{\alpha_2,\beta_2}(A)$.

Definition 4 [12] Let (F, +, ·) be a field. An intuitionistic fuzzy subset K of F is said to be an intuitionistic fuzzy subfield (IFSF) of F if the following conditions are satisfied:

(i) $\mu_{\mathrm{K}}(\mathrm{x} + \mathrm{y}) \ge \min\{\mu_{\mathrm{K}}(\mathrm{x}),\ \mu_{\mathrm{K}}(\mathrm{y})\}$, for all x and y in F.
(ii) $\mu_{\mathrm{K}}(-\mathrm{x}) \ge \mu_{\mathrm{K}}(\mathrm{x})$, for all x in F.
(iii) $\mu_{\mathrm{K}}(\mathrm{xy}) \ge \min\{\mu_{\mathrm{K}}(\mathrm{x}),\ \mu_{\mathrm{K}}(\mathrm{y})\}$, for all x and y in F.
(iv) $\mu_{\mathrm{K}}(\mathrm{x}^{-1}) \ge \mu_{\mathrm{K}}(\mathrm{x})$, for all x in F − {0}.
(v) $\nu_{\mathrm{K}}(\mathrm{x} + \mathrm{y}) \le \max\{\nu_{\mathrm{K}}(\mathrm{x}),\ \nu_{\mathrm{K}}(\mathrm{y})\}$, for all x and y in F.
(vi) $\nu_{\mathrm{K}}(-\mathrm{x}) \le \nu_{\mathrm{K}}(\mathrm{x})$, for all x in F.
(vii) $\nu_{\mathrm{K}}(\mathrm{xy}) \le \max\{\nu_{\mathrm{A}}(\mathrm{x}),\ \nu_{\mathrm{K}}(\mathrm{y})\}$, for all x and y in F.
(viii) $\nu_{\mathrm{K}}(\mathrm{x}^{-1}) \le \nu_{\mathrm{K}}(\mathrm{x})$, for all x in F − {0}.

Definition 5 [10] Let $A, A_1,\ \ldots,\ A_n$ be intuitionistic fuzzy subsets of V and let K be an intuitionistic fuzzy subset of F.

(1) Define the intuitionistic fuzzy subset $A_1 + \cdots + A_n$ of V by the following: for all $x \in V$

$$\mu_{(A_1+A_2+\cdots+A_n)}(x) = \begin{cases} \sup\{\min\{\mu_{A_1}(x_1), \ldots, \mu_{A_n}(x_n)\}\} & if\, x = x_1 + \cdots + x_n, \\ 0 & otherwise \end{cases}$$

$$\nu_{(A_1+A_2+\cdots+A_n)}(x) = \begin{cases} \inf\{\max\{\nu_{A_1}(x_1), \ldots, \nu_{A_n}(x_n)\}\} & if\, x = x_1 + \cdots + x_n, \\ 1 & otherwise \end{cases}$$

(2) Define the intuitionistic fuzzy subset $K \circ A$ of V by for all $x \in V$,

$$\mu_{(K \circ A)}(x) = \sup\{\min\{K(c),\ A(y)\} | c \in F,\ y \in V,\ x = cy\}.$$

$$\nu_{(K \circ A)}(x) = \inf\{\max\{K(c),\ A(y)\} | c \in F,\ y \in V,\ x = cy\}.$$

Definition 6 [3, 7] Let $\alpha, \beta \in [0, 1]$ with $\alpha + \beta \leq 1$. An intuistionistic fuzzy point written as $x_{(\alpha,\beta)}$, is defined to be an intuitionistic fuzzy subset of A. given by

$$x_{(\alpha,\beta)}(y) = \begin{cases} (\alpha,\ \beta) \text{ if } x = y, \\ (0, 1) \text{ otherwise} \end{cases}$$

An intuitionistic fuzzy point $x_{(\alpha,\beta)}$ is said to belong in IFS $A = (\mu_A, \nu_A)$ denoted by $x_{(\alpha,\beta)} \in (\mu_A, \nu_A)$ if $\mu_A(x) \geq \alpha$ and $\nu_A(x) \leq \beta$.

3 Intuitionistic Fuzzy Spanning

Definition 7 Let $A = (\mu_A, \nu_A)$ be an intuitionistic fuzzy subset of vector space V. A is called an intuitionistic fuzzy subspace of V over an intuitionistic fuzzy subfield K of F if $\forall x, y \in V, c \in F$ the following conditions are satisfied:

(i) $\mu_A(0) > 0$.
(ii) $\mu_A(x - y) \geq \min\{\mu_A(x), \mu_A(y)\}$ and $\nu_A(x - y) \leq \max\{\nu_A(x),\ \nu_A(y)\}$.
(iii) $\mu_A(cx) \geq \min\{\mu_K(c), \mu_A(x)\}$, and $\nu_A(cx) \leq \max\{\nu_K(c), \nu_A(x)\}$.

Proposition 2 *If K is an intuitionistic fuzzy subfield of F and if $x \in F, x \neq 0$, then $\mu_K(0) = \mu_K(1) \geq \mu_K(x) = \mu_K(-x) = \mu_K(x^{-1})$ and $\nu_K(0) = \nu_K(1) \leq \nu_K(x) = \nu_K(-x) = \nu_K(x^{-1})$* [15].

In the following we let $\mathscr{K}$ denote the set of all intuitionistic fuzzy subfields of F and we let $\mathscr{A}_K$ denote the set of all intuitionistic fuzzy subspaces of V over $K \in \mathscr{K}$. For S a subset of F, we let $K_S = (\delta_S, 1 - \delta_S)$ where δ_s denote the characteristic function of S.

Proposition 3 *Let $K_F = (\delta_F, 1 - \delta_F)$ and let $A \in \mathscr{A}_{K_F}$. Then for all $0 \leq \alpha + \beta \leq 1$, $C_{\alpha,\beta}(A)$ is a subspace of V.*

Proof Let $\alpha, \beta \in [0, 1]$ with $\alpha + \beta \leq 1$ we have:

$$\begin{aligned} C_{\alpha,\beta}(A) &= \{x \in V \ : \mu_A(x) \geq \alpha, \ \nu_A(x) \leq \beta\} \\ &= \{x \in V \ : \ \mu_A(x) \geq \alpha, 1 - \nu_A(x) \geq 1 - \beta\} \\ &= \{x \in V \ : \ \mu_A(x) \geq \alpha\} \bigcap \{x \in V \ : \ 1 - \nu_A(x) \geq 1 - \beta\} \text{ or} \\ &\{x \in V \ : \ \mu_A(x) \geq \alpha\} \text{ and } \{x \in V \ : \ 1 - \nu_A(x) \geq 1 - \beta\} \end{aligned}$$

are a subspaces of V (see [9]). So $C_{\alpha,\beta}(A)$ is a subspace of V.

Proposition 4 *Let A be an intuitionistic fuzzy subset of V. if $C_{\alpha,\beta}(A)$ is a subspace of V for for all $(\alpha, \beta) \in Im(A)$, then $A \in \mathscr{A}_{K_F}$.*

Proposition 5 *Let $V_1 \subset V_2 \subset \ldots \subset V_i \subset \ldots$ be a strictly ascending chain of subspaces of V. Define the intuitionistic fuzzy subset $A = (\mu_A, \nu_A)$ of V by $(\mu_A(x), \nu_A(x)) = (t_i, s_i)$ if $x \in V_i \backslash V_{i-1}$, where $t_i > t_{i+1}, s_i < s_{i+1}$ for $i = 1, 2, \ldots$ and $V_0 = \phi$; and $(\mu_A(x), \nu_A(x)) = (0, 1)$ if $x \in V \backslash \bigcup_{i=1}^{\infty} V_i$. Then $A \in \mathscr{A}_{K_F}$.*

Proof Let $c \in F$. If $x \in V \backslash \bigcup_{i=1}^{\infty} V_i$, then $\mu_A(cx) \geq 0 = \mu_A(x) \geq \min\{\delta_F(c), \mu_A(x)\}$. and $\nu_A(cx) \leq 1 = \nu_A(x) \leq \max\{(1 - \delta_F)(c), \nu_A(x)\}$. Suppose that $x \in V_i \backslash V_{i-1}$. Then $cx \in V_i$. Thus $\mu_A(cx) \geq t_i = \mu_A(x) \geq \min\{\delta_F(c), \mu_A(x)\}$. and $\nu_A(cx) \leq s_i = \nu_A(x) \leq \max\{(1 - \delta_F)(c), \nu_A(x)\}$.

Proposition 6 *Let $V = V_0 \supset V_1 \supset \ldots \supset V_i \supset \ldots$ be a strictly descending chain of subspaces of V. Define the intuitionistic fuzzy subset $A = (\mu_A, \nu_A)$ of V by $\mu_A(x) = t_{i-1}$ and $\nu_A(x) = s_{i-1}$ if $x \in V_{i-1} \backslash V_i$. where $t_{i-1} < t_i < 1$,$s_{i-1} > s_i > 0$ for $i = 1, 2, ...$ and $(\mu_A(x), \nu_A(x)) = (1, 0)$ if $x \in \cap_{i=1}^{\infty} V_i$. Then $A \in \mathscr{A}_{K_F}$.*

Proof Let $c \in F$. If $x \in \bigcap_{i=1}^{\infty} V_i$, then $cx \in \bigcap_{i=1}^{\infty} V_i$ and so $\mu_A(cx) = 1 \geq \min \{\delta_F(c), \mu_A(x)\}$and $\nu_A(cx) = 0 \leq \max\{(1 - \delta_F)(c), \nu_A(x)\}$. Suppose that $x \in V_{i-1} \backslash V_i$. Then $cx \in V_{i-1}$. Thus, $\mu_A(cx) \geq t_{i-1} = A(x) \geq \min\{\delta_F(c), \mu_A(x)\}$. and $\nu_A(cx) \leq s_{i-1} = \nu_A(x) \leq \max\{(1 - \delta_F)(c), \nu_A(x)\}$.

Theorem 1 *V is finite dimensional (over F) if and only if $A \in \mathscr{A}_{K_F}$ is finite-valued.*

Proof Suppose $A \in \mathscr{A}_{K_F}$ and A is infinite-valued. Then $Im(A)$ contains either a strictly increasing infinite sequence or a strictly decreasing infinite sequence of real numbers. Thus V contains either a strictly descending infinite sequence or a strictly ascending infinite sequence of subspaces, respectively, by Propositions 1 and 3. Hence V is infinite-dimensional. Conversely, suppose V is infinite-dimensional. Then V contains both strictly ascending and strictly descending chains of subspaces. Hence, by either Proposition 5 or Proposition 6 there exists $A \in \mathscr{A}_{K_F}$ such that A is infinite-valued.

Proposition 7 *Let $A = (\mu_A, \nu_A)$ be an intuitionistic fuzzy subset of V and K be an intuitionistic fuzzy subset of F. Let $d \in F$ and $x \in V$. Suppose that $0 \leq \alpha, \beta, \alpha', \beta' \leq 1$ and $\alpha + \beta \leq 1, \alpha' + \beta' \leq 1$. Then for all $z \in V$,*

(1) $(d_{(\alpha,\beta)} \circ A)(z) = (\alpha \wedge \mu_A(\frac{1}{d}z), \beta \vee \nu_A(\frac{1}{d}z))$, *if* $d \neq 0$.
(2) $(0_{(\alpha,\beta)} \circ A)(z) = (\sup_{y \in V}\{\alpha \wedge \mu_A(y)\}, \inf_{y \in V}\{\beta \vee \nu_A(y)\})$ *if* $z = 0$
$(0_{(\alpha,\beta)} \circ A)(z) = (0, 1)$ *if* $z \neq 0$.
(3) $(K \circ x_{(\alpha,\beta)})(z) = (\sup_{c \in F, z=cx}\{\alpha \wedge \mu_K(c)\}, \inf_{c \in F, z=cx}\{\beta \vee \nu_A(y)\})$ *if* $x \neq 0$ $z \in sp(x)$ *;and*
$(K \circ x_{(\alpha,\beta)})(z) = (0, 1)$ *if* $x \neq 0$ *and* $z \notin sp(x)$.
(4) $(K \circ 0_{(\alpha,\beta)})(z) = (\sup_{c \in F}\{\alpha \wedge \mu_K(c)\}, \inf_{c \in F}\{\beta \vee \nu_K(c)\})$ *if* $z = 0$*;*
$(K \circ 0_{(\alpha,\beta)})(z) = (0, 1)$ *if* $z \neq 0$.

Proof (1) Let $(d_{(\alpha,\beta)} \circ A)(z) = (\mu(z), \nu(z))$, we have:

$$\begin{aligned}\mu(z) &= \sup\{\min\{d_\alpha(c), \mu_A(y)\} | c \in F,\ y \in V,\ z = cy\}\\ &= \sup\{\min\{\alpha, \mu_A(y)\} | y \in V,\ z = dy\} = \min\{\alpha, \mu_A((1/d)z)\}.\end{aligned}$$

and

$$\begin{aligned}\nu(z) &= \inf\{\max\{d_\beta(c), \nu_A(y)\} | c \in F,\ y \in V,\ z = cy\}\\ &= \inf\{\max\{\beta, \nu_A(y)\} | y \in V,\ z = dy\} = \max\{\beta, \nu_A((1/d)z)\}.\end{aligned}$$

(2) Let $(0_{(\alpha,\beta)} \circ A)(z) = (\mu(z), \nu(z))$, we have:

$$\begin{aligned}\mu(z) &= \sup\{\min\{0_\alpha(c), \mu_A(y)\} | c \in F, y \in V, z = cy\}\}\\ &= \sup\{\min\{\alpha, \mu_A(y)\} | y \in V,\ z = 0y\}\end{aligned}$$

and

$$\begin{aligned}\nu(z) &= \inf\{\max\{0_\beta(c), \nu_A(y)\} | c \in F, y \in V, z = cy\}\}\\ &= \inf\{\max\{\alpha, \mu_A(y)\} | y \in V,\ z = 0y\} \text{ if } z = 0.\end{aligned}$$

if $z \neq 0$, then $0_{(\alpha,\beta)}(c) = (0, 1)$, since $c \neq 0$ when $z = cy$
(3) Let $(K \circ x_{(\alpha,\beta)})(z) = (\mu(z), \nu(z))$, we have:

$$\begin{aligned}\mu(z) &= \sup\{\min(\mu_K(c),\ x_\alpha(y)\} | c \in F,\ y \in V, z = cy\}\\ &= \sup\{\min\{\mu_K(c), \alpha\} | c \in F,\ z = cx\}.\end{aligned}$$

and

$$\begin{aligned}\nu(z) &= \inf\{\max(\nu_K(c),\ x_\beta(y)\} | c \in F,\ y \in V, z = cy\}\\ &= \inf\{\max\{\mu_K(c), \beta\} | c \in F,\ z = cx\} \text{ if } z \in sp(x).\end{aligned}$$

and $(\mu(z), \nu(z)) = (0, 1)$ otherwise.

(4) Let Let $(K \circ 0_{(\alpha,\beta)})(z) = (\mu(z), \nu(z))$, we have:

$$\begin{aligned}\mu(z) &= \sup\{\min(\mu_K(c),\ 0_\alpha(y)\}|c \in F,\ y \in V, z = cy\} \\ &= \sup\{\min\{\mu_K(c), \alpha\}|c \in F,\ z = c0\}\end{aligned}$$

and

$$\begin{aligned}\nu(z) &= \inf\{\max(\nu_K(c), 0_\beta(y)\}|c \in F,\ y \in V, z = cy\} \\ &= \inf\{\max\{\mu_K(c), \beta\}|c \in F,\ z = c0\} \text{ if } z = 0,\end{aligned}$$

if $z \neq 0$ then $(\mu(z), \nu(z)) = (0, 1)$, since $y \neq 0$ when $z = cy$.

Proposition 8 *Let* $x, y \in V$ *,* $c, d \in F$ *and* $0 \leq \alpha, \beta, \alpha^{'}, \beta^{'}, s, t, s', t' \leq 1$ *and* $\alpha + \beta \leq 1, s + t \leq 1, \alpha^{'} + \beta^{'} \leq 1, s' + t' \leq 1.$

(1) $x_{(t,s)} + y_{(\alpha,\beta)} = (x + y)_{(t\wedge\alpha, s\vee\beta)}$.
(2) $x_{(t,s)}y_{(\alpha,\beta)} = (xy)_{(t\wedge\alpha, s\vee\beta)}$.
(3) $c_{(t,s)} \circ x_{(\alpha,\beta)} = (cx)_{(t\wedge\alpha, s\vee\beta)}$.
(4) $c_{(t,s)} \circ x_{(\alpha,\beta)} + d_{(t',s')} \circ y_{(\alpha',\beta')} = (cx + dy)_{(\min(t,\alpha,\alpha',t'),\max(s,\beta,\beta',s'))}$.

Proof For (1), (2) and (3) see [4].
(3) The result follows from conditions (1), (2) and (3).

If $c_{1(\alpha_1,\beta_1)}, \ldots, c_{n(\alpha_n,\beta_n)}, x_{1(\lambda_1,\gamma_1)}, \ldots, x_{n(\lambda_n,\gamma_n)}$ are intuitionistic fuzzy points, where $c_i \in F$ and $x_i \in V$, $i = 1, 2, \ldots, n$, then $\Sigma_{i=1}^n c_{i(\alpha_i,\beta_i)} \circ x_{i(\lambda_i,\gamma_i)}$ is called an intuitionistic fuzzy linear combination of intuitionistic fuzzy points. By (4) of Proposition 8, it follows that an intuitionistic fuzzy linear combination of intuitionistic fuzzy points is an intuitionistic fuzzy point in V.

Proposition 9 *Let* $A \in \mathscr{A}_K$ *and let* B, C *be intuitionistic fuzzy subsets of* V*. Let* $b, c \in F$*. If* $B \subseteq A$ *and* $C \subseteq A$*, then* $b_{(\alpha,\beta)} \circ B + c_{(\alpha',\beta')} \circ C \subseteq A$*, where* $0 \leq \alpha \leq \mu_K(b)$ *,* $0 \leq \alpha^{'} \leq \mu_K(c), \nu_k(b) \leq \beta \leq 1$ *and* $\nu_k(c) \leq \beta^{'} \leq 1$*.*

Proof Since $b_{(\alpha,\beta)} \circ B$ and $c_{(\alpha',\beta')} \circ C$ are intuitionistic fuzzy subsets of V, it suffices to show that $b_{(\alpha,\beta)} \circ B \subseteq A$ and $B + C \subseteq A$.
Suppose that $b \neq 0$ and $(b_{(\alpha,\beta)} \circ B)(z) = (\mu, \nu)(z)$. Let $z \in V$, Then $\mu(z) = \min\{\alpha, \mu_B((\frac{1}{b})z)\} \leq \min\{\mu_K(b), \mu_A((\frac{1}{b})z)\} \leq \mu_A(b(\frac{1}{b})z)) = \mu_A(z)$. and $\nu(z) = \max\{\beta, \nu_B((\frac{1}{b})z)\} \geq \max\{\nu_K(b), \nu_A((\frac{1}{b})z)\} \geq \nu_A(b(\frac{1}{b})z)) = \nu_A(z)$ by Proposition 7(1).
Suppose that $b = 0$ and $z = 0$, then $\mu(z) = \sup\{\min\{\alpha, \mu_B(y)|y \in V\}\} \leq \mu_A(0)$ and $\nu(z) = \inf\{\max\{\beta, \nu_B(y)|y \in V\}\} \geq \nu_A(0)$ by Proposition 7(2).
Now

$$\begin{aligned}\mu_{B+C}(z) &= \sup\{\min\{\mu_B(x), \mu_C(y)\}|z = x + y\} \\ &\leq \sup\{\min\{\mu_A(x), \mu_A(y)\}|z = x + y\} \\ &\leq \mu_A(z).\end{aligned}$$

and

$$\begin{aligned}\nu_{B+C}(z) &= \inf\{\max\{\nu_B(x), \nu_C(y)\}|z = x + y\} \\ &\geq \inf\{\max\{\nu_A(x), \nu_A(y)\}|z = x + y\} \\ &\geq \nu_A(z).\end{aligned}$$

Proposition 10 *$\{A_i|A_i \in \mathscr{A}_K,\ i \in I\}$ is nonempty, then $\bigcap_{i\in I} A_i \in \mathscr{A}_K$.*

Proof Let $c \in F$, $x \in V$ and $(\bigcap_{i\in I} A_i)(cx) = (\mu, \nu)(cx)$. Then:

$$\begin{aligned}\mu(cx) &= \inf\{A_i(cx)|i \in I\} \\ &\geq \inf\{\min\{\mu_K(c), \mu_{A_i}(x)|i \in I\} \\ &= \text{either}\mu_K(c) \text{ or } \inf\{\mu_{A_i}(x)|i \in I\}.\end{aligned}$$

Hence, $\mu(cx) \geq \min\{\mu_K(c), \mu(x)\}$.
The same we show that $\nu(cx) \leq \max\{\nu_K(c), \nu(x)\}$.

Definition 8 Let $A \in \mathscr{A}_K$ and let X be an intuitionistic fuzzy subset of V such that $X \subseteq A$.Let $\langle X\rangle$ denote the intersection of all intuitionistic fuzzy subspaces of V (over K) that contain X and are contained in A. Then $\langle X\rangle$ is called the intuitionistic fuzzy subspace of A intuitionistic fuzzily spanned (or generated) by X.

Proposition 11 *Let $c_{1(\alpha_1,\beta_1)}, \ldots, c_{n(\alpha_n,\beta_n)}, x_{1(\alpha_1',\beta_1')}, \ldots, x_{n(\alpha_n',\beta_n')}$ are intuitionistic fuzzy points ,$A \in \mathscr{A}_K$ and let X be an intuitionistic fuzzy subset of V that $X \subseteq A$. Define the subset S of V by the following: for all $x \in V$,*

$$\mu_S(x) = \sup\{\mu_{\left(\sum_{i=1}^n c_{i(\alpha_i,\beta_i)} \circ x_{i(\alpha_i',\beta_i')}\right)}(x)|c_i \in F,\ x_i \in V,\ K(c_i) = (\alpha_i, \beta_i),$$
$$X(x_i) = (\alpha_i', \beta_i'),\ i = 1, \ldots, n,\ n \geq 1\}.$$
$$\nu_S(x) = \inf\{\nu_{\left(\sum_{i=1}^n c_{i(\alpha_i,\beta_i)} \circ x_{i(\alpha_i',\beta_i')}\right)}(x)|c_i \in F,\ x_i \in V,\ K(c_i) = (\alpha_i, \beta_i),$$
$$X(x_i) = (\alpha_i', \beta_i'),\ i = 1, \ldots, n,\ n \geq 1\}.$$

Then S is an intuitionistic fuzzy subset of V.

Theorem 2 *Let $c_{1(\alpha_1,\beta_1)}, \ldots, c_{n(\alpha_n,\beta_n)}, x_{1(\alpha_1',\beta_1')}, \ldots, x_{n(\alpha_n',\beta_n')}$ are intuitionistic fuzzy points ,$A \subset \mathscr{A}_K$ and let X be an intuitionistic fuzzy subset of V that $X \subseteq A$. Define the intuitionistic fuzzy subset S of V the following : for all $x \in V$,*
$\mu_S(x) = \sup\{\mu_{\left(\sum_{i=1}^n c_{i(\alpha_i,\beta_i)} \circ x_{i(\alpha_i',\beta_i')}\right)}(x)|c_i \in F,\ x_i \in V,\ \mu_K(c_i) = (\alpha_i, \beta_i),\ X(x_i) = (\alpha_i', \beta_i'),\ i = 1, \ldots, n,\ n \geq 1\}$.
$\nu_S(x) = \inf\{\nu_{\left(\sum_{i=1}^n c_{i(\alpha_i,\beta_i)} \circ x_{i(\alpha_i',\beta_i')}\right)}(x)|c_i \in F,\ x_i \in V,\ K(c_i) = (\alpha_i, \beta_i),\ X(x_i) = (\alpha_i', \beta_i'),\ i = 1, \ldots, n,\ n \geq 1\}$.
Then $\langle X\rangle = S$ and $S \in \mathscr{A}_K$.

Proof We have $x_{i(\alpha_i',\beta_i')} \subseteq X \subseteq \langle X \rangle$. Thus by Proposition 8(4) and 9, $S \subseteq \langle X \rangle$. In order to show that $S \supseteq \langle X \rangle$, it suffices to show that $S \in \mathscr{A}_K$ and $S \supseteq X$. Let $x \in V$ and let $X(x) = (\alpha, \beta)$.Then $\mu_{x_{(\alpha,\beta)}}(x) \leq \mu_S(x)$ and $\nu_{x_{(\alpha,\beta)}}(x) \geq \nu_S(x)$ and so $x_{(\alpha,\beta)} \subseteq S$. Thus $X \subseteq S$.

Let $u, v \in V$. Then $\mu_S(u)$ and $\mu_S(v)$ are supremums of the numbers of the forms, $(\sum_{i=1}^m c_{i(\alpha_i,\beta_i)} \circ y_{i(\alpha_i',\beta_i')})(u)$ and $\sum_{i=1}^m (d_{i(\lambda_i,\gamma_i)} \circ z_{i(\lambda_i',\gamma_i')})(v)$respectively, $\nu_S(u)$ and $\nu_S(v)$ are infimums of the numbers of the forms, $(\sum_{i=1}^m c_{i(\alpha_i,\beta_i)} \circ y_{i(\alpha_i',\beta_i')})(u)$ and $\sum_{i=1}^m (d_{i(\lambda_i,\gamma_i)} \circ z_{i(\lambda_i',\gamma_i')})(v)$respectively.

Suppose that $\mu_S(u) > 0$ and $\mu_S(v) > 0$. Then there exist sequences $(\alpha_j^*, \beta_j^*) = \big(\min\{\alpha_{1j}, \ldots, \alpha_{mj}, \alpha'_{1j}, \ldots, \alpha'_{mj}\}, \max\{\beta_{1j}, \ldots, \beta_{mj}, \beta'_{1j}, \ldots, \beta'_{mj}\}\big)$ and $(\lambda_j^*, \gamma_j^*) = \big(\min\{\lambda_{1j}, \ldots, \lambda_{qj}, \lambda'_{1j}, \ldots, \lambda'_{qj}\}, \max\{\gamma_{1j}, \ldots, \gamma_{qj}, \gamma'_{1j}, \ldots, \gamma'_{qj}\}\big)$ such that $(\alpha_j^*, \beta_j^*) \to (\mu_S(u), \nu_S(u))$ and $(\lambda_j^*, \gamma_j^*) \to (\mu_S(v), \nu_S(v))$.

Now if $u \in sp\{y_1, \ldots, y_m\}$ and $v \in sp\{z_1, \ldots, z_q\}$, then $u + v \in sp(y_1, \ldots, y_m, z_1, \ldots, z_q\}$.

Thus, for $j = 1, 2, \ldots,$ $\mu_S(u + v) \geq \min\{\alpha_{ij}, \alpha'_{ij}, \lambda_{kj}, \lambda'_{kj}, |i = 1, \ldots, m; k = 1, \ldots, q\} = \min\{\alpha_j^*, \lambda_j^*\}$.

Since $\min\{\alpha_j^*, \lambda_j^*\} \to \min\{\mu_S(u), \mu_S(v)\}$,$\mu_S(u + v) \geq \min\{\mu_S(u), \mu_S(v)\}$.

$\nu_S(u + v) \leq \max\{\beta_{ij}, \beta'_{ij}, \gamma_{kj}, \gamma'_{kj}, |i = 1, \ldots, m; k = 1, \ldots, q\} = \max\{\beta_j^*, \gamma_j^*\}$.

Since $\max\{\beta_j^*, \gamma_j^*\} \to \max\{\nu_S(u), \nu_S(v)\}$,$\nu_S(u + v) \leq \max\{\nu_S(u), \nu_S(v)\}$.

If either $\mu_S(u) = 0$ or $\mu_S(v) = 0$, then clearly $\mu_S(u + v) \geq \min\{\mu_S(u), \mu_S(v)\}$.

Clearly $(\mu_S(x), \nu_S(x)) = (\mu_S(-x), \nu_S(-x))$ for all $x \in V$. Let $c \in F, x \in V$ and $(\mu, \nu) = (\min\{\alpha_1, \ldots, \alpha_n, \alpha'_1, \ldots, \alpha'_n\}, \max\{\beta_1, \ldots, \beta_n, \beta'_1, \ldots, \beta'_n\})$.

Suppose $c \neq 0$. Now $cx = \sum_{i=1}^n c_i x_i$ if and only if $x = \sum_{i=1}^n c^{-1} c_i x_i$ where $c_i \in F$ and $x_i \in V$ $i = 1, \ldots, n$. Also

$$\begin{aligned}
(\mu_{(\sum_{i=1}^n c_{i\mu_i} \circ x_{i\lambda_i})}(cx) = \mu &= \min\{\mu_K(c_1), \ldots \mu_K(c_n), \alpha'_1, \ldots, \alpha'_n\} \\
&\geq \min\{\mu_K(c), \mu_K(c^{-1}c_1), \ldots \mu_K(c^{-1}c_n), \alpha'_1, \ldots, \alpha'_n\} \\
&\geq \min\{\mu_K(c), \min\{\mu_K(c^{-1}c_1), \ldots \mu_K(c^{-1}c_n), \alpha'_1, \ldots, \alpha'_n\}\}
\end{aligned}$$

Thus

$$\begin{aligned}
\mu_S(cx) &= \sup\{\min\{\mu_K(c_1), \ldots \mu_K(c_n), \alpha'_1, \ldots, \alpha'_n\} | cx = \sum\nolimits_{i=1}^n c_i x_i\} \\
&\geq \sup\{\min\{\mu_K(c), \min\{\mu_K(c^{-1}c_1), \ldots \mu_K(c^{-1}c_n), \alpha'_1, \ldots, \alpha'_n\}\} | x = \sum\nolimits_{i=1}^n c^{-1} c_i x_i\} \\
&\geq \min\{\{\mu_K(c), \sup\{\mu_K(c^{-1}c_1), \ldots \mu_K(c^{-1}c_n), \alpha'_1, \ldots, \alpha'_n\}\} | x = \sum\nolimits_{i=1}^n c^{-1} c_i x_i\} \\
&= \min\{\mu_K(c), \mu_S(x)\}.
\end{aligned}$$

Suppose that $c = 0$. Then

$$\mu_S(cx) = \mu_S(0) \geq \sup\{(\mu_{\left(0_{(\alpha,\beta)} \circ y_{(\alpha',\beta')}\right)}(0) | 0 \in F, y \in V, \mu_K(0) = \alpha = 1, \mu_X(y) = \alpha'\}$$
$$= \sup\{\min\{1, \alpha'\} | y \in V, \mu_X(y) = \alpha'\}$$
$$= \sup\{\mu_X(y) | y \in V\} \geq \mu_S(x) = \min\{\mu_K(0), \mu_S(x)\}.$$

The same we show that

$$\nu_S(cx) \leq \max\{\nu_K(c), \nu_S(x)\}$$

References

1. K.T. Atanassov, Intuitionistic fuzzy sets. Fuzzy Sets Syst. **20**(1), 87–96 (1986)
2. K.T. Atanassov, New operations defined over intuitionistic fuzzy sets. Fuzzy Sets Syst. **61**(2), 137–142 (1994)
3. I. Bakhadach, S. Melliani, M. Oukessou, S.L. Chadli, Intuitionistic fuzzy ideal and intuitionistic fuzzy prime ideal in a ring. Notes Intuit. Fuzzy Sets **22**(2), 59–63 (2016)
4. D. Coker, On intuitionistic fuzzy point. NIFSI **2**, 79–84 (1995)
5. P. Das, Fuzzy vector spaces under triangular norms. Fuzzy Sets Syst. **3**, 73-G (1988)
6. K. Hur, S.Y. Jans, H.W. Kang, Intuitionistic fuzzy subgroupoids. Int. J. Fuzzy Log. Intell. Syst. **3**(1), 72–77 (2003)
7. Y.B. Jun, S.Z. Song, Intuitionistic fuzzy semi preopen sets and Intuitionistic fuzzy semi precontinuous mappings. J. Appl. Math. Comput. **19**(1–2), 467–474 (2005)
8. A.K. Katsaras, D.B. Liu, Fuzzy vector spaces and topological vector spaces. J. Math. Anal. Appl. **58**, 135–146 (1977)
9. D.S. Malik, J.N. Mordeson, Fuzzy vector spaces. Inf. Sci. **55**, 271–281 (1991)
10. M.J. Mohammed, G.A. Ataa, On intuitionistic fuzzy topological vector space. J. Coll. Educ. Pure Sci. **4**, 32–51 (2014)
11. C. Moumita, S.K. Samanta, Intuitionistic fuzzy basis of an intuitionistic fuzzy vector space **23**(4), 62–74 (2017)
12. M. Muthusamy, N. Palaniappan, K. Arjunan, Study on intuitionistic fuzzy subfields. Int. J. General Topol **4**, 57–65 (2011)
13. S. Nanda, Fuzzy fields and fuzzy linear spaces. Fuzzy Sets Syst. **19**, 89–94 (1986)
14. P.M. Pu, Y.M. Mandal, Fuzzy topology, neibghbourhood structure of a fuzzy point and Moore-Smith convergence. J. Math. Anal. Appl. **76**(2), 571–599 (1980)
15. A. Rosenfeld, Fuzzy groups. J. Math. Anal. Appl. **35**, 512–517 (1971)
16. L.A. Zadeh, Fuzzy sets. Inf. Control **8**, 338–353 (1965)

Time-Dependent Neutral Stochastic Delay Partial Differential Equations Driven by Rosenblatt Process in Hilbert Space

E. Lakhel and A. Tlidi

Abstract In this paper, we investigate a class of time-dependent neutral stochastic functional differential equations with finite delay driven by Rosenblatt process in a real separable Hilbert space. We prove the existence of unique mild solution by the well-known Banach fixed point principle. At the end we provide a practical example in order to illustrate the viability of our result.

1 Introduction

The theory of the stochastic evolution equations have attracted great interest due to its many real applications in several areas such as biology, medicine, physics, finance, electrical engineering, telecommunication networks. For further details, the reader may refer to the works of [5]. As many phenomena exhibit a memory effect or aftereffect, there has been a real need for developing stochastic evolution systems with delay which incorporate the effect of delay on state equations. The neutral functional differential equations are often used to fulfill this aim, specially in natural phenomena such as extreme weather and natural disasters which often display long-term memory as well as in many stochastic dynamical systems which depend not only on present and past, but also contain the derivatives with delays.

Recently, there has been a growing interest on the stochastic functional differential equations driven by fractional Brownian motion (here after, fBm). The reader is referred to the works of [3, 4, 6], among others. The literature concerning the existence and qualitative properties of solutions of time-dependent functional stochastic differential equations is very restricted.

The fBm has several properties such as self-similarity, stationarity of increments and long-range dependence. Due to these nice properties, the fBm is of interest in real application and it is generally prefered among other processes because it is Gaussian

E. Lakhel · A. Tlidi (✉)
Cadi Ayyad university, National School of Applied Sciences, 46000 Safi, Morocco
e-mail: mtlidi2010@gmail.com

E. Lakhel
e-mail: e.lakhel@uca.ma

S. Melliani and O. Castillo (eds.), *Recent Advances in Intuitionistic Fuzzy Logic Systems*, Studies in Fuzziness and Soft Computing 372,
https://doi.org/10.1007/978-3-030-02155-9_24

and the calculus is easier. However, in some situations specially when the gaussianity property is not satisfied, the Rosenblatt process is often used instead. Although introduced during the 60 and 70 s [13, 15] in the literature, the Rosenblatt processes has been developed only during the last decade due to their appearance in the Non-Central Limit Theorem and to its desirable properties cited above i.e self-similarity, stationarity of increments and long-range dependence. The Rosenblatt processes can also be an input in models where self-similarity is observed in empirical data which appears to be non-Gaussian. In the literature, there exists a numerous studies that focuses on different theoretical aspects of the Rosenblatt processes. Leonenko and Ahn [8] studied the rate of convergence to the Rosenblatt process in the Non Central Limit Theorem. Tudor [16] analysed the Rosenblatt process. Maejima and Tudor [9] gave the distribution of the Rosenblatt process. Lakhel [7] established the existence of the unique solution for a class of neutral stochastic differential equation with delay and Poisson jumps driven by Rosenblatt process in Hilbert space.

To the best of our knowledge, there are no studies on time-dependent neutral stochastic functional differential equations with delays driven by Rosenblatt process. The aim of this paper is to fill this gap by providing the existence and uniqueness of mild solutions for a class of time-dependent neutral functional stochastic differential equations driven by non-Gaussian noises. This class is described as follow:

$$\begin{cases} d[x(t) + g(t, x(t-\tau))] = [A(t)x(t) + f(t, x(t-\tau))]dt + \sigma(t)dZ_H(t), \ 0 \le t \le T, \\ x(t) = \varphi(t), \ -\tau \le t \le 0, \end{cases} \tag{1}$$

In a real Hilbert space X with inner product $< ., . >$ and norm $\|.\|$, where $\{A(t),\ t \in [0, T]\}$ is a family of linear closed operators from a space X into X that generates an evolution system of operators $\{U(t, s),\ 0 \le s \le t \le T\}$. Z_H is a Rosenblatt process on a real and separable Hilbert space Y, and $f, g : [0, +\infty) \times X \to X,\ \sigma : [0, +\infty) \to \mathscr{L}_2^0(Y, X)$, are appropriate functions. Here $\mathscr{L}_2^0(Y, X)$ denotes the space of all Q-Hilbert-Schmidt operators from Y into X.

The rest of the paper is structured as follows: Section 2 is devoted to basic notations and concepts and results about Rosenblatt process as well as Wiener integral with respect to Hilbert space and recall some results about evolution operator. New technical lemma for the $\mathbb{L}^2-$estimate of stochastic convolution integral is proved. Section 3 gives sufficient conditions for the existence and uniqueness of the problem (1). Section 4 gives an example to illustrate the efficiency of the obtained result. Section 5 concludes.

2 Preliminaries

In this section we recall some basic results about evolution family, and we introduce the Rosenblatt process as well as the Wiener integral with respect to it. We also establish some important results which will be needed throughout the paper. At first, we introduce the notion of evolution family.

2.1 Evolution Families

Definition 1 A set $\{U(t,s) : 0 \leq s \leq t \leq T\}$ of bounded linear operators on a Hilbert space X is called an *evolution family* if

(a) $U(t,s)U(s,r) = U(t,r)$, $U(s,s) = I$ if $r \leq s \leq t$,
(b) $(t,s) \to U(t,s)x$ is strongly continuous for $t > s$.

Let $\{A(t),\ t \in [0,T]\}$ be a family of closed densely defined linear unbounded operators on the Hilbert space X under a domain $D(A(t))$ which is independent of t and satisfies the following conditions introduced by [1].

There exist constants $\lambda_0 \geq 0, \theta \in (\frac{\pi}{2}, \pi), L, K \geq 0$, and $\mu, \nu \in (0,1]$ with $\mu + \nu > 1$ such that

$$\Sigma_\theta \cup \{0\} \subset \rho(A(t) - \lambda_0), \quad \|R(\lambda, A(t) - \lambda_0)\| \leq \frac{K}{1 + |\lambda|} \tag{2}$$

and

$$\|(A(t) - \lambda_0)R(\lambda, A(t) - \lambda_0)\big[R(\lambda_0, A(t)) - R(\lambda_0, A(s))\big]\| \leq L|t-s|^\mu |\lambda|^{-\nu}, \tag{3}$$

for $t, s \in \mathbb{R}, \lambda \in \Sigma_\theta$ where $\Sigma_\theta := \big\{\lambda \in \mathbb{C} - \{0\} : |\arg \lambda| \leq \theta\big\}$.

It is well known, that this assumption implies that there exists a unique evolution family $\{U(t,s) : 0 \leq s \leq t \leq T\}$ on X such that $(t,s) \to U(t,s) \in \mathscr{L}(X)$ is continuous for $t > s$, $U(\cdot, s) \in \mathscr{C}^1((s,\infty), \mathscr{L}(X))$, $\partial_t U(t,s) = A(t)U(t,s)$, and

$$\|A(t)^k U(t,s)\| \leq C(t-s)^{-k} \tag{4}$$

for $0 < t - s \leq 1$, $k = 0, 1$, $0 \leq \alpha < \mu$, $x \in D((\lambda_0 - A(s))^\alpha)$, and a constant C depending only on the constants in (2)-(3). Moreover, $\partial_s^+ U(t,s)x = -U(t,s)A(s)x$ for $t > s$ and $x \in D(A(s))$ with $A(s)x \in \overline{D(A(s))}$. We say that $A(\cdot)$ generates $\{U(t,s) : 0 \leq s \leq t \leq T\}$. Note that $U(t,s)$ is exponentially bounded by (4) with $k = 0$.

Remark 1 If $\{A(t),\ t \in [0,T]\}$ is a second order differential operator A, that is $A(t) = A$ for each $t \in [0,T]$, then A generates a C_0-semigroup $\{e^{At}, t \in [0,T]\}$.

For further details on evolution system and their properties, the reader may refer to [11].

2.2 Rosenblatt Process

In this section, we collect some definitions and lemmas on Wiener integrals with respect to an infinite dimensional Rosenblatt process and we recall some basic results about analytical semi-groups and fractional powers of their infinitesimal generators, which will be used throughout the whole of this paper.

For details of this section, we refer the reader to [11, 16] and references therein.

Let $(\Omega, \mathscr{F}, \mathbb{P})$ be a complete probability space. Selfsimilar processes are invariant in distribution under suitable scaling. They are of considerable interest in practice since aspects of the selfsimilarity appear in different phenomena like telecommunications, turbulence, hydrology or economics. A self-similar processes can be defined as limits that appear in the so-called Non-Central Limit Theorem (see [15]). We briefly recall the Rosenblatt process as well as the Wiener integral with respect to it.

Let us recall the notion of Hermite rank. Denote by $H_j(x)$ the Hermite polynomial of degree j given by $H_j = (-1)^j e^{\frac{x^2}{2}} \frac{d^j}{dx^j} e^{\frac{-x^2}{2}}$ and let g be a function on $\mathbb{R}$ such that $\mathbb{E}[g(\zeta_0)] = 0$ and $\mathbb{E}[g(\zeta_0)^2] < \infty$. Assume that g has the following expansion in Hermite polynomials

$$g(x) = \sum_{j\geq 0} c_j H_j(x),$$

where $c_j = \frac{1}{j!}\mathbb{E}(g(\zeta_0 H_j(\zeta_0)))$. The Hermite rank of g is defined by

$$k = min\{j | c_j \neq 0\}.$$

Since $\mathbb{E}[g(\zeta_0)] = 0$, we have $k \geq 1$. Consider $(\zeta_n)_{n\in\mathbb{Z}}$ a stationary Gaussian sequence with mean zero and variance 1 which exhibits long range dependence in the sense that the correlation function satisfies

$$r(n) = \mathbb{E}(\zeta_0\zeta_n) = n^{\frac{2H-2}{k}} L(n),$$

with $H \in (\frac{1}{2}, 1)$ and L is a slowly varying function at infinity. Then the following family of stochastic processes

$$\frac{1}{n^H} \sum_{j=1}^{[nt]} g(\zeta_j)$$

converges as $n \longrightarrow \infty$, in the sense of finite dimensional distributions, to the self-similar stochastic process with stationary increments

$$Z_H^k(t) = c(H, k) \int_{\mathbb{R}^k} \left(\int_0^t \prod_{j=1}^{k} (s - y_j)_+^{-(\frac{1}{2}+\frac{1-H}{k})} ds \right) dB(y_1)...dB(y_k), \tag{5}$$

where $x_+ = max(x, 0)$. The above integral is a Wiener-Itô multiple integral of order k with respect to the standard Brownian motion $(B(y))_{y\in\mathbb{R}}$ and the constant $c(H, k)$ is a normalizing constant that ensures $\mathbb{E}(Z_H^k(1))^2 = 1$.

The process $(Z_H^k(t))_{t\geq 0}$ is called the Hermite process. When $k = 1$ the process given by (5) is nothing else that the fractional Brownian motion (fBm) with Hurst

parameter $H \in (\frac{1}{2}, 1)$. For $k = 2$ the process is not Gaussian. If $k = 2$ then the process (5) is known as the Rosenblatt process. It was introduced by Rosenblatt in [13] and was given its name by Taqqu in [14]. The fractional Brownian motion is of course the most studied process in the class of Hermite processes due to its significant importance in modelling. A stochastic calculus with respect to it has been intensively developed in the last decade. The Rosenblatt process is, after fBm, the most well known Hermite process.
We also recall the following properties of the Rorenblatt process:

- The process Z_H^k is H-selfsimilar in the sense that for any $c > 0$,

$$(Z_H^k(ct)) =^{(d)} (c^H Z_H^k(t)), \tag{6}$$

where "$=^{(d)}$" means equivalence of all finite dimensional distributions. It has stationary increments and all moments are finite.

- From the stationarity of increments and the self-similarity, it follows that, for any $p \geq 1$

$$\mathbb{E}|Z_H(t) - Z_H(s)|^p \leq |\mathbb{E}(Z_H(1))|^p |t-s|^{pH}.$$

As a consequence the Rosenblatt process has Hölder continuous paths of order γ with $0 < \gamma < H$.

Self-similarity and long-range dependence make this process a useful driving noise in models arising in physics, telecommunication networks, finance and other fields. Consider a time interval $[0, T]$ with arbitrary fixed horizon T and let $\{Z_H(t), t \in [0, T]\}$ the one-dimensional Rosenblatt process with parameter $H \in (1/2, 1)$. By Tudor [16], it is well known that Z_H has the following integral representation:

$$Z_H(t) = d(H) \int_0^t \int_0^t \left[\int_{y_1 \vee y_2}^t \frac{\partial K^{H'}}{\partial u}(u, y_1) \frac{\partial K^{H'}}{\partial u}(u, y_2) du \right] dB(y_1) dB(y_2), \tag{7}$$

where $B = \{B(t) : t \in [0, T]\}$ is a Wiener process, $H' = \frac{H+1}{2}$ and $K^H(t, s)$ is the kernel given by

$$K^H(t, s) = c_H s^{\frac{1}{2}-H} \int_s^t (u-s)^{H-\frac{3}{2}} u^{H-\frac{1}{2}} du,$$

for $t > s$, where $c_H = \sqrt{\frac{H(2H-1)}{\beta(2-2H, H-\frac{1}{2})}}$ and $\beta(,)$ denotes the Beta function. We put $K^H(t, s) = 0$ if $t \leq s$ and $d(H) = \frac{1}{H+1}\sqrt{\frac{H}{2(2H-1)}}$ is a normalizing constant.

The covariance of the Rosenblatt process $\{Z_H(t), t \in [0, T]\}$ satisfies, for every $s, t \geq 0$,

$$R_H(s, t) := \mathbb{E}(Z_H(t) Z_H(s)) = \frac{1}{2}(t^{2H} + s^{2H} - |t-s|^{2H}).$$

The basic observation is the fact that the covariance structure of the Rosenblatt process is similar to the one of the fractional Brownian motion and this allows the use of the same classes of deterministic integrands as in the fractional Brownian motion case whose properties are known.

Now, we introduce Wiener integrals with respect to the Rosenblatt process. We refer to [16] for additional details on the Rosenblatt process .

By formula (7) we can write

$$Z_H(t) = \int_0^t \int_0^t I(\mathbf{1}_{[0,t]})(y_1, y_2) dB(y_1) dB(y_2),$$

where by I we denote the mapping on the set of functions $f : [0, T] \longrightarrow \mathbb{R}$ to the set of functions $f : [0, T]^2 \longrightarrow \mathbb{R}$

$$I(f)(y_1, y_2) = d(H) \int_{y_1 \vee y_2}^T f(u) \frac{\partial K^{H'}}{\partial u}(u, y_1) \frac{\partial K^{H'}}{\partial u}(u, y_2) du.$$

Let us denote by $\mathcal{E}$ the class of elementary functions on R of the form

$$f(.) = \sum_{j=1}^n a_j \mathbf{1}_{(t_j, t_{j+1}]}(.), \qquad 0 \leq t_j < t_{j+1} \leq T, \quad a_j \in \mathbb{R}, \quad i = 1, ..., n.$$

For $f \in \mathcal{E}$ as above, it is natural to define its Wiener integral with respect to the Rosenblatt process Z_H by

$$\begin{aligned} \int_0^T f(s) dZ_H(s) &:= \sum_{j=1}^n a_j \left[Z_H(t_{j+1}) - Z_H(t_j) \right] \\ &= \int_0^T \int_0^T I(f)(y_1, y_2) dB(y_1) dB(y_2). \end{aligned} \tag{8}$$

Let $\mathcal{H}$ be the set of functions f such that

$$\mathcal{H} = \left\{ f : [0, T] \longrightarrow \mathbb{R} : \quad \|f\|_{\mathcal{H}} := \int_0^T \int_0^T (I(f)(y_1, y_2))^2 \, dy_1 dy_2 < \infty \right\}.$$

It hold that (see Maejima and Tudor [10])

$$\|f\|_{\mathcal{H}} = H(2H-1) \int_0^T \int_0^T f(u) f(v) |u-v|^{2H-2} du dv,$$

and, the mapping

$$f \longrightarrow \int_0^T f(u) dZ_H(u) \tag{9}$$

provides an isometry from $\mathscr{E}$ to $L^2(\Omega)$. On the other hand, it has been proved in [12] that the set of elementary functions $\mathscr{E}$ is dense in $\mathscr{H}$. As a consequence the mapping (9) can be extended to an isometry from $\mathscr{H}$ to $L^2(\Omega)$. We call this extension as the Wiener integral of $f \in \mathscr{H}$ with respect to Z_H.

Let us consider the operator K_H^* from $\mathscr{E}$ to $\mathbb{L}^2([0,T])$ defined by

$$(K_H^*\varphi)(y_1, y_2) = \int_{y_1 \vee y_2}^{T} \varphi(r)\frac{\partial K}{\partial r}(r, y_1, y_2)dr,$$

where $K(.,.,.)$ is the kernel of Rosenblatt process in representation (7)

$$K(r, y_1, y_2) = \mathbf{1}_{[0,t]}(y_1)\mathbf{1}_{[0,t]}(y_2)\int_{y_1 \vee y_2}^{t} \frac{\partial K^{H'}}{\partial u}(u, y_1)\frac{\partial K^{H'}}{\partial u}(u, y_2)du.$$

We refer to [16] for the proof of the fact that K_H^* is an isometry between $\mathscr{H}$ and $L^2([0,T])$. It follows from [16] that $\mathscr{H}$ contains not only functions but its elements could be also distributions. In order to obtain a space of functions contained in $\mathscr{H}$, we consider the linear space $|\mathscr{H}|$ generated by the measurable functions ψ such that

$$\|\psi\|^2_{|\mathscr{H}|} := \alpha_H \int_0^T \int_0^T |\psi(s)||\psi(t)||s-t|^{2H-2}dsdt < \infty,$$

where $\alpha_H = H(2H-1)$. The space $|\mathscr{H}|$ is a Banach space with the norm $\|\psi\|_{|\mathscr{H}|}$ and we have the following inclusions (see [16]).

Lemma 1

$$\mathbb{L}^2([0,T]) \subseteq \mathbb{L}^{1/H}([0,T]) \subseteq |\mathscr{H}| \subseteq \mathscr{H},$$

and for any $\psi \in \mathbb{L}^2([0,T])$, *we have*

$$\|\psi\|^2_{|\mathscr{H}|} \leq 2HT^{2H-1}\int_0^T |\psi(s)|^2 ds.$$

Let X and Y be two real, separable Hilbert spaces and let $\mathscr{L}(Y,X)$ be the space of bounded linear operator from Y to X. For the sake of convenience, we shall use the same notation to denote the norms in X, Y and $\mathscr{L}(Y,X)$. Let $Q \in \mathscr{L}(Y,Y)$ be an operator defined by $Qe_n = \lambda_n e_n$ with finite trace $tr\,Q = \sum_{n=1}^{\infty} \lambda_n < \infty$. where $\lambda_n \geq 0$ $(n = 1, 2...)$ are non-negative real numbers and $\{e_n\}$ $(n = 1, 2...)$ is a complete orthonormal basis in Y. We define the infinite dimensional $Q-$Rosenblatt process on Y as

$$Z_H(t) = Z_Q(t) = \sum_{n=1}^{\infty} \sqrt{\lambda_n} e_n z_n(t), \tag{10}$$

where $(z_n)_{n\geq 0}$ is a family of real independent Rosenblatt process.

Note that the series (10) is convergent in $L^2(\Omega)$ for every $t \in [0,T]$, since

$$\mathbb{E}|Z_Q(t)|^2 = \sum_{n=1}^{\infty} \lambda_n \mathbb{E}(z_n(t))^2 = t^{2H} \sum_{n=1}^{\infty} \lambda_n < \infty.$$

Note also that Z_Q has covariance function in the sense that

$$E\langle Z_Q(t), x\rangle\langle Z_Q(s), y\rangle = R(s,t)\langle Q(x), y\rangle \quad \text{for all } x, y \in Y \text{ and } t, s \in [0, T].$$

In order to define Wiener integrals with respect to the Q-Rosenblatt process, we introduce the space $\mathcal{L}_2^0 := \mathcal{L}_2^0(Y, X)$ of all Q-Hilbert-Schmidt operators $\psi : Y \to X$. We recall that $\psi \in \mathcal{L}(Y, X)$ is called a Q-Hilbert-Schmidt operator, if

$$\|\psi\|^2_{\mathcal{L}_2^0} := \sum_{n=1}^{\infty} \|\sqrt{\lambda_n}\psi e_n\|^2 < \infty,$$

and that the space $\mathcal{L}_2^0$ equipped with the inner product $\langle \varphi, \psi\rangle_{\mathcal{L}_2^0} = \sum_{n=1}^{\infty}\langle \varphi e_n, \psi e_n\rangle$ is a separable Hilbert space.

Now, let $\phi(s)$; $s \in [0, T]$ be a function with values in $\mathcal{L}_2^0(Y, X)$, such that $\sum_{n=1}^{\infty} \|K^*\phi Q^{\frac{1}{2}} e_n\|^2_{\mathcal{L}_2^0} < \infty$. The Wiener integral of ϕ with respect to Z_Q is defined by

$$\begin{aligned}\int_0^t \phi(s)dZ_Q(s) &= \sum_{n=1}^{\infty} \int_0^t \sqrt{\lambda_n}\phi(s)e_n dz_n(s)\\ &= \sum_{n=1}^{\infty} \int_0^t \int_0^t \sqrt{\lambda_n} K_H^*(\phi e_n)(y_1, y_2)dB(y_1)dB(y_2). \quad (11)\end{aligned}$$

Now, we end this subsection by stating the following result which is fundamental to prove our result.

Lemma 2 *If* $\psi : [0, T] \to \mathcal{L}_2^0(Y, X)$ *satisfies* $\int_0^T \|\psi(s)\|^2_{\mathcal{L}_2^0} ds < \infty$ *then the above sum in (11) is well defined as a* X*-valued random variable and we have*

$$\mathbb{E}\|\int_0^t \psi(s)dZ_H(s)\|^2 \leq 2Ht^{2H-1}\int_0^t \|\psi(s)\|^2_{\mathcal{L}_2^0} ds.$$

Proof By Lemma 1, we have

$$\begin{aligned}\mathbb{E}\|\int_0^t \psi(s)dZ_H(s)\|^2 &= \sum_{n=1}^{\infty} \mathbb{E}\|\int_0^t\int_0^t \sqrt{\lambda_n}K_H^*(\psi e_n)(y_1, y_2)dB_n(y_1)dB_n(y_2)\|^2\\ &\leq \sum_{n=1}^{\infty} 2Ht^{2H-1}\int_0^t \lambda_n \|\psi(s)e_n\|^2 ds\\ &= 2Ht^{2H-1}\int_0^t \|\psi(s)\|^2_{\mathcal{L}_2^0} ds.\end{aligned}$$

2.3 Definition and Assumption

Henceforth we will assume that the family $\{A(t),\ t \in [0, T]\}$ of linear operators generates an evolution system of operators $\{U(t, s),\ 0 \leq s \leq t \leq T\}$.

Definition 2 An X-valued stochastic process $\{x(t),\ t \in [-\tau, T]\}$, is called a mild solution of Eq. (1) if

(i) $x(.) \in \mathscr{C}([-\tau, T], \mathbb{L}^2(\Omega, X))$,
(ii) $x(t) = \varphi(t),\ -\tau \leq t \leq 0$.
(iii) For arbitrary $t \in [0, T]$, $x(t)$ satisfies the following integral equation:

$$\begin{aligned} x(t) &= U(t,0)(\varphi(0) + g(0, \varphi(-\tau))) - g(t, x(t-\tau)) \\ &\quad - \int_0^t U(t,s)A(s)g(s, x(s-\tau))ds + \int_0^t U(t,s)f(s, x(s-\tau))ds \\ &\quad + \int_0^t U(t,s)\sigma(s)dZ_Q(s) \quad \mathbb{P} - a.s \end{aligned}$$

We introduce the following assumptions:
($\mathscr{H}$.1)

(i) The evolution family is exponentially stable, that is, there exist two constants $\beta > 0$ and $M \geq 1$ such that

$$\|U(t,s)\| \leq Me^{-\beta(t-s)}, \qquad for\ all \quad t \geq s,$$

(ii) There exist a constant $M_* > 0$ such that

$$\|A^{-1}(t)\| \leq M_* \qquad for\ all \quad t \in [0, T].$$

($\mathscr{H}$.2) The maps $f, g : [0, T] \times X \to X$ are continuous functions and there exist two positive constants C_1 and C_2, such that for all $t \in [0, T]$ and $x, y \in X$:

(i) $\|f(t,x) - f(t,y)\| \vee \|g(t,x) - g(t,y)\| \leq C_1\|x - y\|$.

(ii) $\|f(t,x)\|^2 \vee \|A^k(t)g(t,x)\|^2 \leq C_2(1 + \|x\|^2), \quad k = 0, 1.$

($\mathscr{H}$.3)

(i) There exists a constant $0 < L_* < \frac{1}{M_*}$ such that

$$\|A(t)g(t,x) - A(t)g(t,y)\| \leq L_*\|x - y\|,$$

for all $t \in [0, T]$ and $x, y \in X$.

(ii) The function g is continuous in the quadratic mean sense: for all $x(.) \in \mathscr{C}([0, T], L^2(\Omega, X))$, we have

$$\lim_{t \longrightarrow s} \mathbb{E}\|g(t, x(t)) - g(s, x(s))\|^2 = 0.$$

($\mathscr{H}$.4)

(i) The map $\sigma : [0, T] \longrightarrow \mathscr{L}_2^0(Y, X)$ is bounded, that is : there exists a positive constant L such that $\|\sigma(t)\|_{\mathscr{L}_2^0(Y,X)} \leq L$ uniformly in $t \in [0, T]$.

(ii) Moreover, we assume that the initial data $\varphi = \{\varphi(t) : -\tau \leq t \leq 0\}$ satisfies $\varphi \in \mathscr{C}([-\tau, 0], \mathbb{L}^2(\Omega, X))$.

3 Existence and Uniqueness of Mild Solutions

In this section we study the existence and uniqueness of mild solutions of Eq. (1). First, it is of great importance to establish the basic properties of the stochastic convolution integral of the form

$$X(t) = \int_0^t U(t, s)\sigma(s)dZ_Q(s), \qquad t \in [0, T],$$

where $\sigma(s) \in \mathscr{L}_2^0(Y, X)$ and $\{U(t, s),\ 0 \leq s \leq t \leq T\}$ is an evolution system of operators.

The properties of the process X are crucial when regularity of the mild solution to stochastic evolution equation is studied, see [5] for asystematic account of the theory of mild solutions to infinite-dimensional stochastic equations. Unfortunately, the process X is not a martingale, and standard tools of the martingale theory, yielding e.g. continuity of the trajectories or $\mathbb{L}^2$–estimates are not available. The following result on the stochastic convolution integral X holds.

Lemma 3 *Suppose that* $\sigma : [0, T] \to \mathscr{L}_2^0(Y, X)$ *satisfies* $\sup_{t\in[0,T]} \|\sigma(t)\|^2_{\mathscr{L}_2^0} < \infty$, *and suppose that* $\{U(t, s),\ 0 \leq s \leq t \leq T\}$ *is an evolution system of operators satisfying* $\|U(t, s)\| \leq Me^{-\beta(t-s)}$, *for some constants* $\beta > 0$ *and* $M \geq 1$ *for all* $t \geq s$. *Then, we have*

1. *The stochastic integral* $X :\ t \longrightarrow \int_0^t U(t, s)\sigma(s)dZ_Q(s)$ *is well-defined and we have*

$$\mathbb{E}\|\int_0^t U(t, s)\sigma(s)dZ_Q(s)\|^2 \leq C_H M^2 t^{2H} (\sup_{t\in[0,T]} \|\sigma(t)\|_{\mathscr{L}_2^0})^2.$$

2. *The stochastic integral* $X :\ t \longrightarrow \int_0^t U(t, s)\sigma(s)dZ_Q(s)$ *is continuous.*

Proof 1. Let $\{e_n\}_{n\in\mathbb{N}}$ be the complete orthonormal basis of Y and $\{z_n\}_{n\in\mathbb{N}}$ is a sequence of independent, real-valued Rosenblatt process each with the same parameter $H \in (\frac{1}{2}, 1)$. Thus, using isometry property one can write

$$\begin{aligned}
\mathbb{E}\|\int_0^t U(t,s)\sigma(s)dZ_Q(s)\|^2 &= \sum_{n=1}^{\infty}\mathbb{E}\|\int_0^t U(t,s)\sigma(s)e_n dz_n(s)\|^2\\
&= H(2H-1)\int_0^t \{\|U(t,s)\sigma(s)\|\\
&\times \int_0^t \|U(t,r)\sigma(r)\||s-r|^{2H-2}dr\}ds\\
&\leq H(2H-1)M^2\int_0^t \{e^{-\beta(t-s)}\|\sigma(s)\|_{\mathcal{L}_2^0}\\
&\times \int_0^t e^{-\beta(t-r)}|s-r|^{2H-2}\|\sigma(r)\|_{\mathcal{L}_2^0}dr\}ds.
\end{aligned}$$

Since σ is bounded, one can then conclude that

$$\begin{aligned}
\mathbb{E}\|\int_0^t U(t,s)\sigma(s)dZ_H(s)\|^2 \leq H(2H-1)M^2(\sup_{t\in[0,T]}\|\sigma(t)\|_{\mathcal{L}_2^0})^2\int_0^t \{e^{-\beta(t-s)}\\
\times \int_0^t e^{-\beta(t-r)}|s-r|^{2H-2}dr\}ds.
\end{aligned}$$

Make the following change of variables, $v = t - s$ for the first integral and $u = t - r$ for the second. One can write

$$\begin{aligned}
\mathbb{E}\|\int_0^t U(t,s)\sigma(s)dZ_H(s)\|^2 &\leq H(2H-1)M^2(\sup_{t\in[0,T]}\|\sigma(t)\|_{\mathcal{L}_2^0})^2\int_0^t \{e^{-\beta v}\\
&\times \int_0^t e^{-\beta u}|u-v|^{2H-2}du\}dv\\
&\leq H(2H-1)M^2(\sup_{t\in[0,T]}\|\sigma(t)\|_{\mathcal{L}_2^0})^2\int_0^t\int_0^t |u-v|^{2H-2}dudv.
\end{aligned}$$

By using the equality,

$$R_H(t,s) = H(2H-2)\int_0^t\int_0^s |u-v|^{2H-2}dudv,$$

we get that

$$\mathbb{E}\|\int_0^t U(t,s)\sigma(s)dZ_Q(s)\|^2 \leq C_H M^2 t^{2H}(\sup_{t\in[0,T]}\|\sigma(t)\|_{\mathcal{L}_2^0})^2.$$

2. Let $h > 0$ small enough, we have

$$\begin{aligned}
\mathbb{E}\|\int_0^{t+h} U(t+h,s)\sigma(s)dZ_Q(s) - \int_0^t U(t,s)\sigma(s)dZ_Q(s)\|^2 &\leq 2\|\int_0^t (U(t+h,s)-U(t,s))\sigma(s)dZ_Q(s)\|^2\\
&+ 2\|\int_t^{t+h} U(t+h,s)\sigma(s)dZ_H(s)\|^2\\
&\leq 2[\mathbb{E}\|I_1(h)\|^2 + \mathbb{E}\|I_2(h)\|^2].
\end{aligned}$$

By Lemma 2, we get that

$$E\|I_1(h)\|^2 \leq 2Ht^{2H-1}\int_0^t \|[U(t+h,s)-U(t,s)]\sigma(s)\|^2_{\mathscr{L}_2^0}ds.$$

Since

$$\lim_{h\to 0}\|[U(t+h,s)-U(t,s)]\sigma(s)\|^2_{\mathscr{L}_2^0}=0,$$

and

$$\|(U(t+h,s)-U(t,s))\sigma(s)\|_{\mathscr{L}_2^0} \leq MLe^{-\beta(t-s)}e^{-\beta h+1} \in \mathbb{L}^1([0,T],\, ds),$$

we conclude, by the dominated convergence theorem that,

$$\lim_{h\to 0}\mathbb{E}\|I_1(h)\|^2=0.$$

Again by Lemma 2, we get that

$$\mathbb{E}\|I_2(h)\|^2 \leq \frac{2Ht^{2H-1}LM^2(1-e^{-2\beta h})}{2\beta}.$$

Thus,

$$\lim_{h\to 0}\mathbb{E}\|I_2(h)\|^2=0.$$

Remark 2 Thanks to Lemma 3, the stochastic integral $X(t)$ is well-defined and it belongs to the space $\mathscr{C}([-\tau,0],\mathbb{L}^2(\Omega,X))$.

We have the following theorem on the existence and uniqueness of mild solutions of Eq. (1).

Theorem 1 *Suppose that* $(\mathscr{H}.1)$*-*$(\mathscr{H}.4)$ *hold. Then, for all* $T>0$*, the Eq. (1) has a unique mild solution on* $[-\tau,T]$*.*

Proof Fix $T>0$ and let $B_T:=\mathscr{C}([-\tau,T],\mathbb{L}^2(\Omega,X))$ be the Banach space of all continuous functions from $[-\tau,T]$ into $\mathbb{L}^2(\Omega,X)$, equipped with the supremum norm

$$\|x\|^2_{B_T}=\sup_{-\tau\leq t\leq T}\mathbb{E}\|x(t,\omega)\|^2.$$

Let us consider the set

$$S_T(\varphi)=\{x\in B_T : x(s)=\varphi(s),\ \text{for}\ s\in[-\tau,0]\}.$$

$S_T(\varphi)$ is a closed subset of B_T provided with the norm $\|.\|_{B_T}$.
We transform (1) into a fixed-point problem. Consider the operator ψ on $S_T(\varphi)$ defined by $\psi(x)(t)=\varphi(t)$ for $t\in[-\tau,0]$ and for $t\in[0,T]$

$$\begin{aligned}
\psi(x)(t) &= U(t,0)(\varphi(0) + g(0, \varphi(-\tau))) - g(t, x(t-\tau)) \\
&\quad - \int_0^t U(t,s)A(s)g(s, x(s-\tau))ds + \int_0^t U(t,s)f(s, x(s-\tau))ds \\
&\quad + \int_0^t U(t,s)\sigma(s)dZ_Q(s) \\
&= \sum_{i=1}^{5} I_i(t).
\end{aligned}$$

Clearly, the fixed points of the operator ψ are mild solutions of (1). The fact that ψ has a fixed point will be proved in several steps. We will first prove that the function ψ is well defined.

Step 1: For arbitrary $x \in S_T(\varphi)$, we are going to show that each function $t \to I_i(t)$ is continuous on $[0, T]$ in the $\mathbb{L}^2(\Omega, X)$-sense.

For the first term $I_1(h)$, by Definition 1, we obtain

$$\lim_{h \longrightarrow 0} (U(t+h, 0) - U(t, 0))(\varphi(0) + g(0, \varphi(-\tau))) = 0.$$

From $(\mathscr{H}.1)$, we have

$$\begin{aligned}
&\|(U(t+h,0) - U(t,0))(\varphi(0) + g(0, \varphi(-\tau)))\| \le Me^{-\beta t}(e^{-\beta h} + 1)\|\varphi(0) \\
&\quad + g(0, \varphi(-\tau))\| \in L^2(\Omega).
\end{aligned}$$

Then we conclude by the Lebesgue dominated theorem that

$$\lim_{h \longrightarrow 0} \mathbb{E}\|I_1(t+h) - I_1(t)\|^2 = 0.$$

For the second term $I_2(h)$, assumption $(\mathscr{H}.2)$ ensures that

$$\lim_{h \longrightarrow 0} \mathbb{E}\|I_2(t+h) - I_2(t)\|^2 = 0.$$

To show that the third term $I_3(h)$ is continuous, we suppose $h > 0$ (similar calculus for $h < 0$). We have

$$\begin{aligned}
\|I_3(t+h) - I_3(t)\| &\le \left\| \int_0^t (U(t+h, s) - U(t,s))A(s)g(s, x(s-\tau))ds \right\| \\
&\quad + \left\| \int_t^{t+h} U(t,s)g(s, x(s-\tau))ds \right\| \\
&\le I_{31}(h) + I_{32}(h).
\end{aligned}$$

By Hölder's inequality, we have

$$\mathbb{E}\|I_{31}(h)\| \leq t\mathbb{E}\int_0^t \|(U(t+h,s) - U(t+h,s))A(s)g(s,x(s-\tau)\|^2 ds.$$

By Definition 1, we obtain

$$\lim_{h\longrightarrow 0}(U(t+h,s) - U(t,s))A(s)g(s,x(s-\tau)) = 0.$$

From ($\mathcal{H}$.1) and ($\mathcal{H}$.2), we have

$$\begin{aligned}&\|(U(t+h,s) - U(t,s))A(s)g(s,x(s-\tau))\| \\ &\leq C_2 M e^{-\beta(t-s)}(e^{-\beta h}+1)\|A(s)g(s,x(s-\tau))\| \in L^2(\Omega).\end{aligned}$$

Then we conclude by the Lebesgue dominated theorem that

$$\lim_{h\longrightarrow 0}\mathbb{E}\|I_{31}(h)\|^2 = 0.$$

So, estimating as before. By using ($\mathcal{H}$.1) and ($\mathcal{H}$.2), we get

$$\mathbb{E}\|I_{32}(h)\|^2 \leq \frac{M^2C_2(1-e^{-2\beta h})}{2\beta}\int_t^{t+h}(1+\mathbb{E}\|x(s-\tau)\|^2)ds.$$

Thus,

$$\lim_{h\longrightarrow 0}\mathbb{E}\|I_{32}(h)\|^2 = 0.$$

For the fourth term $I_4(h)$, we suppose $h > 0$ (similar calculus for $h < 0$). We have

$$\begin{aligned}\|I_4(t+h) - I_4(t)\| &\leq \left\|\int_0^t (U(t+h,s) - U(t,s))f(s,x(s-\tau))ds\right\| \\ &\quad + \left\|\int_t^{t+h} U(t,s)f(s,x(s-\tau))ds\right\| \\ &\leq \|I_{41}(h)\| + \|I_{42}(h)\|.\end{aligned}$$

By Hölder's inequality, we have

$$\mathbb{E}\|I_{41}(h)\| \leq t\mathbb{E}\int_0^t \|(U(t+h,s) - U(t,s))f(s,x(s-\tau))\|^2 ds.$$

Again exploiting properties of Definition 1, we obtain

$$\lim_{h\longrightarrow 0}(U(t+h,s) - U(t,s))f(s,x(s-\tau)) = 0,$$

and

$$\begin{aligned}&\|(U(t+h,s)-U(t,s))f(s,x(s-\tau))\|\\&\leq Me^{-\beta(t-s)}(e^{-\beta h}+1)\|f(s,x(s-\tau))\|\in L^2(\Omega).\end{aligned}$$

Then we conclude by the Lebesgue dominated theorem that

$$\lim_{h\longrightarrow 0}\mathbb{E}\|I_{41}(h)\|^2=0.$$

On the other hand, by ($\mathscr{H}$.1) , ($\mathscr{H}$.2), and the Hölder's inequality, we have

$$\mathbb{E}\|I_{42}(h)\|\leq\frac{M^2C_2(1-e^{-2\beta h})}{2\beta}\int_t^{t+h}(1+\mathbb{E}\|x(s-\tau)\|^2)ds.$$

Thus

$$\lim_{h\to 0}\mathbb{E}\|I_{42}(h)\|^2=0.$$

Now, for the term $I_5(h)$, we have

$$\begin{aligned}\mathbb{E}\|I_5(t+h)-I_5(t)\|^2\leq 2\mathbb{E}\|&\int_0^t(U(t+h,s)-U(t,s))\sigma(s)dZ_Q(s)\|^2\\&+2\mathbb{E}\|\int_t^{t+h}U(t+h,s)\sigma(s)dZ_Q(s)\|^2.\end{aligned}$$

By Lemma 3 we get

$$\lim_{h\to 0}\|I_5(t+h)-I_5(t)\|^2=0.$$

The above arguments show that $\lim_{h\to 0}\mathbb{E}\|\psi(x)(t+h)-\psi(x)(t)\|^2=0$. Hence, we conclude that the function $t\to\psi(x)(t)$ is continuous on $[0,T]$ in the $\mathbb{L}^2$-sense.
Step 2: Now, we are going to show that ψ is a contraction mapping in $S_{T_1}(\varphi)$ with some $T_1\leq T$ to be specified later. Let $x,y\in S_T(\varphi)$, by using the inequality

$$(a+b+c)^2\leq\frac{1}{\nu}a^2+\frac{2}{1-\nu}b^2+\frac{2}{1-\nu}c^2,$$

where $\nu:=L_*M_*<1$, we obtain for any fixed $t\in[0,T]$

$$\begin{aligned}&\|\psi(x)(t)-\psi(y)(t)\|^2\\&\qquad\leq\frac{1}{\nu}\|g(t,x(t-\tau))-g(t,y(t-\tau))\|^2\\&\qquad+\frac{2}{1-\nu}\|\int_0^t U(t,s)A(s)(g(s,x(s-\tau))-g(s,y(s-\tau)))ds\|^2\end{aligned}$$

$$+\frac{2}{1-\nu}\|\int_0^t U(t,s)(f(s,x(s-\tau))-f(s,y(s-\tau)))ds\|^2$$
$$=\sum_{k=1}^{3} J_k(t).$$

By using the fact that the operator $\|(A^{-1}(t))\|$ is bounded, combined with the condition ($\mathscr{H}$.3), we obtain that

$$\begin{aligned}
\mathbb{E}\|J_1(t)\| &\le \frac{1}{\nu}\|A^{-1}(t)\|^2\mathbb{E}|A(t)g(t,x(t-\tau))-A(t)g(t,y(t-\tau))\|^2\\
&\le \frac{L_*^2M_*^2}{\nu}\mathbb{E}\|x(t-\tau)-y(t-\tau)\|^2\\
&\le \nu \sup_{s\in[-\tau,t]} \mathbb{E}\|x(s)-y(s)\|^2.
\end{aligned}$$

By hypothesis ($\mathscr{H}$.3) combined with Hölder's inequality, we get that

$$\begin{aligned}
\mathbb{E}\|J_2(t)\| &\le \mathbb{E}\|\int_0^t U(t,s)\left[A(t)g(t,x(t-\tau))-A(t)g(t,y(t-\tau))\right]ds\|\\
&\le \frac{2}{1-\nu}\int_0^t M^2e^{-2\beta(t-s)}ds\int_0^t \mathbb{E}\|x(s-\tau)-y(s-\tau)\|^2ds\\
&\le \frac{2M^2L_*^2}{1-\nu}\frac{1-e^{-2\beta t}}{2\beta}t \sup_{s\in[-\tau,t]} \mathbb{E}\|x(s)-y(s)\|^2.
\end{aligned}$$

Moreover, by hypothesis ($\mathscr{H}$.2) combined with Hölder's inequality, we can conclude that

$$\begin{aligned}
E\|J_3(t)\| &\le E\|\int_0^t U(t,s)\left[f(s,x(s-\tau))-f(s,y(s-\tau))\right]ds\|^2\\
&\le \frac{2C_1^2}{1-\nu}\int_0^t M^2e^{-2\beta(t-s)}ds\int_0^t \mathbb{E}\|x(s-\tau)-y(s-\tau)\|^2ds\\
&\le \frac{2M^2C_1^2}{1-\nu}\frac{1-e^{-2\beta t}}{2\beta}t \sup_{s\in[-\tau,t]} \mathbb{E}\|x(s)-y(s)\|^2.
\end{aligned}$$

Hence

$$\sup_{s\in[-\tau,t]} \mathbb{E}\|\psi(x)(s)-\psi(y)(s)\|^2 \le \gamma(t) \sup_{s\in[-\tau,t]} \mathbb{E}\|x(s)-y(s)\|^2,$$

where

$$\gamma(t) = \nu + [L_*^2 + C_1^2]\frac{2M^2}{1-\nu}\frac{1-e^{-2\beta t}}{2\beta}t$$

By condition ($\mathscr{H}$.3), we have $\gamma(0) = \nu = L_* M_* < 1$. Then there exists $0 < T_1 \leq T$ such that $0 < \gamma(T_1) < 1$ and ψ is a contraction mapping on $S_{T_1}(\varphi)$ and therefore has a unique fixed point, which is a mild solution of Eq. (1) on $[-\tau, T_1]$. This procedure can be repeated in order to extend the solution to the entire interval $[-\tau, T]$ in finitely many steps. This completes the proof.

4 An Example

In recent years, the interest in neutral systems has been growing rapidly due to their successful applications in practical fields such as physics, chemical technology, bioengineering, and electrical networks. We consider the following stochastic partial neutral functional differential equation with finite delay τ $(0 \leq \tau < \infty,)$, driven by a Rosenblatt process

$$\begin{cases} d\,[u(t,\zeta) + G(t, u(t-\tau,\zeta))] = \Big[\frac{\partial^2}{\partial^2\zeta}u(t,\zeta) + b(t,\zeta)u(t,\zeta) + F(t, u(t-\tau,\zeta))\Big]dt \\ \qquad\qquad + \sigma(t)dZ_H(t),\ 0 \leq t \leq T,\ 0 \leq \zeta \leq \pi, \\ u(t,0) = u(t,\pi) = 0, \qquad 0 \leq t \leq T \\ u(t,\zeta) = \varphi(t,\zeta), \qquad t \in [-\tau, 0],\ 0 \leq \zeta \leq \pi, \end{cases} \tag{12}$$

where Z_H is a Rosenblatt process, $b(t,\zeta)$ is a continuous function and is uniformly Hölder continuous in t, $F, G : \mathbb{R}^+ \times \mathbb{R} \longrightarrow \mathbb{R}$ are continuous functions.
To study this system, we consider the space $X = L^2([0,\pi])$, $Y = \mathbb{R}$ and the operator $A : D(A) \subset X \longrightarrow X$ given by $Ay = y''$ with

$$D(A) = \{y \in X : y'' \in X,\quad y(0) = y(\pi) = 0\}.$$

It is well known that A is the infinitesimal generator of an analytic semigroup $\{T(t)\}_{t\geq 0}$ on X. Furthermore, A has discrete spectrum with eigenvalues $-n^2$, $n \in \mathbb{N}$ and the corresponding normalized eigenfunctions given by

$$e_n := \sqrt{\frac{2}{\pi}}\sin nx,\ n = 1, 2, \ldots.$$

In addition $(e_n)_{n\in\mathbb{N}}$ is a complete orthonormal basis in X and

$$T(t)x = \sum_{n=1}^{\infty} e^{-n^2 t} < x, e_n > e_n,$$

for $x \in X$ and $t \geq 0$.

Now, we define an operator $A(t) : D(A) \subset X \longrightarrow X$ by

$$A(t)x(\zeta) = Ax(\zeta) + b(t, \zeta)x(\zeta).$$

By assuming that $b(., .)$ is continuous and that $b(t, \zeta) \leq -\gamma$ $(\gamma > 0)$ for every $t \in \mathbb{R}$, $\zeta \in [0, \pi]$, it follows that the system

$$\begin{cases} u'(t) = A(t)u(t), & t \geq s, \\ u(s) \; = x \in X, \end{cases}$$

has an associated evolution family given by

$$U(t, s)x(\zeta) = \left[T(t-s) \exp^{\int_s^t b(\tau,\zeta)d\tau} x \right](\zeta).$$

From this expression, it follows that $U(t, s)$ is a compact linear operator and that for every $s, t \in [0, T]$ with $t > s$

$$\|U(t, s)\| \leq e^{-(\gamma+1)(t-s)}$$

In addition, $A(t)$ satisfies the assumption $\mathscr{H}_1$ (see [2]).

To rewrite the initial-boundary value problem (12) in the abstract form we assume the following:

(i) The substitution operator $f : [0, T] \times X \longrightarrow X$ defined by $f(t, u)(.) = F(t, u(.))$ is continuous and we impose suitable conditions on F to verify assumption $\mathscr{H}_2$.
(ii) The substitution operator $g : [0, T] \times X \longrightarrow X$ defined by $g(t, u)(.) = G(t, u(.))$ is continuous and we impose suitable conditions on G to verify assumptions $\mathscr{H}_2$ and $\mathscr{H}_3$.
(iii) The function $\sigma : [0, T] \longrightarrow \mathscr{L}_2^0(L^2([0, \pi]), \mathbb{R})$ is bounded, that is, there exists a positive constant L such that $\|\sigma(t)\|_{\mathscr{L}_2^0} \leq L < \infty$, uniformly in $t \in [0, T]$, where $L := sup_{t \in [0,T]} e^{-t}$.

If we put

$$\begin{cases} u(t)(\zeta) = u(t, \zeta), \; t \in [0, T], \; \zeta \in [0, \pi] \\ u(t, \zeta) = \varphi(t, \zeta), \; t \in [-\tau, 0], \; \zeta \in [0, \pi], \end{cases} \tag{13}$$

then, the problem (12) can be written in the abstract form

$$\begin{cases} d[x(t) + g(t, x(t-\tau))] = [A(t)x(t) + f(t, x(t-\tau))]dt + \sigma(t)dZ_H(t), \; 0 \leq t \leq T, \\ x(t) = \varphi(t), \; -\tau \leq t \leq 0. \end{cases}$$

Furthermore, if we assume that the initial data $\varphi = \{\varphi(t) : -\tau \leq t \leq 0\}$ satisfies $\varphi \in \mathscr{C}([-\tau, 0], \mathbb{L}^2(\Omega, X))$, thus all the assumptions of Theorem 1 are fulfilled. Therefore, we conclude that the system (12) has a unique mild solution on $[-\tau, T]$.

5 Conclusion

In this paper, we have studied a class of time-dependent neutral stochastic functional differential equations with finite delay driven by Rosenblatt process. We have provided sufficient conditions ensuring the existence of the unique mild solution of neutral stochastic functional differential equations by using stochastic analysis and a fixed-point strategy. We have also proposed an illustrative example to show the effectiveness of our theoretical result.

Finally, it may be interesting for further studies to consider the qualitative behavior of neutral stochastic differential equations with infinite delay driven by Rosenblatt process by studying for instance the transportation inequalities for the law of the mild solution and invariant measures.

References

1. P. Acquistapace, B. Terreni, A unified approach to abstract linear parabolic equations. Tend. Sem. Mat. Univ. Padova **78**, 47–107 (1987)
2. D. Aoued, S. Baghli, Mild solutions for Perturbed evolution equations with infinite state-dependent delay. Electron. J. Qual. Theory Differ. Equ. **59**, 1–24 (2013)
3. B. Boufoussi, S. Hajji, E. Lakhel, Functional differential equations in Hilbert spaces driven by a fractional Brownian motion. Afrika Matematika **23**(2), 173–194 (2011)
4. T. Caraballo, M.J. Garrido-Atienza, T. Taniguchi, The existence and exponential behavior of solutions to stochastic delay evolution equations with a fractional Brownian motion. Nonlinear Anal. **74**, 3671–3684 (2011)
5. G. Da Prato, J. Zabczyk, *Stochastic Equations in Infinite Dimension* (Cambridge University Press, Cambridge, 1992)
6. S. Hajji, E. Lakhel, Existence and uniqueness of mild solutions to neutral SFDEs driven by a fractional Brownian motion with non-Lipschitz coefficients. J. Numer. Math. Stochast. **7**(1), 14–29 (2015)
7. E. Lakhel, Exponential stability for stochastic neutral functional differential equations driven by Rosenblatt process with delay and Poisson jumps. Random Oper. Stoch. Equ. **24**(2), 113–127 (2016)
8. N.N. Leonenko, V.V. Ahn, Rate of convergence to the Rosenblatt distribution for additive functionals of stochastic processes with long-range dependence. J. Appl. Math. Stoch. Anal. **14**, 27–46 (2001)
9. M. Maejima, C.A. Tudor, On the distribution of the Rosenblatt process. Statist. Probab. Lett. **83**, 1490–1495 (2013)
10. M. Maejima, C.A. Tudor, Wiener integrals with respect to the Hermite process and a non central limit theorem. Stoch. Anal. Appl. **25**, 1043–1056 (2007)
11. A. Pazy, *Semigroups of Linear Operators and Applications to Partial Differential Equations, Applied Mathematical Sciences* (Springer, New York, 1983)

12. V. Pipiras, M.S. Taqqu, Integration questions related to the fractional Brownian motion. Probab. Theory Relat. Fields **118**, 251–281 (2001)
13. M. Rosenblatt, Independence and dependence, in Proceedings of the Fourth Berkeley Symposium on Mathematical Statistics and Probability, Volume 2: Contributions to Probability Theory, pp. 431-443. (University of California Press, Berkeley, CA, 1961)
14. M.S. Taqqu, Weak convergence to fractional Brownian motion and the Rosenblatt Process. Z. Wahr. Geb. **31**, 287–302 (1975)
15. M. Taqqu, Convergence of integrated processes of arbitrary Hermite rank. Z. Wahrscheinlichkeitstheor, Verw. Geb., **50**, 53–83 (1979)
16. C.A. Tudor, Analysis of the Rosenblatt process. ESAIM Probab. Statist. **12**, 230–257 (2008)

Printed in the United States
By Bookmasters